DATE DUE

GAYLORD			PRINTED IN U.S.A.

Intelligent Life in the Universe

Advances in Astrobiology and Biogeophysics

springer.com

This series aims to report new developments in research and teaching in the interdisciplinary fields of astrobiology and biogeophysics. This encompasses all aspects of research into the origins of life – from the creation of matter to the emergence of complex life forms – and the study of both structure and evolution of planetary ecosystems under a given set of astro- and geophysical parameters. The methods considered can be of theoretical, computational, experimental and observational nature. Preference will be given to proposals where the manuscript puts particular emphasis on the overall readability in view of the broad spectrum of scientific backgrounds involved in astrobiology and biogeophysics.
The type of material considered for publication includes:

- Topical monographs

- Lectures on a new field, or presenting a new angle on a classical field

- Suitably edited research reports

- Compilations of selected papers from meetings that are devoted to specific topics

The timeliness of a manuscript is more important than its form which may be unfinished or tentative. Publication in this new series is thus intended as a service to the international scientific community in that the publisher, Springer, offers global promotion and distribution of documents which otherwise have a restricted readership. Once published and copyrighted, they can be documented in the scientific literature.

Series Editors:

Dr. André Brack
Centre de Biophysique Moléculaire
CNRS, Rue Charles Sadron
45071 Orléans, Cedex 2, France
Brack@cnrs-orleans.fr

Dr. Gerda Horneck
DLR, FF-ME, Radiation Biology
Linder Höhe
51147 Köln, Germany
Gerda.Horneck@dlr.de

Prof. Dr. Michel Mayor
Observatoire de Genève
1290 Sauverny, Switzerland
Michel.Mayor@obs.unige.ch

Dr. Christopher P. McKay
NASA Ames Research Center
Moffet Field, CA 94035, USA

Prof. Dr. H. Stan-Lotter
Institut für Genetik
und Allgemeine Biologie
Universität Salzburg
Hellbrunnerstr. 34
5020 Salzburg, Austria

Peter Ulmschneider

Intelligent Life in the Universe

Principles and Requirements
Behind Its Emergence

Second Edition

With 156 Figures and 24 Tables
Including 37 Color Figures

 Springer

Professor Dr. Peter Ulmschneider
Universität Heidelberg
Institut für Theoretische Astrophysik
Albert-Ueberle-Straße 2
69120 Heidelberg, Germany
e-mail: ulm@ita.uni-heidelberg.de

P. Ulmschneider, Intelligent Life in the Universe, Adv. Astrobiol. Biogeophys. (Springer, Berlin Heidelberg 2006), DOI 10.1007/11614371

Library of Congress Control Number: 2006929538

ISSN 1610-8957

ISBN-10 3-540-32836-X 2nd ed. Springer Berlin Heidelberg New York
ISBN-13 978-3-540-32836-0 2nd ed. Springer Berlin Heidelberg New York
ISBN 3-540-43988-9 1st ed. Springer-Verlag Berlin Heidelberg New York

Springer is a part of Springer Science+Business Media
springer.com

© Springer-Verlag Berlin Heidelberg 2003, 2006
Printed in Germany

Cover design: Erich Kirchner, Heidelberg
Production and Typesetting: LE-TeX Jelonek, Schmidt & Vöckler GbR, Leipzig, Germany

Printed on acid-free paper   55/3100/YL   5 4 3 2 1 0

Preface to the Second Edition

In the first edition of this book it was argued that the uniqueness of human intelligence is the consequence of a very large brain and man's outstanding specializations in communication and tool use. No other life form on Earth is able to communicate in such a detailed manner by both vision and language and is able to handle so many diverse objects and tools. Yet apes, monkeys, dogs, elephants, seals, dolphins and even corvids all show highly intelligent behavior, which in recent years has become increasingly understood and appreciated. Palaeanthropologists argue that exceptional human intelligence arose from keen vision acquired in the rainforest, an upright walk together with a complete freeing of the hands for tool use adopted after our ancestors entered the open savannahs, and from our intimate social interactions in group living. Since our technological intelligence is based on the development of hands it is intimately connected with life on land. This is seen, for instance, by the modification of arms into fins or flippers when vertebrate land animals evolved back to life in the oceans. The development of our type of intelligence therefore is a consequence of the conquest of the land by animals and plants, which by a mutualistic relationship makes animal life on land possible.

For this reason a whole new Chap. 3 "The Earth" on geology has been added, in which the phenomena of plate tectonics and continent formation are discussed. This has resulted in a renumbering of the remaining chapters. In addition, sections on the conquest of the land by plants and animals are greatly expanded to show the enormous difficulties that life encountered before it finally mastered the land 470 million years ago, over 3.5 billion years after it first appeared in the Earth's oceans. The planetological history of the early Earth is discussed in greater detail to give more insight into how Earth-like planets form and how the chemical composition favorable for life arose. Finally, the rapid advances over the last four years in all fields, from the search for planets to the search for the Last Universal Common Ancestor of all life, has been brought up to date.

Acknowledgements

I am greatly indebted to my colleagues Volker Storch (Zoology), Peter Leins (Botany), Mario Trieloff (Mineralogy), Hans-Peter Gail (Astrophysics) from

Heidelberg as well as Uwe Walzer (Earth Sciences) from Jena who critically read parts of the manuscript. If there are remaining errors they are entirely my own. I also thank my children Katharina, Martin and Jakob as well as Philip Salmon for many critical and fruitful discussions about the content and logic of the presentation. The book is dedicated to my wife Helgard who never read any of it.

Heidelberg, January 2006   *Peter Ulmschneider*

Preface to the First Edition

One of the most exciting questions for mankind is whether we are alone in the universe. That intelligent nonhuman beings exist was commonly believed in prehistoric times as well as in antiquity. Creatures such as giants, centaurs, angels, and fairies were essential and universally accepted parts of Greek, Jewish, and Germanic mythologies. Although no fossil traces of such beings have ever been found, most of us firmly believe that nonhuman intelligent beings do indeed exist. This conviction is derived from the staggering size of the universe with roughly 100 billion times 100 billion (10^{22}) stars, which makes it inconceivable that we could be the only intelligent society in the universe. Indeed, modern science has shown that since the Copernican revolution all attempts to define our position as an exceptional one in the universe have failed dismally.

But if other intelligent civilizations do exist, how can we find them? Why is there no terrestrial or astronomical trace of them, despite great technological advances in recent centuries and especially in modern times? Why have we never found artifacts discarded by visiting aliens, which would convincingly prove the existence of nonhuman intelligent beings? Is the number of planets on which life is able to evolve too small, or is the formation of life – and particularly intelligent life – an extremely rare event? Could these intelligent societies face insurmountable difficulties in traveling over large galactic distances, or do they no longer exist?

Recent advances in search techniques for planets, in the theory of planet formation, and particularly in biochemistry, molecular, and cell biology are about to give answers to these questions: how life appeared and how many planets can be expected in the universe on which life, and eventually intelligent life, developed. New in this book is the argument that, by thinking carefully about the future development of mankind, one can gain insight into the nature of extraterrestrial civilizations.

The book consists of three parts: planets, life, and intelligence. In *Part I*, Chaps. 1–3 discuss stars, galaxies, and the origin of chemical elements, our recent planet formation theories, the search methods for extrasolar planets and what has been found so far. Chapter 4, "Planets suitable for life", describes what constitutes an Earth-like planet and how many of them can be expected in the universe. In *Part II*, Chaps. 5 and 6 outline life and its origin on Earth, how it evolved, and how intelligent life developed. Chapter 7 discusses the search for extraterrestrial life and intelligent societies. In *Part III*, Chap. 8, "The future of mankind", gives possible insights into what can be expected about the nature of extraterrestrials. Finally, Chap. 9, on extraterrestrial intelligent life, constructs a likely picture of these beings and attempts to answer the question of why they don't interact with us.

Heidelberg, June 2002   *Peter Ulmschneider*

Table of Contents

Part I Planets

1 Stars, Galaxies, and the Origin of Chemical Elements 3
  1.1 The History of the Universe 3
  1.2 Molecular Clouds 6
  1.3 The Pre-Main Sequence Evolution of Stars 8
  1.4 The Post-Main Sequence Evolution of Stars 11
  1.5 Element Composition and Dating 13
     1.5.1 Population I and Population II Stars 13
     1.5.2 Dating with Radiometric Clocks 15

2 Planet Formation 19
  2.1 Accretion Disks and Planetesimal Formation 19
  2.2 Terrestrial Planets 21
  2.3 Jovian Planets and Kuiper Belt Objects 24
  2.4 The Migration of Jovian Planets 25
  2.5 The T-Tauri Stage 26
  2.6 Asteroids 28
  2.7 Comets 31
  2.8 Meteorites 33
  2.9 Early History of the Solar System 34

3 The Earth 39
  3.1 Planetological History of the Early Earth 39
  3.2 Formation of the Moon 40
  3.3 Ocean-Vaporizing Impacts 42
  3.4 The End of the Heavy Bombardment 44
  3.5 The Environment on the Early Earth 45
  3.6 Seismology and the Earth's Interior Structure 49
  3.7 Volcanism and the Composition of Rocks 52
  3.8 The Earth's Core and Mantle 56
  3.9 The Earth's Magnetic Field and Sea-Floor Spreading 58
  3.10 Convection, Hot Spots and Plate Tectonics 60
  3.11 Mountain Building and the Evolution of Continents 66
  3.12 Plate Tectonics on Mars and Venus? 71

4   The Search for Extrasolar Planets 73
   4.1   The Recently Discovered Planets 73
   4.2   Direct Search Methods for Planets 76
   4.3   Indirect Search Methods 76
   4.4   Circumstellar Disks 79
   4.5   New Search Strategies 80

5   Planets Suitable for Life 87
   5.1   Habitable Zones 87
       5.1.1   The Solar Habitable Zone 88
       5.1.2   Habitable Zones Around Other Stars 90
   5.2   Planetary Mass and the Evaporation of the Atmosphere 91
   5.3   The Lifetimes of the Stars 94
   5.4   Tidal Effects on Planets 95
   5.5   The Increase in Solar Luminosity
       and the Continuously Habitable Zone 97
   5.6   Instabilities of the Planetary Atmosphere 98
       5.6.1   The Greenhouse Effect 99
       5.6.2   The Carbonate Silicate Cycle 99
       5.6.3   The Runaway Greenhouse Effect 100
       5.6.4   Irreversible Glaciation 101
   5.7   Axis Variations of the Planets 103
   5.8   Biogenic Effects on Planetary Atmospheres 105
   5.9   Proterozoic Glaciations and Snowball Earth 107
   5.10   The Requirements for Continuous Habitability 109
   5.11   The Drake Formula 109
   5.12   The Number of Habitable Planets 111

Part II Life

6   Life and its Origin on Earth 117
   6.1   What is Life? 117
   6.2   The Special Role of Organic Chemistry 118
   6.3   The Elements of Biochemistry 118
       6.3.1   Proteins, Carbohydrates, Lipids, and Nucleic Acids ... 119
       6.3.2   The Genetic Code 124
       6.3.3   ATP, the Energy Currency of the Biochemical World . 124
       6.3.4   Synthesizing RNA, DNA, and Proteins 125
   6.4   Cells and Organelles 127
   6.5   Sequencing and the Classification of Organisms 129
       6.5.1   Classification by Sequencing 129
       6.5.2   The Molecular Clock 129
       6.5.3   The Evolutionary Tree of Bacteria 130
       6.5.4   The Timetable of the Evolution of Life 131

      6.5.5   Sequencing and the Complete Genome 133

  6.6   Geological Traces of Life 135

  6.7   The Stage for the Appearance of Life 136

      6.7.1   The Origin of the Genetic Code 137

      6.7.2   The Urey–Miller Experiments..................... 138

      6.7.3   The Search for the Last Universal Common Ancestor . 139

      6.7.4   Summary: The Boundary Conditions 142

  6.8   Abiotic Chemical Evolution
and the Theories of How Life Formed 143

7   Evolution 149

  7.1   Darwin's Theory 149

  7.2   The Development of Eukaryotes and Endosymbiosis........ 151

  7.3   Oxygen as an Environmental Catastrophe 153

  7.4   The Cell Nucleus and Mitosis 154

  7.5   Sexuality and Meiosis 155

  7.6   Genetic Evolution 157

  7.7   Multicellularity, the Formation of Organs,
and Programmed Cell Death 159

  7.8   Problems of Life on Land................................ 162

      7.8.1   Conquest of the Land by Plants 163

      7.8.2   New Organs of Land Plants....................... 166

      7.8.3   Conquest of the Land by Animals 171

  7.9   The Great K/T Boundary Event 173

  7.10 The Tertiary and the Evolution of Mammals 177

  7.11 Primate Evolution 178

  7.12 DNA Hybridization 187

  7.13 Brain Evolution and Tool Use............................ 188

  7.14 Stone Tool Culture 190

  7.15 Diet and Social Life..................................... 192

  7.16 The Logic of the Human Body Plan 193

  7.17 Evolution, Chance, and Information 196

  7.18 Cultural Evolution...................................... 199

8   The Search for Extraterrestrial Life....................... 201

  8.1   Life in the Solar System 201

  8.2   Europa's Ocean 202

  8.3   Life on Mars 204

      8.3.1   Early Searches 204

      8.3.2   The Viking Experiments 206

      8.3.3   Mars Meteorites.................................. 208

  8.4   The Early Atmosphere of Mars........................... 210

  8.5   Future Mars Missions 212

  8.6   Life Outside the Solar System 214

  8.7   UFOs 216

Part III Intelligence

9   The Future of Mankind 221
9.1   Predicting Mankind's Future 221
9.2   Settlement of the Solar System 222
9.2.1   The Space Station 223
9.2.2   Moon and Mars Projects 225
9.2.3   Space Travel 228
9.2.4   Near-Earth Asteroids and the Mining
of the Solar System 230
9.2.5   Space Habitats 231
9.2.6   Cultural Impact of Space Colonization 234
9.3   Interstellar Travel 236
9.4   Mastering the Biological World.......................... 237
9.4.1   Creating Life in the Laboratory 238
9.4.2   The Decoding of the Human Genome 239
9.4.3   Understanding Intelligence 239
9.5   Androids and Miniaturization 240
9.6   Connected Societies..................................... 241
9.7   Fear of the Future 242
9.8   The Dangers for Mankind 242
9.8.1   Bacterial or Viral Infection 243
9.8.2   Episodes of Extreme Volcanism 244
9.8.3   Irreversible Glaciation
and the Runaway Greenhouse Effect 245
9.8.4   Comet or Asteroid Impact 246
9.8.5   Supernova Explosions and Gamma Ray Bursts 248
9.8.6   Irreversible Environmental Damage 250
9.8.7   Uncontrollable Inventions 250
9.8.8   War, Terrorism, and Irrationality 251
9.9   Survival Strategies...................................... 252

10   Extraterrestrial Intelligent Life 255
10.1   Does Extraterrestrial Intelligent Life Exist? 255
10.2   What is the Hypothetical Nature of the Extraterrestrials? ... 257
10.3   The Drake Formula, the Number
of Extraterrestrial Societies 260
10.4   The Lifetime of an Extraterrestrial Civilization 262
10.5   Distances to the Extraterrestrial Societies 263
10.6   SETI, the Search for Extraterrestrial Intelligent Life 265
10.6.1   Radio and Optical Searches
for Extraterrestrial Civilizations 266
10.6.2   Possible Contact in the not too Distant Future 270
10.7   The Fermi Paradox: Where are the Extraterrestrials? 272

10.7.1 They do not Exist 273
10.7.2 Technically, a Visit is not Possible 274
10.7.3 They are Nearby, but have not been Detected 275
10.7.4 They are not Interested in Us 275
10.8 The Zoo Hypothesis 276

References 279

Author Index 297

Subject Index 303

Part I

Planets

1 Stars, Galaxies, and the Origin of Chemical Elements

"That I am mortal I know, and that my days are numbered, but when in my mind I follow the multiply entwined orbits of the stars, then my feet do no longer touch the Earth. At the table of Zeus himself do I eat Ambrosia, the food of the Gods". These words by Ptolemy from around 125 A.D. are handed down together with his famous book *The Almagest*, the bible of astronomy for some 1500 years. They capture mankind's deep fascination with the movements of the heavens, and the miracles of the physical world. After the Babylonians observed the motions of the Sun, Moon, and planets for millennia, the ancient Greeks were the first to speculate about the nature of these celestial bodies. Yet it is only as a consequence of developments in the last 150 years that a much clearer picture of the physical universe has begun to emerge. Among the most important discoveries have been the stellar parallax, confirming Copernicus's heliocentric system, the realization that galaxies are comprised of billions of stars, the awareness of the size of the universe, and the evolutionary nature of living organisms.

Although life is known only from Earth, without doubt here and elsewhere it emerged in close association with planets, stars, and galaxies. The material out of which living organisms are made and the planets on which life formed are composed of chemical elements that have been synthesized in stars. To understand the nature of life and its origin it is therefore necessary to briefly review in this chapter the history of the universe, the formation and development of stars, and how the chemical elements were generated. Planet formation is discussed in Chap. 2 and the emergence of life in Chap. 6.

1.1 The History of the Universe

About 14 billion years ago, our universe made its appearance in the Big Bang. It is currently believed that it was at this starting point that space, time, matter, energy, and the laws of nature all came into being. Evidence for the existence of the Big Bang is the observationally well established Hubble law. Edwin Hubble in 1924 found that galaxies move away from us with a speed that increases with distance. Retracing these motions back in time, one finds not only when the universe came into being (Ferreras et al. 2001) but also that it must have originated from a tiny volume. Another indication

P. Ulmschneider, Stars, Galaxies, and the Origin of Chemical Elements. In: *Intelligent Life in the Universe*, P. Ulmschneider, Adv. Astrobiol. Biogeophys., pp. 3–18 (2006)
DOI 10.1007/11614371_1   Springer-Verlag Berlin Heidelberg 2006

for the Big Bang is the observed 3 K cosmic background radiation, which is believed to be the remnant of the primordial fireball through which the universe made its appearance. In about a million years after the Big Bang, the temperature of this fireball decreased from unbelievably high values of more than 10^{32} K to a few thousand K, and hydrogen and helium gas formed. No other constituents, except for traces of some very light chemical elements, were present at that time. However, after about 100 million years, due to the expansion of the universe, the fireball became so dim, with temperatures dropping below 300 K, that the universe would have become dark to human eyes because its peak radiation had moved into the infrared spectral range. As the universe continued its expansion up to the present time, the temperature of the fireball radiation decreased further, to the mentioned value of 3 K.

The so-called "dark age" of the universe lasted for about a billion years. After this time, the rapid expansion had led to a filamentary distribution of matter with local accumulations in which galaxy clusters, galaxies, and the first stars, the so-called population III stars, formed. These stars brought visible light back to the universe. Figure 1.1 displays the mass distribution of a tiny section of the universe, generated using a computer simulation in a box with a side length of 100 Mpc. Here distances are given in pc (parsec), where 1 pc = 3.26 Ly (light years) = 3.09×10^{18} cm. The red and white regions show areas of high mass concentrations where galaxies and stars form, while the dark regions indicate voids where there is little matter. The largest gravitationally bound objects in the universe are galaxy clusters, which have diameters of about 4 Mpc, while individual galaxies like our Milky Way have sizes of about 30 kpc.

The first detailed models of population III stars, consisting purely of H and He, have recently been constructed. One finds that these stars were very massive, with 100–300 $M_\odot$ (where 1 $M_\odot = 2 \times 10^{33}$ g is the mass of the Sun), and had a short lifetime of a few million years. They ended their lives with a supernova explosion (discussed below). It is important for the chemical element composition in the universe that in the cores of population III stars the elements H and He were transmuted by nuclear fusion into heavier elements, up to Fe. These heavy elements were subsequently ejected into the interstellar medium by the terminal supernova explosion, in which even heavier elements were generated. Mixed together with fresh H and He, the enriched material then accumulated into the next generation of stars, the population II stars. By accretion into massive stars with short lifetimes, this process of enrichment of heavy elements continued over several generations of stars, until finally the metal-rich population I star chemical element mixture formed, which was the material out of which our Sun and the planets were made.

Figure 1.2 shows the spiral galaxy M74, which lies roughly 11 Mpc away from Earth and is very similar to our own galaxy, also containing about 100 billion stars. Here, the conspicuous dark absorption bands indicate the presence of dust, while the luminous emission in the spiral arms shows regions of star formation. Viewed from the side, spiral galaxies have a disk-like shape.

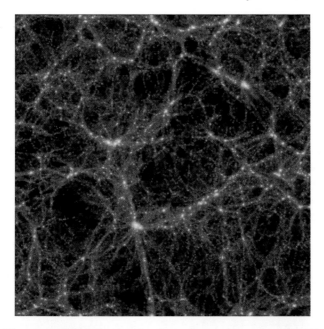

Fig. 1.1. The matter distribution in the universe from the VIRGO simulation (Jenkins et al. 1998). The figure shows a slice out of a cube of side length 100 Mpc, or 3×10^{26} cm

Fig. 1.2. Spiral galaxy M74 in the constellation Pisces, seen face on (courtesy of NASA)

The spiral galaxy NGC891 (Fig. 1.3) represents a good example, which shows that the dust (and gas) layers are concentrated in the central plane of these systems.

Galaxies, stars, and planets form as the result of a gravitational collapse of large amounts of gas and dust from the interstellar medium. Our galaxy originated from a spherically shaped pre-galactic cloud, while stars and planets form readily from giant molecular clouds in spiral and irregular galaxies, because these systems possess abundant amounts of gas and dust. However, a gravitational collapse does not occur easily, as it has to overcome severe obstacles such as differential rotation, turbulence, magnetic fields, and the need to concentrate matter from a large volume.

Fig. 1.3. Spiral galaxy NGC891 in the constellation Andromeda, seen edge on (courtesy of NASA)

1.2 Molecular Clouds

While our galaxy has a mass of about 10^{11} $M_{\odot}$, and typical stars possess masses in the range 0.1–120 $M_{\odot}$, giant molecular clouds have masses of up to 10^6 $M_{\odot}$. They are the most massive objects in our galaxy and there are large numbers of them. Their name comes from the many molecules identified within them by radio astronomy, some of which are listed in Table 1.1. In addition to hydrogen in the form of H_2, the most abundant molecules are OH, H_2O, CO, and NH_3. But they even harbor organic compounds, albeit much less complicated ones than those found in living organisms. Although more than 99% of the mass in molecular clouds is made up of gas, they also contain large quantities of interstellar dust, which cools through infrared radiation and serves to shield the cloud's interior from heating by stellar radiation. As a result, the cloud cores become very cold, with temperatures as low as 5–10 K, and resulting densities as large as 10^3–10^5 particles per cm^3. These cloud cores are the seats of star formation, the separate stages of which are shown in Fig. 1.4.

Table 1.1. Molecules detected in molecular clouds (Wootten 2002)

Simple hydrides, oxides, sulfides, halides, and related molecules				
H_2	CO	NH_3	CS	NaCl
HCl	SiO	SiH_4	SiS	AlCl
H_2O	SO_2	CC	H_2S	KCl
	OCS	CH_4	PN	AlF

Nitriles, acetylene derivatives, and related molecules				
HCN	HC_3N	CH_3C_3N	CH_3CH_2CN	CH_2CH_2
H_3CCN	HC_5N	CH_3CCH	CH_2CHCN	CHCH
CCCO	HC_7N	CH_3C_4H	HNC	
CCCS	HC_9N		HNCO	
HCCCHO	$HC_{11}N$		HNCS	
H_3CNC				

Aldehydes, alcohols, ethers, ketones, amides, and related molecules			
H_2CO	CH_3OH	HCOOH	CH_2NH
CH_2S	CH_3CH_2OH	CH_3COOH	CH_2NH_2
CH_3CHO	CH_3SH	CH_3OCH_3	NH_2CN
NH_2CHO		CH_2CO	

Cyclic molecules		
C_3H_2	SiC_2	C_3H

Ions		
CH	HCO	HCNH
NH_2	HCOO	SO
	HCS	

Radicals				
OH	C_3H	CN	HCO	C_2S
CH	C_4H	C_3N	NO	NS
C_2H	C_5H	CH_2CN	SO	
	C_6H	HSiCC	HSCC	

When the density wave of a galactic spiral arm rushes over a molecular cloud, or a neighboring supernova explosion ejects its material against it, fragments of the cloud become compressed, and start to collapse under their own gravity. Figure 1.4a shows, for example, the molecular cloud L1152 (Hartmann 1998), which has been mapped using the radio lines emitted from its molecules CO, CS, and NH_3. Giant molecular clouds are lengthy objects,

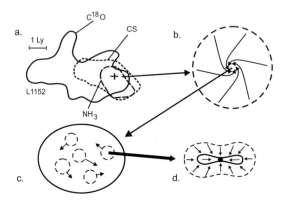

Fig. 1.4. The collapse of a molecular cloud core and the formation of a solar system.
a. The molecular cloud core; the indicated scale is 1 Ly. **b.** Collapsing cloud core.
c. Fragmentation. **d.** Precursor of a solar system with an accretion disk (seen from
the side). There is a large difference in scale between each of these four stages

which contain many cloud cores. Because the collapsing cloud material con-
verges from a very extended volume, the small amount of rotation that it
originally possessed becomes strongly enhanced and, due to conservation
of angular momentum, the collapsing core region starts to rotate rapidly
(Fig. 1.4b). As the collapse proceeds in a chaotic and nonradial fashion, the
rotating regions break up into fragments, and their rotation is converted into
orbital motion (see Fig. 1.4c).

  This process of collapse, increased rotation, and fragmentation repeats
itself several times, until small enough flattened sub-fragments (the predecessors of solar systems) are generated (see Fig. 1.4d), in the center of which
an accretion disk is formed. Note that in going from the stage of Fig. 1.4a
to that of Fig. 1.4d the size is reduced by about a factor of 2000. The flattened shape of the fragment and the accretion disk derives from the fact
that the collapse is much easier parallel to the rotation axis than perpendicular to it, where direct collapse violates the conservation law of angular
momentum. The rate of rotation determines whether a solar system ends
up as a multiple stellar system (70% of the cases), or as a system with
a single central star (30%). It is important to note that this entire process
does not lead to the formation of individual stars. Instead, whole star clusters with many hundreds or even thousands of stars are effectively created
simultaneously.

1.3 The Pre-Main Sequence Evolution of Stars

Figure 1.4d shows that the precursor of a single-star solar system is a flat
cocoon-like object, at the center of which a protostar accumulates, surrounded
by a rotating structure called an *accretion disk*, which feeds the protostar.

Such accretion disks have been observed (see Figs. 4.4 and 4.5). It is within these disks that the formation of planets takes place. Before discussing planet formation, however, it is necessary to consider the development and subsequent evolution of stars in greater detail, especially those which are similar to the Sun, because G-stars are potentially the most promising for extraterrestrial life.

A very useful insight into a star's evolution can be obtained by plotting its total radiated energy per second, the luminosity L, against its effective temperature T_{eff} (which is essentially the surface temperature of a star), to produce what is known as a *Hertzsprung–Russell diagram*. Because for a given T_{eff}, larger stars have higher luminosities, this diagram also displays the dependence of the radius R of a star against its surface temperature. The Hertzsprung–Russell diagram in Fig. 1.5a shows the computed pre-main sequence evolutionary tracks of Sun-like stars (dashed), where L is plotted in units of the solar luminosity $L_{\odot}$, and the dotted lines indicate stellar radii in units of the solar radius $R_{\odot} = 7 \times 10^{10}$ cm.

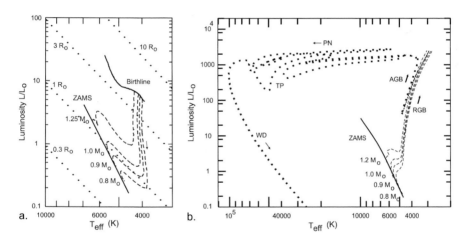

Fig. 1.5. Hertzsprung–Russell diagrams with computed evolutionary tracks of Sun-like stars. **a.** The pre-main sequence evolution (after Bernasconi 1996). **b.** The post-main sequence evolution (after Bressan et al. 1993 *dashed*; Bloecker 1995 *dotted*). ZAMS indicates the zero age main sequence branch, RGB the red giant branch, AGB the asymptotic giant branch, TP the thermal pulsation phase, while PN and WD mark the planetary nebula and white dwarf stages, respectively

Initially, the collapse of a cloud occurs in a rather cold gas; the energy gained by the contraction is radiated away efficiently in the infrared spectrum by the dust, and the temperature in the protostar remains roughly unchanged. Eventually, however, after the collapse has proceeded for about 500 000 years, the accumulated mass makes the core region of the protostar so dense and opaque that radiation can no longer escape, and the released gravitational energy creates a rapid rise of the core temperature. When this temperature reaches about a million K, nuclear reactions set in and deuterium starts to burn (into ^{3}He). Figure 1.5a shows this latter phase, called the *deuterium main sequence* or *birthline*. At this point, Sun-like stars with masses of 0.8–1.25 $M_\odot$ have about six times the solar luminosity and about five times the solar radius. They become optically visible, having previously been hidden behind their dusty parental cocoons.

Because deuterium, an isotope of hydrogen, has an abundance of only about 10^{-5} relative to hydrogen, deuterium burning (with its low-energy yield) is not powerful enough to balance the protostar's high light output. The star therefore continues to collapse, releasing its gravitational energy. This constant decrease of the radius of the collapsing star can be seen from the evolutionary tracks in Fig. 1.5a (dashed) in relation to the dotted lines. The movement along these tracks is the result of a close interaction between the energy-producing core and the overlying stellar envelope. The latter provides weight to compress the core to fusion temperatures and carries the generated energy away to the surface. As is well known, heat can be transported in two ways, both of which are employed in stellar envelopes. First, there is the heat directly radiated from a hot region; and second, there is convection, where energy is carried away by hot moving gas bubbles. When the stellar envelope transports energy mainly by radiation, the tracks in the Hertzsprung–Russell diagram (Fig. 1.5a) slope to the right, and are known as *radiative tracks*, but when convection dominates, they become more or less vertical, and are called *Hayashi tracks*.

Figure 1.5a shows that at low surface temperatures $T_{\rm eff}$, the stellar envelope is dominated by convection, and that the stars move down the Hayashi tracks. After further collapse, however, the surface temperature increases and convection becomes less important, and the stars then move along the radiative tracks. During the entire contraction process, the core temperature continues to rise until a critical limit of around 10 million K is reached, at which hydrogen burning sets in. As hydrogen is the most abundant element (see Table 1.2), and because this particular fusion process (of four hydrogen nuclei to one helium nucleus) liberates the maximum amount of fusion energy of all processes, the star now has a huge reservoir of fuel and is easily able to balance the energy expended by its luminosity. The stars settle at the endpoints of their evolutionary tracks at a line called the *hydrogen zero-age main sequence* (ZAMS), which is shown solid in Fig. 1.5a. It takes 36 million years for a 1.25 $M_\odot$ star to get to this ZAMS phase from the start of its collapse, 38 million years for a 1 $M_\odot$, and 72 million years for a 0.8 $M_\odot$ star. It is in the ensuing main sequence phase, however, that the stars spend most of their lifetime: the 1.25 $M_\odot$ star about 5 billion years, the 1 $M_\odot$ star (Sun) about 11 billion years, and the 0.8 $M_\odot$ star roughly 26 billion years. During this time the luminosity of the stars slowly grows by a factor of four, before increasing rapidly in the post-main-sequence evolution (see Fig. 5.8).

1.4 The Post-Main Sequence Evolution of Stars

An understanding of the post-main sequence stellar evolution is not only important because of what it can tell us about the destiny of the Sun, but also because *supernova explosions* which occur at the endpoints of the evolution of massive stars are essential for the generation of life. The fate of stars depends strongly on their mass, which can be divided into low (0.075–0.5 $M_\odot$), intermediate (0.5–6 $M_\odot$), and massive (6–120 $M_\odot$) categories. We are mainly interested in intermediate and massive stars.

As the hydrogen in the stellar core becomes used up toward the end of the main sequence phase, an inert He core forms around which a shell hydrogen burning zone develops. In order to balance the radiative energy loss of this core in the absence of nuclear energy generation, it contracts, causing the central temperature to rise. The process of contraction and shell burning leads to an expansion of the star's outer envelope, which appears to "roll back" the pre-main sequence evolution. In Fig. 1.5b (dashed) it can be seen that the stars first move backward along the radiative tracks but then, when the stellar surface temperatures become low enough, climb up along the Hayashi tracks. Since stars at this stage have much larger radii than when they were on the main sequence, they are called *red giants* and are said to be on the *red giant branch* of the Hertzsprung–Russell diagram(RGB; see Fig. 1.5b).

During the course of this evolution, the inactive helium core of the star contracts to such a high density that quantum effects come into play. The *Pauli exclusion principle*, that no more than two electrons (with different spins) can occupy the same energy level, leads to an *electron degeneracy pressure*, which strongly resists further compression of the stellar core. At this point the core is said to be *degenerate*. When the core temperature eventually rises to around 100 million K, *helium burning* sets in (three He nuclei fuse to form one C nucleus, and some of the He and C to an O nucleus), which ignites at the top of the red giant branch. In Fig. 1.5b these ignition points (the top ends of the dashed tracks) are reached 29, 12, and 6 billion years after the start of the ZAMS phase, for the 0.8 $M_\odot$, 1.0 $M_\odot$, and 1.2 $M_\odot$ stars, respectively. The subsequent evolutionary phases, which are roughly similar for the stars shown in Fig. 1.5b, are displayed for the 1.0 $M_\odot$ star (dotted track).

Low-mass stars of less than 0.5 $M_\odot$ never reach central temperatures high enough for helium burning. Here nuclear fusion eventually dies, and the stars contract to become fully degenerate stars, called *white dwarfs* (WD). Such stars, with masses up to 1.4 $M_\odot$, the *Chandrasekhar limit*, are very condensed. A 1.0 $M_\odot$ white dwarf, for instance, has roughly the same size as the Earth, and one teaspoonful of white dwarf matter would weigh about 5 tons. Having exploited all available energy sources, white dwarfs simply cool as they age. The very frequent stellar objects with masses below 0.075 $M_\odot$ (or 75 Jupiter masses) never reach even hydrogen fusion, and are known as *brown dwarfs*.

They contract continuously and also end up as degenerate white dwarf-like stars.

For intermediate-mass stars, Fig. 1.5b shows what happens after core helium burning starts in a star with 1.0 $M_\odot$ (dotted). This sequence is also representative of massive stars. At first, the core expands and the outer envelope contracts, placing the stars briefly on a *helium main sequence*. But the core helium burning source quickly expires, and a He-shell source around an inert C/O core develops in addition to the H-shell source. The C/O core of the star then contracts, the outer envelope expands, and the star rises up the *asymptotic giant branch* (AGB, Fig. 1.5b). Some 110 million years elapse from the red giant tip, before a 1.0 $M_\odot$ star reaches this state.

The evolution of massive stars is essential for the formation of life in the universe, because these stars play the primary role in the production of heavy elements. For these stars, with more than 6 $M_\odot$, there are further developments after helium burning. As their contracting C/O core becomes progressively hotter, C burns to Ne, and O to S as well as Si. Then Si fuses to Fe. Thereafter, because elements heavier than Fe can no longer provide energy in nuclear fusion processes, the core of the star contracts until it becomes degenerate. For massive stars above 8 $M_\odot$, it is likely that in their core the Chandrasekhar limit for white dwarfs will be exceeded. At this limit the degeneracy pressure can no longer hold the degenerate core against gravity and it collapses into a *neutron star*, in which electrons and protons combine to produce neutrons in a degenerate neutron core. Such a collapse releases a huge amount of energy and produces a gigantic explosion that can momentarily outshine an entire galaxy. The stellar envelope, containing all the heavier elements produced in the different fusion processes, is ejected into space and enriches the composition (heavy element abundances) of the interstellar medium. In addition, capture of neutrons produces elements heavier than Fe. This event is called a *type II supernova*, an example of which occurred in our galaxy in the year 1054, and was visible in broad daylight for weeks, creating the Crab Nebula. In 1987, another type II supernova occurred in the Large Magellanic Cloud, a satellite galaxy of our Milky Way.

For stars with even larger masses, a neutron star option is no longer possible because there is nothing that can prevent a complete gravitational collapse, and the resulting formation of a *black hole*. Stellar black holes have radii of about 3 km and possess such a strong gravitational field that not even light can escape from them – hence their name. The generation of a black hole involves a supernova explosion. By the way, not only very massive stars produce supernovae. It is also possible, under certain circumstances, that stars with less mass are able to generate this phenomenon. When the stellar wind in a binary system dumps mass from a neighboring massive star on top of a white dwarf, its mass can be pushed over the Chandrasekhar limit and the white dwarf explodes in a similarly energetic event, called a *type I supernova*.

Let us now return to the fate of intermediate-mass stars like our Sun. There is no carbon burning, but near the tip of the AGB phase an interesting phenomenon develops, called *thermal pulsations* (TP; Fig. 1.5b), indicated by extensive loops toward the left side of the diagram. This is because He burning becomes unstable, due to the large mass loss from the stellar wind in the AGB phase, which narrows the layer between the stellar surface and the He burning shell. There are about four pulsations until, on a final occasion, the outer envelope of the star is completely thrown off and the remaining degenerate core becomes visible, with surface temperatures of up to 100 000 K. The ejected mass is observed as a *planetary nebula* (PN). In Fig. 1.5b, the evolution of the core of the planetary nebula is seen by the horizontal tracks to the left (dotted). These very hot white dwarfs, with masses between 0.5–0.9 $M_\odot$, subsequently cool and become feeble degenerate stars, which radiate their stored thermal energy until eventually they fade from visibility. The evolution from the AGB to the leftmost point takes about 170 000 years, and to the bottommost point of the dotted track about 22 million years.

1.5 Element Composition and Dating

1.5.1 Population I and Population II Stars

Before planet formation and the creation of life are discussed, it is necessary to consider the origin and abundances of chemical elements in the universe in more detail. The theories of star formation described above depend on the chemical elements which were supplied by the originalmolecular cloud. Table 1.2 (left column) shows the most abundant elements of cosmic matter out of which the solar system was formed. It can be seen that hydrogen is by far the most frequent element, and that together with helium it makes up essentially all of the cosmic matter. The remaining elements constitute tiny but essential contaminations, consisting mainly of carbon and oxygen.

Table 1.2. Fractional abundances (by mass in percent) of the most frequent chemical elements in the Sun and the Earth's mantle. Also listed for Earth are the infrequent elements C and H (after Cox 2000; Holleman and Wiberg 1995)

Sun element	(%)	Earth element	(%)
H	70.68	O	48.9
He	27.43	Si	26.3
O	0.955	Al	7.7
C	0.306	Fe	4.7
Ne	0.174	Ca	3.4
Fe	0.136	Na	2.7
N	0.110	K	2.4
Si	0.070	Mg	2.0
Mg	0.065	(H	0.74)
S	0.036	(C	0.02)

Compared to this, the composition of the Earth's mantle is very different (see Table 1.2, right column). Note that the Earth's core consists mainly of Fe and Ni. This disparity in composition between the Sun and the Earth is due to the history of planet formation where in the accumulation process of the terrestrial planets mainly the non-volatile heavy elements contributed (see Chap. 2). The Sun still maintains the cosmic element mixture of the initial molecular cloud fragment from which it was formed. This particular mixture is called *population I* stellar abundances, and stars with element mixtures such as those of the Sun are called *population I stars*.

As already mentioned above, the so-called *population II* stars formed at much earlier epochs of our universe, and therefore have a very different mixture of elements. In these stars all elements except H and He are typically 1/10 to 1/1000 less abundant than in the Sun. It is thought that the formation of population II stars has led to much smaller terrestrial planets, which therefore cannot retain a sufficiently dense atmosphere to be seats of life (see Chap. 5). To understand why these different element abundances in stars came about, let us briefly come back to the early history of our universe.

It was mentioned above, that our galaxy formed from a spherically shaped pre-galactic cloud. Traces of that initial configuration are still found by observing long-lived (low-mass) population II stars, created at those early times. They occupy a wide spherical halo around our galaxy. Other fossils of that early epoch, so-called *globular star clusters*, also retain the old spherical distribution of the pre-galactic cloud. Globular clusters are highly concentrated spherical systems of up to 10 million stars, and are among the oldest population II objects, with a maximum age of around 13.5 billion years (Sandage et al. 2003). Figure 1.6 shows an example of a system of globular clusters around the elliptical galaxy M87. These clusters are seen as faint star-like objects in the halo around M87. Today's disk-shaped spiral galaxies, in which the population I stars were created, were formed by the collapse of the pre-galactic gas and dust cloud. From this history, it is understandable that there are still many long-lived population II stars around, and that they amount to about 39% of the stars in our galaxy.

To determine the age of star clusters, one uses the fact that their member stars have all been born at the same time. One computes the evolution of a large number of cluster stars and plots the results in a diagram such as Fig. 1.5. By connecting the points that the stars reach after fixed times, so-called *isochrones* can be constructed. The comparison of the isochrones with the observed cluster stars gives the age of the cluster.

The most conspicuous population I star objects in our galaxy are *open star clusters* such as M67 and NGC6451 that have ages up to 10 billion years (Salaris et al. 2004). Open clusters are much less massive than globular clusters and usually contain at most several thousand stars. Population I stars such as our Sun were all created in open clusters and loose associations, that long ago have dissolved, because compared to globular clusters they are

Fig. 1.6. The elliptical galaxy M87 in Virgo. Note the spherical system of globular clusters in the halo of the galaxy, which look like very distant bright stars (courtesy of NASA)

much less tightly bound. The members escape in typical time-scales of 100 million years and become *field stars* like the Sun.

From studies of the enrichment with time of elements heavier than He (the heavier elements are called metals in astronomy) in our and other spiral galaxies, one finds that population I type abundances were present 10 billion years ago (Lineweaver 2001), consistent with the age of the oldest open clusters. One also finds that the amount of metals is somewhat higher in the inner parts of the galaxy than in the outer parts, probably because of a higher supernova rate due to the higher matter density in the inner parts. Based on this decrease of metal content with distance from the galactic center Gonzalez et al. (2001) proposed the existence of a *Galactic Habitable Zone* in which complex life does primarily flourish and about which Lineweaver et al. (2004) have elaborated. However, because new results show that the galactic metal content does not decrease in the outer parts of the galaxy between 10 and 23 kpc (Carney et al. 2005) the existence of such a Galactic Habitable Zone is not well established.

1.5.2 Dating with Radiometric Clocks

Different from establishing the ages of star clusters and stars, the methods used to date the history of our solar system are much more accurate and direct. *Radiometric clocks* (also called radioactive clocks) are particularly effective in the dating of meteorites as well as lunar and terrestrial rocks. They function by counting the decay products of radioactive isotopes. Isotopes are atoms of the same element with different numbers of neutrons. For example, the radioactive carbon isotope ^{14}C has two additional neutrons compared to the regular carbon atom ^{12}C, which has six neutrons and six protons in its nucleus. Carbon occurs on Earth as a mixture of 98.9% ^{12}C and 1.1% ^{13}C, together with tiny amounts of ^{14}C. The radioactive parent isotope decays into a daughter isotope following an exponential law in which, after a certain time-span known as the *half-life time*, half of the parent isotope atoms have decayed. This decay goes on over many half-life times until all of the parent isotopes have vanished. The measurement of the isotopes is carried out with a mass spectrometer, in which, for example, a small amount of material from a meteorite is vaporized and ionized. The ions are then accelerated in an electric potential and the different isotopes separated by a magnetic field. In this separation individual atoms can be counted and the abundance ratios of the different particles accurately determined.

The major isotopic systems used to date meteorites and rocks are shown in Table 1.3. It can be seen for instance, that the parent isotope ^{14}C decays into the daughter ^{14}N with a half-life time of 5730 years. This radiometric clock becomes unreliable when used over time spans longer than about ten half-life times, or about 60 000 years, because then only 1/1000 of the original ^{14}C nuclei are still around and can no longer be measured accurately because of contaminations of the sample. As typical meteorites surviving from the early history of the solar system are billions of years old, one would not expect to find that any trace remains of the parent isotopes of the first four systems in Table 1.3. Nevertheless, some of these isotopes are often found in meteorites, because they have been newly created in space by cosmic rays. This is also true of the ^{14}C found in biological materials, by which historical and prehistoric terrestrial objects can be dated. The ^{14}C found in these materials has been created by cosmic rays in the Earth's atmosphere.

Table 1.3. Major isotopes used as radiometric clocks

Parent isotope	Daughter isotope	Half-life time (years)
^{14}C	^{14}N	5730
^{26}Al	^{26}Mg	720 000
^{182}Hf	^{182}W	9 million
^{129}I	^{129}Xe	17 million
^{235}U	^{207}Pb	704 million
^{40}K	^{40}Ar	1.3 billion
^{238}U	^{206}Pb	4.5 billion
^{232}Th	^{208}Pb	14 billion
^{87}Rb	^{87}Sr	49 billion

Figure 1.7 shows examples of dating with radiometric clocks. The Efremovka meteorite fell in Kazakhstan in 1962, while the Acfer 059 meteorite was found in the Sahara in southwestern Algeria. Both are carbonaceous chondrites of the CV and CR types, respectively, which contain calcium-aluminum-rich inclusions (CAIs) and chondrules (see Sect. 2.8) that could be precisely dated with the Pb-Pb method to give a likely age of the solar system of 4.5672 ± 0.0006 billion years (Fig. 1.7a). Another very old meteorite dating from the earliest solar system is the Allende CV chondrite (Fig. 2.12), which fell in Mexico in 1969 and has an age of 4.566 billion years (Allègre et al. 1995).

The ^{207}Pb/^{206}Pb isotopic chronometer is based on the radioactive decay of two long-lived radionucleides ^{235}U (into ^{207}Pb) and ^{238}U (into ^{206}Pb, see Table 1.3) where ^{235}U decays about six times faster than the latter and started with a high abundance in the early solar system. In the method developed by Amelin et al. (2002) particularly precise age determinations are achieved because only the three Pb isotopes ^{204}Pb, ^{206}Pb and ^{207}Pb need to be measured. Here, the stable non-radiogenic and non-radioactive ^{204}Pb serves as a reference isotope. It can be seen in Fig. 1.7a that the age is determined by the slope of the line going through the Pb ratio measurements from several chondrules of the Acfer 059 meteorite and several regions of two CAIs of the Efremovka meteorite. The slope of the plotted lines depends only on the age and the known decay laws.

The other example (Fig. 1.7b) shows dating of terrestrial rocks that contain remnants from the earliest continental crust found in the Mount Narryer and Jack Hills regions of Western Australia. These rocks contain tiny zircon crystals that are the remains from disintegrated older rocks. In these detrital zircons the four isotopes ^{206}Pb, ^{207}Pb, ^{235}U, and ^{238}U can be measured together with the isotopes of other elements. Plotting the ratio ^{206}Pb/^{238}U against that of ^{207}Pb/^{235}U with the known decay laws the gray line shown in Fig. 1.7b is found. The numbers on that line give the age in billions of

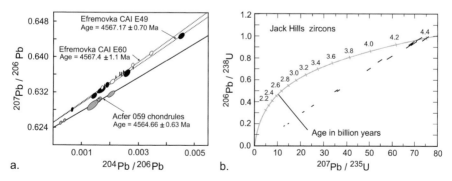

Fig. 1.7a. Radioactive isotope dating of two meteorites Efremovka and Acfer 059 (after Amelin et al. 2002), the ages are in million years, and **b.** of zircons that are remnants of the Earth's early continental crust (after Wilde et al. 2001)

years. The measurements (solid) show that the Jack Hills zircons are up to 4.4 billion years old.

As mafic crust (see Sect. 3.10) gets returned into the Earth's mantle by which process the zircons are destroyed, the survival of these early zircons proves that some continental crust must have been present since 4.4 billion years ago. Moreover, based on high $^{18}O/^{16}O$ oxygen isotope ratios measured in the zircons, Valley (2005) and Cavosie et al. (2005) showed that there must have been oceans 4.2 billion years ago. Finally, it was suggested on the basis of $^{176}Hf/^{177}Hf$ isotope ratios in zircons (Harrison et al. 2005) that continents and oceans could have existed as early as 4.4–4.5 billion years ago. When considering zircon ages it should be recalled that these crystals are merely debris of rock that long ago disintegrated by weathering and were subsequently included in younger rocks. The oldest known intact (but metamorphic, see Sect. 3.7) rock, the so-called Acasta gneisses of the western Slave Province, Canada are 4.00–4.03 billion years old (Bowring and Williams 1999). Similar extremely old rocks are found in Labrador, western Greenland (Isua Formation), and in the Barberton Greenstone Belt of eastern South Africa (see Fig. 3.24).

Dating lunar and terrestrial rocks using the $^{182}Hf–^{182}W$ chronometry (see Table 1.3) shows that the Earth underwent early and rapid accretion and core formation, with most of the accumulation occurring in about 10 million years and reaching its final mass after the giant collision of Proto-Earth with a Mars-sized impactor arriving after 30–40 million years (Kleine et al. 2005, Jacobsen 2005) that produced our Moon 4.527 ± 0.010 billion years ago (see Sects. 2.9, 3.2).

2 Planet Formation

The previous chapter briefly outlined the history of the universe, the formation of galaxies and stars, as well as how the stars develop and how, as a consequence of stellar evolution, the present cosmic abundances of chemical elements came about. Planets do not form independently; rather, their generation is an inevitable byproduct of star formation. This is because the gravitational collapse of large amounts of gas and dust from the interstellar medium unavoidably results in rotating accretion disks. These disks not only feed the growing protostars but also give rise to planets. As we are interested in the existence of extraterrestrial life and the question, whether Earth-like planets exist elsewhere in our galaxy, we now describe the present views about how planets are created. The unique properties of a terrestrial planet such as the Earth, the formation of the Moon, the conditions on the early Earth as the stage on which life made its appearance are discussed in Chap. 3. Another way to learn about planets is to conduct detailed search programs, which recently have become very successful, although they have not yet found an Earth-like planet. These planet-search methods are discussed in Chap. 4 and the question of what constitutes a planet suitable for life is addressed in Chap. 5.

2.1 Accretion Disks and Planetesimal Formation

The formation of a single star like our Sun results from the collapse of an interstellar molecular cloud core, which finally produces a rotating fragment that contains a protostar and its surrounding accretion disk (see Fig. 1.4d). In the case of our solar system, the accretion disk is called the solar nebula. As can be seen from the cross-section in Fig. 2.1, it has a fan-shaped structure that extends away from the center to several 100 AU. The astronomical unit

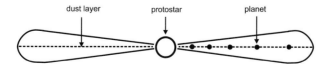

Fig. 2.1. The cross-section of a stellar accretion disk

P. Ulmschneider, Planet Formation. In: *Intelligent Life in the Universe*, P. Ulmschneider, Adv. Astrobiol. Biogeophys., pp. 19–38 (2006)
DOI 10.1007/11614371_2   Springer-Verlag Berlin Heidelberg 2006

(1 AU $= 1.5 \times 10^{13}$ cm) is the mean distance between the Earth and the Sun. The collapsing cloud continues to deposit matter onto the accretion disk, and from there feeds the protostar. Like the planets, the accretion disk rotates around the protostar, with matter orbiting more rapidly in its inner parts than in the outer regions. This is because closer to the protostar, the gravitational attraction is larger. In this so-called Keplerian disk, the centrifugal acceleration caused by the rotation balances the gravitational attraction of the star. In order for matter to move toward the protostar, therefore, its rotational motion must be slowed to diminish the centrifugal acceleration. This is achieved through friction. Since the inner material in the disk moves more rapidly than the outer material, friction arising from trying to make the motions equal slows down the inner, and accelerates the outer, material. The heat created by this friction is radiated away as infrared light and can be observed (see Figs. 4.4 and 4.5). After slowly migrating through the disk to its inner boundary, most of the disk material is then captured by the protostar, while some of it forms planets (Lissauer 1993).

The temperature in the solar nebula at the location where the planets form is of great importance, as it determines which types of material get accumulated into the planets. Resulting from frictional heating, the temperature decreases with the distance from the central star. Figure 2.2 shows the situation for the solar nebula. From the observed mean densities of the planets and moons, the materials involved in their formation can be derived. As the identified materials can only form at certain temperatures, the formation temperatures and the distance over which these materials have formed can be determined (crosses in Fig. 2.2). These empirical values fit well with a theoretical viscous planetary accretion disk model, shown by the solid line. These theoretical models assume that there is a certain mass infall rate $\dot{M}$, in grams per second. Figure 2.2 shows that accumulating 10 times more or 10 times less material (the two dashed lines) would not agree with the observations, as these rates provide too much or too little frictional heating. Note that jovian planets are not listed in the figure, as they consist mainly of hydrogen and helium, and the densities of their rocky cores cannot be measured.

During the collapse of a molecular cloud to an accretion disk, not only gas, but also large amounts of dust accumulates. Sedimenting down, the dust particles suffer friction and rapidly (in about 10 000 years) collect into a thin layer in the midplane of the accretion disk (Fig. 2.1). In this dense dust layer, static electrical forces bring the particles together: they stick to each other and form extended fluffy grain-like conglomerates. By this process, the diameter of the dust grains grows rapidly from less than 1 μm to sizes of 1 mm. From this size, larger grains grow even more rapidly as a result of electrical, magnetic and gas – solid surface interactions. It has been estimated that bodies as large as 10 m are formed in 1000 years, and 10 km comet size *planetesimals* in 10 000 years. Collisions of planetesimals lead to further growth, but also to fragmentation. After about 100 000 years, planetesimals of Ceres size (100–1000 km) are formed.

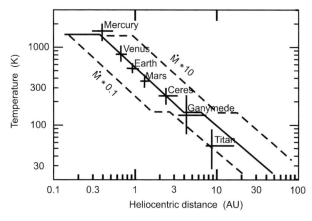

Fig. 2.2. Temperature distribution in the solar nebula (after Delsemme 1997)

As the planetesimals ultimately accumulate from neighboring grains, it is the temperature in the solar nebula that determines their chemical composition. Close to the Sun, in the relatively warm regions where the terrestrial planets form, the accumulating material consists mainly of silicate and iron grains, while very few volatile gases such as CO, H_2O, and hydrogen become incorporated into the grains. This means that most planetesimals that form the terrestrial planets contain essentially no carbon or water.

At a distance of about 3 AU from the Sun in the solar nebula, however, at what is known as the *ice-formation boundary* or *snow-line*, the temperature becomes low enough (150 K) for ice grains to form. This is important because H_2O is a very abundant molecule, as the elements hydrogen and oxygen are among the most frequent elements found in the interstellar medium (see Table 1.2). Beyond this boundary, large quantities of ice grains can easily form and are rapidly accumulated into large planetesimals. This rapid accumulation is also favored by the slow relative speed of the neighboring material orbiting the star at these distances.

2.2 Terrestrial Planets

Once the planetesimals have reached the size of Ceres, with diameters of around 1000 km, gravitational effects begin to play an additional role in their growth. Small planetesimals collide by chance, when they happen to be in each other's way, but larger planetesimals attract each other gravitationally and enforce collisions from a much wider volume around their flight path. In addition, heat created by impacts and by the decay of radioactive isotopes from the interstellar material melts the interior of some of the planetesimals and produces sedimentation of the heavy material, such as iron, into their cores. Eventually, planetesimals accumulate into planets.

Simulations have shown that the development from planetesimals to planets occurs in time spans of several 10 million years (Wetherill 1990). In a time-dependent calculation by Wetherill (1986), the motion of 500 planetesimals in their orbit around the Sun was modeled (see Fig. 2.3). These initially had masses between that of Ceres and half that of the Moon (Table 2.1), and were distributed across a distance range between 0.4 and 2 AU. In the course of the calculation, the number of planetesimals decreases due to three types of catastrophic events. First, some of them collide to form larger bodies; second, some fall into the Sun; and third, some are thrown far out to the outer boundaries of the solar system or escape from it altogether. In the latter scenario, a close encounter between two planetesimals gives one of them sufficient energy to overcome solar gravity. In the two phases of the calculation shown in Fig. 2.3, the eccentricity of the planetesimals is plotted against the semimajor axis of their elliptical orbit around the Sun.

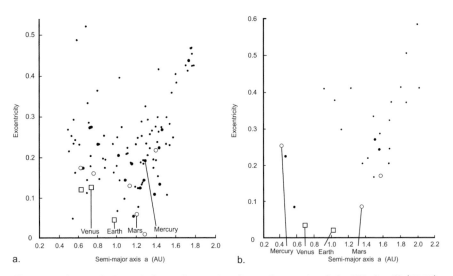

Fig. 2.3. A simulation of planet formation from planetesimals by Wetherill (1986). **a.** After 10.9 million years. **b.** After 64 million years

Table 2.1. The masses of the terrestrial planets and the Moon

	Mass (g)	Mass in Earth masses
Earth	6×10^{27}	1
Mercury	3×10^{26}	1/20
Venus	5×10^{27}	0.8
Moon	7×10^{25}	1/81
Mars	6×10^{26}	1/10
Ceres	1×10^{24}	1/6000

The significance of the eccentricity is shown in Fig. 2.4, where three orbits of eccentricity, 0.6, 0.3, and 0, are shown. The eccentricity tells us how far the focal point (occupied by the Sun) is away from the center of the ellipse. Eccentricity 0 signifies a circular orbit. Very eccentric orbits are long ellipses and, since this means crossing the orbits of many other planetesimals, they are destined to suffer frequent collisions. Eccentricity, therefore, provides a measure of how unstable the orbit is. Note that in Fig. 2.4 the semimajor axis a (one half the largest diameter) of all three orbits is the same. This quantity determines the energy of the planetesimal in the solar gravitational field.

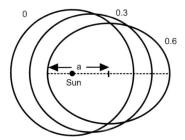

Fig. 2.4. Orbits with the same semimajor axis a but different indicated eccentricities

In Fig. 2.3 the symbols marking the individual planetesimals indicate different amounts of mass. Objects smaller than the Moon are indicated by small dots, those between the size of the Moon and that of Mercury by large dots, those between the size of Mercury and Mars by rings, and masses greater than $1/6$ of the Earth by squares. The figure shows that the number of bodies has decreased markedly with time and that their mass has become concentrated into very few protoplanets with stable orbits of very low eccentricity.

Figure 2.3a shows the calculation after 10.9 million years, by which time 60% of the mass has accumulated into the four protoplanets, which eventually develop into terrestrial planets. But note also that one additional Venus-size and five Mercury to Mars-size bodies are present, which will later be eliminated. Proto-Mercury is still almost at the Mars distance. Proto-Venus has a high-mass neighbor that will later collide with it to form Venus. Figure 2.3b shows the scenario 64 million years after the start of the calculation. Here, about 90% of the mass is in the final protoplanets. Mercury has now assumed its correct position. But there is still a Mars-size planetesimal close to the Mars orbit which has yet to find its final place. In addition to suffering the three catastrophic events mentioned above, this object could also collide with the proto-Earth to form a massive Moon.

2.3 Jovian Planets and Kuiper Belt Objects

In principle, the formation of the jovian (Jupiter-like) planets occurs in the same way as that of terrestrial planets. But the greater numbers and larger size of the planetesimals beyond the ice-formation boundary leads to protoplanets of a much larger size at those distances. Jupiter, for instance, has a silicate and ice core of about 15 Earth masses. With such massive protoplanetary cores, the gravitational attraction becomes so strong that they also accumulate gas from the solar nebula, in particular hydrogen (H_2) and helium. That increases the mass even more, which in turn magnifies the gravitation of these protoplanets, and so on. The resulting runaway growth, which took about 1–2 million years for Jupiter and similar times for the other giant planets (Alibert et al. 2005), is limited by the amount of available gas near the planet and by the time that the T-Tauri phase (see below) permits. As a consequence of these runaway processes, Jupiter, Saturn, Uranus, and Neptune grew to 318, 95, 15, and 17 Earth masses, and consist mainly of liquid and even solid hydrogen as well as helium. Using Table 2.1, Jupiter thus has a mass of $M_J = 2 \times 10^{30}$ g, or $1/1000$ $M_\odot$. Incidentally, it should be noted that the tidal perturbation from the strong gravitational attraction of Jupiter prevented the asteroids (which orbit between Mars and Jupiter) from accumulating into a single planet, and was also responsible for the low mass of Mars.

Since the cloud fragment out of which the solar system forms has only a limited amount of mass, it is clear that the material in the solar nebula must decrease with greater distance from the Sun. The planets further out are therefore smaller, as there is less material to accumulate. Eventually, at Pluto's distance, only small planets and cometary nuclei form. These so-called *Kuiper belt objects* consist mainly of ice and, due to their small mass, cannot accrete hydrogen and helium gas. Accumulation beyond the ice-formation boundary also accounts for the composition of the three large outer moons of Jupiter (Europa, Ganymede, and Callisto).

A typical calculation by Hughes (1992), in Fig. 2.5, shows the distances and types of planets that develop out of the planetary accretion disks. Here the big dots indicate jovian planets, and the small ones terrestrial planets and Kuiper belt objects. The generation of this sequence of planets with distance has also been found in simulations of Ida and Lin (2004, 2005). From Fig. 2.5, for example, it is seen that 5 terrestrial planets form up to Jupiter distances around G-type stars. That such a result is typical has recently been concluded by Jacobsen (2005) in a review of the planetesimal and planet embryo phase as well as the main planet-building stage. He reported that computer simulations of the final stages of accretion have succeeded in producing systems containing 2–4 terrestrial planets between 0 and 2 AU from the star.

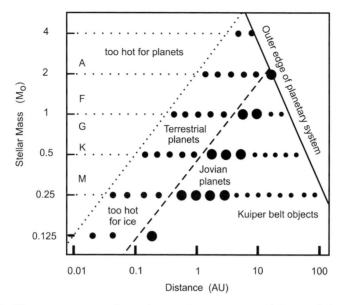

Fig. 2.5. Planetary systems formed around various types of stars (after Hughes 1992)

2.4 The Migration of Jovian Planets

Recent discoveries of many Jupiter-size planets at very short distances from their central star (see Table 4.1) seem to strongly contradict the current theories of giant planet formation. Favoring a generation beyond the ice-formation boundary at roughly 3 AU (e.g. Lissauer 1993), these theories claim that Jupiter-like planets cannot form close to the star because the terrestrial planets there would not be massive enough to initiate a runaway growth. This discrepancy between observations and theory can be resolved by assuming planet migration.

It was found by detailed simulations (Murray et al. 1998), that from their initial distance of formation, jovian planets can migrate to practically any position down to a limit of about 0.01–0.3 AU. Whether a planet migrates depends on the mass in the accretion disk and, in particular, on the amount of mass within its orbital radius, which the jovian planet must speed up in order to migrate inward. There appears to be a critical disk mass above which one has migration, while below it there is little or none.

At a distance of 0.01–0.3 AU the migration of the planet stops, since there are no more planetesimals inside its orbit. This is because at those distances the magnetic fields of the protostar and the high disk temperatures do not allow planetesimal formation. In addition, at closer distances the jovian planets probably get destroyed from photo-evaporation by the central star. Another simulation of a sample of planets with initial masses between 1 M_J

and 5 M_J by Trilling et al. (1998) suggests that migrations are frequent. They find that 50% of the giant planets are destroyed by loss of all mass due to the evaporation of their atmospheres, 33% survive and migrate to distances of less than 1 AU, while about 17% stay at distances larger than 1 AU. In one third of these, or a total of 6% of all cases, the Jupiter-like planets are not expected to migrate appreciably. In the cases where giant planets migrate into the inner parts of the system, the terrestrial planets get destroyed and there is not enough time for them to form a second time (Armitage 2003).

In recent simulations of giant planet formation Alibert et al. (2005) found that of their sample of 1000 planets all migrated. Yet for 13 planets the migration stopped at distances of more than 4.4 AU. This result is not greatly at odds with the previous estimate of 6% of planetary systems where giant planet migration does not disturb the formation of terrestrial planets. Such uncertainties in theoretical work must be expected, because the simulations depend on many model assumptions and on still poorly understood processes such as viscosity, turbulence and magnetic fields as well as the effect caused by a vigorous T-Tauri wind (Fig. 2.6c) that makes the migration more difficult by sweeping away the remaining gas and dust in the disk.

Moreover, the fact that only about 4% of the investigated systems are observed to have Jupiter-like planets in close orbits (Chap. 4) seems to indicate that systems with nonmigrating Jupiters might be more frequent. This could be experimentally tested if twins of our solar system were discovered where Jupiter-size planets occurred only at great distances. As detection of Jupiter-like planets at 5 AU requires observation times of at least 10 years, it is not surprising that so far only one such outlying planet (55 Cancri d, at 5.26 AU) has been found. However, this system is quite different from the solar system because it has an additional closely orbiting Jupiter at 0.12 AU. Recently 4 systems with giant planets at distances of 4.15–4.3 AU were discovered of which two seem to have only one jovian planet (Schneider 2005). There is a consensus that in a few years more systems of this type will be detected.

2.5 The T-Tauri Stage

The final stages of the accretion disks occur when the star reaches the end of the so-called *T-Tauri stage*. In our earlier discussion of the pre-main sequence evolution (Chap. 1), it was mentioned that on its way to the main sequence a protostar on the Hayashi track develops a deep convection zone. A characteristic property of stars in this phase of their development is that, by accumulating matter from the parental nebula, they rotate rapidly. It is the combination of rapid rotation and convection that generates strong and very extended *magnetic fields* in a protostar, with two important consequences. First, the fields produce very energetic jets with massive gas flows away from the star along its rotation axis. Second, they generate a massive stellar wind that exits from isolated patches all over the star.

In the earliest T-Tauri stage, jets break out at the polar regions of the accretion cocoon around the protostar and form oppositely directed highly collimated outflows (see Fig. 2.6a). Because of these jets, and with the additional help of the magnetic fields, about 2% of the infalling matter is ejected right back into interstellar space, without ever reaching the stellar surface. But these jets also act as a "brake" on the star, decreasing not only the rapid rotation produced by the large amount of infalling mass from the innermost region of the accretion disk, but also the rotation resulting from the star's contraction during its pre-main sequence evolution.

In addition, there is a strong stellar wind, called the T-Tauri wind, which is stalled by the accretion disk and forced to flow out at the polar regions alongside these jets. In the later phases of the T-Tauri state (e.g. Kitamura et al. 1997), the accretion slows, the jets decline, and the wind becomes more important. At this point the outflow region widens (Fig. 2.6b), as the accretion disk has essentially collected all of the matter from the original cloud fragment. Deprived of further mass infall, and depleted from continued planetesimal growth, the disk becomes thin. Eventually, the energetic T-Tauri wind sweeps the remaining gas and dust out of the system, leaving behind only a protoplanetary disk with planetesimals, planets, and Kuiper belt objects (Fig. 2.6c). It is thought that at this stage the protoplanetary disk had of the order of 10^{12} planetesimals of 1 km size or larger. The time scale for the final disappearance of the accretion disk is roughly 3–4 million years (Haisch et al. 2001, Briceño et al. 2001), starting with the clearing of gas and dust in the inner regions and then progressively expanding to the outer parts of the system.

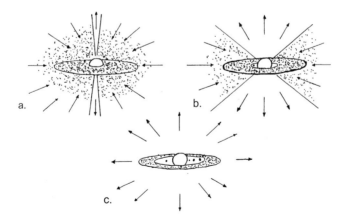

Fig. 2.6. Jets and T-Tauri stages. **a.** Cocoon with jets along the rotation axis. **b.** T-Tauri wind generates a bipolar outflow. **c.** T-Tauri wind has blown away the gas and dust of the accretion disk leaving behind the protoplanetary disk

2.6 Asteroids

Since close to the ice-formation boundary there were many more ice plane-tesimals (Fig. 2.5) one might expect larger terrestrial planets with increasing distance from the Sun. This picture is wrong, as demonstrated by Mars that has only 1/10 of the Earth's mass (see Table 2.1). Even stronger counterexamples are the asteroids, or minor planets – small planetary bodies – that circle the Sun in a belt mainly between the orbits of Mars and Jupiter (Fig. 2.8). A typical asteroid, Ida, with its very irregular shape is shown in Fig. 2.7. Covered with dust and craters it has a size of 56 km × 24 km × 21 km. Surprisingly, it also has a small moon, Dactyl, with a diameter of about 1 km (Fig. 2.7). Starting with the first asteroid, Ceres, found in 1801, followed by Juno, Pallas, and Vesta, we know now more than 120 000 asteroids with determined orbital parameters (see Minor Planets 2005).

It is believed, however, that these constitute only a small fraction of all asteroids. Probably many millions of bodies exist with diameters larger than 100 m. Table 2.2 displays the diameters, masses, rotation rates, and orbital parameters for a sample of asteroids. This list contains the four largest objects and several others that have recently been mentioned in space missions. A plot of the instantaneous position of 19 945 asteroids at distances less than 7 AU from the Sun is shown in Fig. 2.8. As can be seen, the vast majority of the orbits are in the distance range of 1.5–3 AU.

Also shown in Fig. 2.8 are the so-called *near-Earth asteroids* (NEAs). These approach the Sun to distances closer than 1.3 AU. For a list of the three types of near-Earth asteroids, the *Aten asteroids*, *Apollo asteroids* and *Amor asteroids*, see Minor Planets (2005). As of December 2005, more than

Table 2.2. Selected data for the largest asteroids and those mentioned in recent space missions. Columns a and Ecc. give, respectively, the semimajor axis and the eccentricity of the asteroid orbits around the Sun (after Minor Planets 2005)

No.	Name	Diameter (km)	Mass (10^{15} kg)	Rotation period (h)	Orbital period (y)	a (AU)	Ecc.
1	Ceres	960 × 932	870 000	9.08	4.60	2.77	0.08
2	Pallas	570 × 525 × 482	318 000	7.81	4.61	2.78	0.23
3	Juno	240	20 000	7.21	4.36	2.67	0.26
4	Vesta	530	300 000	5.34	3.63	2.36	0.09
243	Ida	56 × 24 × 21	100	4.63	4.84	2.86	0.05
253	Mathilde	66 × 48 × 46	103.3	417.7	4.31	2.65	0.27
433	Eros	40 × 14 × 14	5	5.27	1.76	1.46	0.22
951	Gaspra	19 × 12 × 11	10	7.04	3.29	2.21	0.17
1566	Icarus	1.4	0.001	2.27	1.12	1.08	0.83
1862	Apollo	1.6	0.002	3.06	1.81	1.47	0.56
2060	Chiron	180	4000	5.9	50.7	13.63	0.38
4179	Toutatis	4.6 × 2.4 × 1.9	0.05	130.0	1.10	2.51	0.63

Fig. 2.7. The asteroid Ida with its moon Dactyl (courtesy of NASA)

3700 NEAs are known. It is estimated that most of the NEAs with diameters greater than 1 km are now known, but that there are more than 500 000 with sizes larger than 100 m (see Chap. 9).

That there should be a planet at the distance of the asteroids has been assumed already before 1801. In 1766 J. D. Titius discovered a simple arithmetic progression of the mean distance, a, of the planets from the Sun that, published by J. E. Bode in 1772, was henceforth called the Titius–Bode law. This law, written today with $k = 0, 1, 2, 4, 8, 16, 32, 64, 128$ as

$$a(\text{AU}) = 0.4 + (0.3 \times k) \, ,$$

predicts the distances of all planets (except Neptune) rather well, including even that of Pluto discovered in 1930. A planet at a distance $a = 2.8$ AU was therefore expected, but not such a multitude of minor planets. The reason why there are so many asteroids and why Mars has such a low mass is attributed to the tidal forces exerted by Jupiter. The asteroids can be seen as parts of a failed planet, where the accretion process was severely disturbed because the contributing planetesimals were deflected by Jupiter.

It is significant that the chemical and mineralogical composition of asteroids varies strongly. By measuring the reflected sunlight in three spectral bands (ultraviolet U at 350 nm, blue B at 480 nm, and visual V at 530 nm), one can classify asteroids by their color indices; that is, by the color differences U-B and B-V. Plotting these differences in a color diagram (see Fig. 2.9, left panel), the asteroids can be divided into different groups, the C, E, S, and M types. In addition, there are D and P asteroids which, like comets, are very dark. Most of our knowledge of the chemical composition

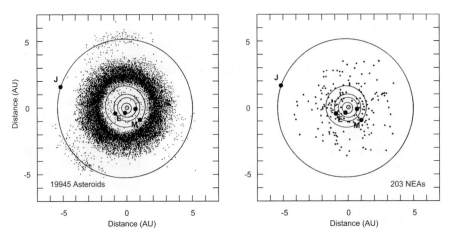

Fig. 2.8. The positions of 19 945 asteroids (*left*), together with Jupiter (J), Mars (M), Earth (E), and the inner planets at a given instant of time. *Right:* positions of 203 asteroids that cross or approach the Earth's orbit, the so-called Near-Earth Asteroids (see Minor Planets 2005)

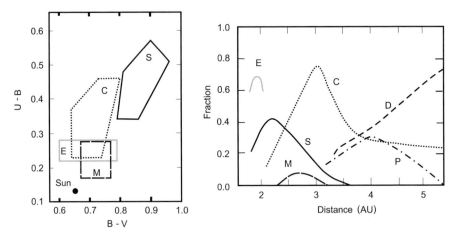

Fig. 2.9. A color diagram of asteroids (*left*) and the fractional distribution of the asteroid types with distance (*right*) (after Kowal 1988)

of asteroids comes from the study of their messengers sent to Earth, the meteorites.

By comparing the observed spectral characteristics of asteroids with those of meteorites (meteorite types are discussed in Sect. 2.8), one can determine the presumed chemical composition of asteroids. C asteroids have very similar reflection properties to chondritic meteorites, and some of them probably have considerable amounts of ice. These asteroids are very dark and are closely related in their composition to other dark types, such as the D and P asteroids that are carbonaceous, covered by soot and tar-like (organic) compounds. As shown in Fig. 2.9, (right panel), the C, D, and P asteroids mostly occur at great distance, near the ice-formation boundary. The S asteroids are probably made of the same material as achondritic meteorites, and are rich in olivine. The M asteroids very likely consist of metal similar to the iron meteorites, while the E asteroids appear to be rich in enstatite, which occurs in ordinary chondritic meteorites. The three types, S, M, and E, populate mainly the inner asteroid belt. Other objects that come to the inner parts of the solar system and could be of commercial interest are comets. They are thought to be composed mainly of ice mixed with rocks and covered with carbon compounds.

2.7 Comets

Unlike asteroids, comets have been known since antiquity (see Comets 2005). The earliest written records of comets in China and Mesopotamia date from around 1000 BC. In Greece, the Pythagoreans around 550 BC considered comets as a kind of migrating planet seen mostly near the horizon in the morning or evening sky, yet in 350 BC Aristotle (Meteorology 344a) took them for some upper-atmosphere phenomenon. The number of detected comets is in the thousands and grows by more than 100 per year, mostly found with space probes such as SOHO (see Minor Planets 2005). By calculating their orbits one finds that some of them (if not most) are periodic, of which 164 are short-period comets with orbital periods of up to 200 years, and about 1500 have longer periods. A well-known short-period example is Comet Halley (Fig. 2.10b), with a 76-year period, of which Chinese records go back to 240 BC. An outstanding feature of comets is their extended tail (see Fig. 2.10a) that results from outgassing caused by heating when the comet comes close to the Sun. These tails are composed of ions, gas and dust and are driven away from the comet body by radiation pressure from the Sun and by the solar wind.

The comets are thought to be composed mainly of dust and rocks baked together with considerable quantities of ice, covered with carbon compounds that are left behind after the evaporation of water. This explains the very dark surface of Comet Halley observed from space craft *Giotto* in 1986 (Fig. 2.10b). With a surface darker than coal, this comet is actually one of the darkest objects in the solar system. Comet Halley has a size of $16 \times 8 \times 8$ km, which is about 1/3 that of the asteroid Ida (Fig. 2.7).

The history of comets proceeded in three phases. In the first, the comet nuclei or ice planetesimals were formed in the solar nebula, beyond the ice-formation boundary at about 3 AU and extending out to the Kuiper belt, which ends at about 300 AU (Fig. 2.11, dark gray). As mentioned earlier, the outer limit of this region is determined by the lack of accreting material, and might additionally be truncated from a stellar encounter.

Fig. 2.10. a. Comet Hyakutake in 1996 with its extended dust tail. **b.** Comet Halley observed with space craft Giotto in 1986 from a distance of 596 km (courtesy of NASA and ESA)

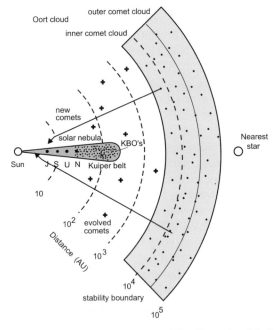

Fig. 2.11. The origin and history of comets. The Oort cloud is indicated as well as the solar nebula with the jovian planets Jupiter, Saturn, Uranus, Neptune, and the Kuiper belt together with the Kuiper belt objects (KBO)

In the next phase, the ice planetesimals collided with each other to form the cores of the jovian planets that subsequently grew to the present-size bodies by runaway accretion of H_2 and He. In this process not all the planetesimals got absorbed by these planets or fell into the Sun, but a large number of them were ejected toward the outer edge of the solar system into the *Oort cloud*, which is a spherical system at a distance of 5000–100 000 AU and is loosely bound to the solar system (Fig. 2.11, light gray). Many comets, however, were also deflected into the inner solar system, where they collided with the terrestrial planets, and – as a result of their impact and breakup in the atmosphere – enriched the surface composition of these planets. The final phase of comet history happens much later when the occasional passage of a star at distances larger than 200 000 AU (essentially the distance to our nearest stellar neighbor α Centauri) disturbs the Oort cloud and sends comets into the inner solar system, where they can be observed and eventually decay by outgassing, fragmentation, impact on planets or the Sun (see Fig. 2.11, crosses).

Until very recently it was thought that the terrestrial planets got most of their carbon and water from comets (Delsemme 1997). Despite the fact that H_2, He, H_2O, CO and N_2 were the most frequent molecules in the solar nebula, their presence on Earth is not trivial, because they are highly volatile and did not get incorporated easily into planetesimals of the inner regions of the solar system. New spectroscopic observations of comet tails show that the deuterium/hydrogen ratio is a factor of two higher than in sea water, which excludes comets from a main role in the delivery of water to the terrestrial planets. Water apparently derived from the release of trapped H_2O in the crystal structure of accreted silicates and as ice from those 10–15% of accreted planetesimals with eccentric orbits that had formed near the ice-formation boundary (Rodgers and Charnley 2002).

2.8 Meteorites

Meteorites are stone-like or metal objects that drop from space to the Earth's surface and have various origins. They constitute primarily rubble created from collisions between asteroids, sometimes surviving debris from the formation of the solar system, and very seldom material ejected from impact events on the Moon or Mars. That asteroid collision is the predominant origin of meteorites has been confirmed by the fact that when it is possible to reconstruct the meteorite orbit before the impact on Earth it invariably extends to the asteroid belt. Typically, therefore, the composition of meteorites is the same as the material found in asteroids. The roughly 30 000 meteorites collected so far can be divided into two groups: "finds", which are unusual rocks or clumps of metal collected on the surface (recently particularly in deserts and on snowfields, where they are easily recognized) and "falls" when the objects are retrieved after an observed fall has been traced to the impact site. Meteorites of the "find" category are noticed only by their unusual color, consistency, or weight, while meteorites of the "fall" group comprise all occurring types.

As displayed in Fig. 2.12, there are three main types of meteorites. First, and easily recognized due to their dark color and great weight, are the *iron meteorites* (top left), which make up 5% of the falls and 40% of the finds. They typically consist of up to 99% of an Fe–Ni–Co alloy, with iron being predominant. Second, *stony-iron meteorites* (top right), which make up 1% of the falls and 4% of the finds, consist of about 50% metal (Fe, Ni, and Co). The third type, the *stone meteorites* (bottom left) make up the largest fraction. They appear in two sub-types: *chondrites*, constituting 86% of the falls and 52% of the finds; and *achondrites*, comprising 8% of the falls and 4% of the finds. The achondrites are very similar in appearance to terrestrial igneous rocks with various amounts of calcium-rich minerals, and possess a glass-like baked consistency. In contrast, chondrites contain different numbers of *chondrules*, silicate spheres of about 1 mm size, set in a very fine-grained (μm sized) background material called the *matrix* that has various compositions.

There are three types of chondrites. Ordinary chondrites make up 93% of our chondrite meteorite inventory with the sub-types H (high), L (low) and LL (low iron, low metal), classified for their iron content. About 2% are enstatite chondrites of types EH and EL, named after the Fe-free Mg-silicate enstatite, a pyroxene mineral (see Sect. 3.7). Most interesting are the 5% carbonaceous chondrites, because their matrix contains large amounts of highly volatile elements like water and carbon with up to 6% carbon compounds such as hydrocarbons, amino acids and tars from which they get a black appearance. The sub-types CI, CM, CO, CK, CV and CR are named after prominent members of their group. Figure 2.12d shows a cut through the CV-type Allende meteorite displaying a high proportion of chondrules. Other carbonaceous chondrites like the CM and CK types have fewer chondrules and the CI type has none. Figure 2.12d shows that in addition to the chondrules and the matrix, these meteorites have a third distinct constituent called *calcium-aluminum-rich inclusions* (CAIs), they appear as white irregular-shaped specks (center and top left).

2.9 Early History of the Solar System

The reason why this bewildering variety of meteorites and their constituents are discussed here is that these details allow us to unravel the sequence of events of the early history of the solar system. They tell us about the formation of planetesimals, asteroids and planets, although some parts of this picture are still poorly understood. From dating with the Pb-Pb method and other radiometric clocks (see Sect. 1.5) it is found that the CAIs are the oldest objects in the solar system (Fig. 2.14). During their formation, temperatures close to 2000 K must have occurred and it is presently still not clear how they were formed (Trieloff and Palme 2005).

Of similar age but somewhat younger are the chondrules. Their origin is also still in discussion, although it appears to be increasingly certain that chondrules and CAIs are produced by melting in the high-temperature zones

Fig. 2.12. The three main types of meteorites, irons (*top left*), stony-irons (*top right*) and stones (*bottom left*). A cut through the Allende meteorite (*bottom right*) shows the three main constituents of a chondrite, a type of stone meteorite: the big white specks are so-called calcium-aluminum-rich inclusions (CAIs), the sand corn like chondrules and a background of darker fine-grained matrix (courtesy of NASA/JPL)

behind powerful shock waves that occurred in the solar nebula. Shock waves are known on Earth from nuclear explosions and on the Sun from *flares* that produce so-called *coronal mass ejections* (CMEs) by which huge amounts of solar material are thrown into the interplanetary space.

A reasonable picture of the formation of chondrules has recently been given by Nakamoto et al. (2005) who pointed out that from T-Tauri stars flares are observed in X-rays that are not only 100 times more energetic than solar flares but also occur 1000 times more frequently (roughly one per day, Feigelson et al. 2002). Flares occur when highly stressed magnetic fields almost instantly release their energy by reconnection of the magnetic field lines. Stars (or planets) possess magnetic fields when appreciable rotation and convection are both present. As mentioned in Sect. 2.5, this is particularly true for pre-main sequence stars on the Hayashi track, such as T-Tauri stars, that have deep convection zones and rapid rotation.

Figure 2.13 shows the computer modeling of such a shock wave, generated by a powerful flare in the magnetosphere of a rapidly rotating protostar that has an accretion disk. This disk is indicated in Fig. 2.13a at the bot-

tom of the picture as an undisturbed region that increases in thickness with
distance from the protostar. The star itself sits at the lower left corner of
the figure and is too small to be drawn. From the flare that occurs close to
the star a powerful shock wave travels primarily upward into the thin region above the accretion disk where dust and gas accretion still goes on. It
is seen that in this blast wave, temperatures as high as 2 million degrees K
are reached but that the accretion disk is not much affected due to its high
matter density.

Figure 2.13b represents a somewhat later cut-out from Fig. 2.13a at the
distance of 1 AU that shows the interaction of the shock wave with the accretion disk. The accretion disk is displayed at the bottom together with layers
of decreasing density toward the top. The shock wave is seen as a discontinuity that is slanted because it runs more quickly in the thinner medium
above. In the high-temperature region behind the shock wave, dust particles
caught in their downfall to the accretion disk from the surrounding nebula
suffer various fates. Small particles of less than 1 µm size are completely
evaporated, while grains of greater size up to 1 mm get melted and become
chondrules. But particles deeper in the accretion disk are shielded and do
not get affected. These different fates, the short duration of the shock events
and the fact that the T-Tauri stage (where an accretion disk is still present)
lasts only for about 3–4 million years, allows to reproduce the age and size
distribution measured for chondrules rather well.

Important for the history of planetesimal formation was the admixture
of the short-lived radioactive isotope ^{26}Al (and less importantly ^{60}Fe) to the

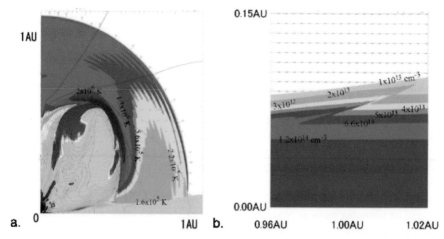

Fig. 2.13. Flare-generated interplanetary shock wave around a protostar with an
accretion disk. **a.** Temperature distribution 6.665 days after the flare event. The
wave has reached distances of 1 AU from the star. **b.** Small cut-out near 1 AU of
the former figure showing the density layering near the accretion disk and the shock
(after Nakamoto et al. 2005)

material. The origin of this isotope, with a half-life time of about 720 000 years is still debated. Its source is thought to be either a close supernova, or nearby massive stars in the late phases of the AGB evolution (see Sect. 1.4), where the winds bringing material from the nuclear-burning zones into the interstellar medium. That a supernova could have happened close to the solar system is not surprising because the system formed together with others in an open star cluster, where massive cluster members could have evolved to a supernova stage within a million years. However, in view of the short decay time scales involved, it is also likely that ^{26}Al was produced by high-energy flare events in the magnetosphere of the protostar itself.

The radioactive decay of ^{26}Al inside the planetesimals generated heat. While dust and small planetesimals quickly radiated this heat into space, it got trapped in larger planetesimals. In bodies with sizes of several 100 km or larger, the heat produced strong melting, whereby due to gravity the heavy metallic iron moved to the center, creating a shell structure with a metallic core and a mantle consisting of silicate rocks. *Differentiated planetesimals* with molten cores are therefore among the earliest objects in the solar system (Fig. 2.14), which is confirmed by recent Pb-Pb dating (Baker et al. 2005).

In such large molten bodies the composition of the core and mantle was severely changed from the original material, while in small planetesimals melting did not occur due to the heat loss to space. From Pb-Pb, Al-Mg dating

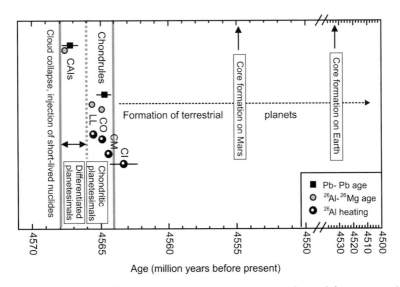

Fig. 2.14. Time scale of events in the early solar system derived from meteorites. CAI and differentiated planetesimal formation was followed by chondrule and chondritic planetesimal formation providing the parent bodies of the LL and other ordinary chondrite meteorites as well as of the CO, CM and CI carbonaceous chondrite meteorites. Also shown are the times of formation of the cores of Mars and Earth (modified after Trieloff and Palme 2005)

and from ^{26}Al heating effects (Fig. 2.14) one finds that together with the chondrules, the *chondritic planetesimals*, the main parent bodies of the chondritic meteorites, that is, of the LL and other ordinary chondrites as well as the CO, CM and CI carbonaceous chondrites, formed about 2–4 million years after the CAIs (Trieloff and Palme 2005). Therefore a swarm of many hundreds of planetesimals of different sizes and in various stages of melting, differentiation and chemical fractionation constituted the parent bodies of the meteorites.

Subsequent collisions between these parent bodies created huge numbers of smaller bodies down to the size of grains. Iron meteorites derive from debris of a metallic core. Stony-iron meteorites with their high metal content originate from the core–mantle regions of large bodies or from the center of smaller planetesimals, where due to a lower gravity, the differentiation was incomplete. Of the stone-meteorites the achondrites with their pronounced melting also derive from inner-mantle regions, while the chondrites originate from smaller or later planetesimals.

As the planets are formed by accreting predominantly the local planetesimals we ask which meteorites are particularly suited to tell us about the original composition of the solar nebula in the region where terrestrial planets formed? Clearly meteorites that came from highly modified regions of their parent bodies cannot show the original state. This leads us to disregard iron- and stony-iron meteorites as well as achondrites. Only chondrites, where the composition of the matrix is relatively unchanged are good sources of information. Measuring the abundances of elements such as Al, Ca, Si, Mg, Fe, Na, O, S and Zn in these meteorites one can classify the different types of chondrites by using individual abundance ratios. These ratios are then compared with the solar ratios (Palme and Jones 2005). Because the Sun retains the element abundances of the solar nebula, those chondrites that show the best agreement with the solar ratios are closest to the original nebula composition. It turns out that only the CI chondrites give a good agreement for all ratios.

Therefore the CI carbonaceous chondrites are thought to be our best indicators of the composition of the bodies that originally occupied the inner regions of the solar nebula. This composition is displayed below in Table 3.2. However, this composition is not close to that of the rocks of the Earth's mantle because of the high proportion of volatile elements. These volatile elements were lost in the accretion of the terrestrial planets.

3 The Earth

The previous chapter briefly outlined how from an accretion disk – the end product of a collapsing interstellar molecular cloud core – a protostar and a planetary system develops. Three types of planets are formed: in the inner regions of the system medium-sized terrestrial planets develop that are differentiated in a rocky mantle and an iron core, further out starting at about 3 AU one has giant-size jovian planets composed mainly of H_2 and He but with a rock and ice core much larger than the terrestrial planets, and finally at great distance there are small Kuiper-belt objects composed mainly of ice mixed with dust. In addition, the collision of planetesimals creates huge numbers of asteroids and meteorites.

As very likely life is closely tied to terrestrial planets it is necessary in the present chapter to look more closely at the basic structure and composition of the singular planet Earth where life abounds. Important properties of our planet are essential for our type of life. Plate tectonics, for instance, the phenomenon of continental drift, helped to produce continents and the division of land and sea. It is very likely that on a planet covered completely by a deep ocean, our type of intelligent life that developed on land and is capable of conducting interstellar communication and travel to the bodies of the solar system would not be possible. In addition, the formation of the Moon, and the conditions on the early Earth are discussed, because this was the stage on which life made its appearance. The conditions necessary for life on terrestrial planets are summarized in Chap. 5, while the question of how life and our type of intelligence came into being is outlined in Chaps. 6 and 7, respectively.

3.1 Planetological History of the Early Earth

Chapter 2 described how the collapse of the core of an interstellar giant molecular cloud eventually resulted in the formation of a protostar with an accretion disk in which a large number of planetesimals of sizes between a few m and thousands of km developed. In the inner regions of the accretion disk planetesimals with diameters of 100 km and more, due to the radioactive decay of isotopes such as ^{26}Al, developed differentiated interiors with a silicate mantle and a liquid-iron core. The collisions that led to the formation of

P. Ulmschneider, The Earth. In: *Intelligent Life in the Universe*, P. Ulmschneider, Adv. Astrobiol. Biogeophys., pp. 39–72 (2006)
DOI 10.1007/11614371_3   Springer-Verlag Berlin Heidelberg 2006

protoplanets accentuated this differentiation by combining cores and mantle of the planetesimals but also due to the strong heating that resulted from the release of the impact energy.

Since the accretion disk lasted only for about 3–4 million years the formation of the terrestrial planets, as shown in Fig. 2.14, was mainly due to the development in the planetesimal disk (Sect. 2.2). As discussed in Sect. 1.5 the accumulation of the Earth by planetesimals essentially ended about 30–40 million years after the start of the solar system 4.567 billion years ago. However, the subsequent phase of heavy bombardment lasted for at least 600 million years from the time where the planet embryo formed around 4.56 billion years ago until the end of the heavy bombardment phase at about 3.9 billion years ago. This time was characterized by the collisions of very large Ceres-, Moon- and even Mars-sized bodies in the first few tens of millions of years (Agnor et al. 1999, Jacobsen 2005) and later by many giant ocean-vaporizing impacts (Figs. 2.3, 3.1, 3.4).

3.2 Formation of the Moon

As Table 2.1 shows, it takes about 6000 Ceres-sized planetesimals to form a planet of Earth-like dimensions. Of course, the vast majority of the planetesimals involved in the Earth's formation were much smaller than Ceres, but occasional impacts with large planetesimals must have occurred. Such very large impacts may also account for the obliquity of the rotation angle of Uranus, and may help to explain the existence of the Moon.

A central question surrounding the Moon's formation is why it has a density half that of the Earth and possesses no iron core. The most likely explanation is that the Moon was formed by a collision of a Mars-size planetesimal with the proto-Earth. Figure 3.1 shows a simulation of such a collision, which is thought to have occurred 4.527 billion years ago (see Sect. 1.5). In this gi-

 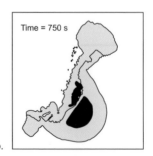

Fig. 3.1a,b. The initial phases of a collision of the proto-Earth with a Mars-size planetesimal (after Benz et al. 1989)

gantic collision, the two metallic cores eventually merged and the light mantle material was thrown out to form the Moon.

Work by Ida et al. (1997) on the later stages of this event (see Fig. 3.2) suggests that after impact a hot torus-shaped silicate debris cloud could have formed around the Earth, and that most material must have fallen back toward the planet. At the same time, a disk consisting of a large number of planetesimals would have formed, out of which one or two moons could have developed by accretion. Depending on the location at which the impact took place, it would either have led to an enhanced total rotation, leading to the formation of two moons (Fig. 3.2b), or a decreased rotation, leaving only one moon (Fig. 3.2a). The work of Benz et al. (1989) and Ida et al. (1997) was largely confirmed by recent more detailed simulations of Canup (2004) who showed that the composition of the Moon is in good agreement with an off-center impact and that on average more than 80% of the Moon's composition comes from the part of the impactor's mantle that escaped the direct hit with the Proto-Earth.

It can be seen in Fig. 3.2a that the Moon formed rather close by, at a distance of about 22 000 km or 3.6 Earth radii from the Earth's center. It was due to tidal interactions (see Chap. 5) that the Earth's rotation rate subsequently decreased from a 5-hour to a 24-hour day, and the Moon became tidally locked and moved to today's distance of 63 Earth radii. Moreover, the Moon itself shows indications of later giant impacts, which were caused by Ceres-size bodies.

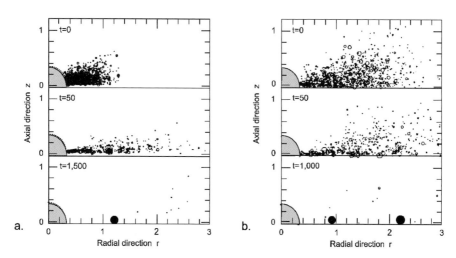

Fig. 3.2. Later phases of formation of the Moon. **a.** A reduced total rotation of the disk leads to the formation of one moon, **b.** an enhanced total rotation to two moons. The distances r and z are given in multiples of 20 000 km, and the times t in multiples of 7 h (after Ida et al. 1997)

3.3 Ocean-Vaporizing Impacts

In the very early phases of the formation of the Earth, when the Moon was created and very large Ceres-type impacts still occurred, liquid water did not yet exist on Earth due to the high surface temperature. But later on oceans developed, and with them the possibility of life. However, there always remained the danger of another giant impact, which could lead to their complete vaporization. This requires bodies with a diameter of at least 500 km; that is, with masses larger than 1/10 that of Ceres. A simulation by Zahnle and Sleep (1997) of an ocean being vaporized through such an impact is shown in Fig. 3.3. Both the impacting body and a large amount of terrestrial rock get vaporized, leading to the formation of a dense hot (100 atm, 2000 °C) rock vapor atmosphere. Its heat radiation causes the oceans to vaporize in a few months (see Fig. 3.3), adding superheated steam to the atmosphere. This heavy atmosphere, however, would not be able to escape into space. The denuded surface of the Earth is subsequently heated to a temperature of 1500 °C for about 100 years. In this situation, all previously formed organic compounds or simple life forms would be destroyed. There is the possibility, however, that by such impacts, rocks with traces of life might be ejected into

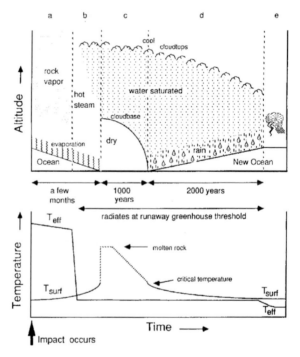

Fig. 3.3. The time development of an ocean-vaporizing impact (after Zahnle and Sleep 1997). T_{eff} indicates the temperature of the cloud covered Earth as seen from space while T_{surf} displays the terrestrial surface temperature

orbit and might re-seed the Earth after the event. In its topmost regions, the atmosphere cools by infrared radiation into space. By atmospheric convection and rainfall the surface slowly cools and over the next 900 years the cloud base sinks down, first slowly but then rapidly (Fig. 3.3). Heavy rains form new oceans that within 2000 years attain their old depth.

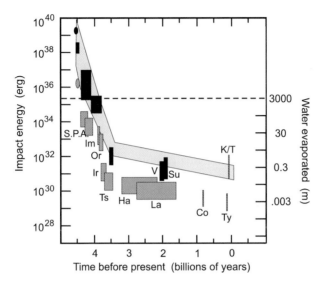

Fig. 3.4. Impact events on Earth (*solid*) and the Moon (*dark gray*) as a function of time (after Zahnle and Sleep 1997). The boxes marked S.P.A., Im, Or, and Ir indicate impacts that formed the lunar mares South Pole Aitken, Mare Imbrium, Orientale, and Iridium. The other symbols denote impacts that formed the craters Tsiolkovsky, Hausen, Langrenus, Copernicus, Tycho, Vredevort, Sudbury, and Chicxulub (responsible for the K/T boundary event)

Figure 3.4 shows the energy of impacts on Earth and the Moon, the black boxes representing terrestrial and the gray ones lunar events. The size of the boxes indicates the uncertainty of the estimated date and impact energy of the events. Ovals mark the energies associated with the formation of the Earth and the Moon. Because the Earth has 81 times the mass and 13 times the cross-section of the Moon, it has a much higher chance (96%) of sustaining an impact by comparison with the lunar surface (4%). In addition, the impact energy on Earth is about a hundred times greater than for similar events on the Moon. As can be seen by the amount of evaporated water in Fig. 3.4, where the dashed line gives the mean depth of the oceans, terrestrial impacts similar to the mare-forming events on the Moon must have been ocean-evaporating. And although such giant events ceased about 4.2 billion years ago, subsequent impacts on a smaller scale still evaporated considerable fractions of the oceans. Note that the last of these major impacts, which was responsible for the K/T boundary event (Chap. 7), evaporated 30 cm of the Earth's oceans.

3.4 The End of the Heavy Bombardment

Apart from the jovian planets (Jupiter, Saturn, Uranus, and Neptune), which have liquid surfaces, the planets and moons in our solar system are all covered by impact craters, with the single exception of Io, whose surface has been heavily reworked by volcanism. One has craters of all sizes, but the small ones predominate. The crater sizes follow a so-called power law. In a wide range of sizes one finds that there are eight times more craters every time the diameter is halved. Almost all of them were produced from planetesimals dating from the early phases of our solar system. Thus the face of the early Earth is preserved in graphic detail on the Moon, where the extreme diversity of the cratering events can be easily observed (see Fig. 3.5). Lunar counts and dating results, based upon the Apollo missions (see Fig. 3.6a), reveal that between 4.1 and 3.8 billion years ago a sharp decrease of the heavy bombardments occurred. Recently, there is mounting evidence that the planetary accretion phase was completed so rapidly that a low impact rate between 4.4–4.0 million years ago occurred, which was followed around 3.9 billion years ago by a phase of late heavy bombardment by impactors that were probably created by collisions in the asteroid belt. Figure 3.6b after Valley et al. (2002) shows this behavior (see also Strom et al. 2005, Kleine et al. 2005 and Jacobsen 2005). This indicates that by 3.8 billion

Fig. 3.5. The Moon showing the Mare Nubium, top is south

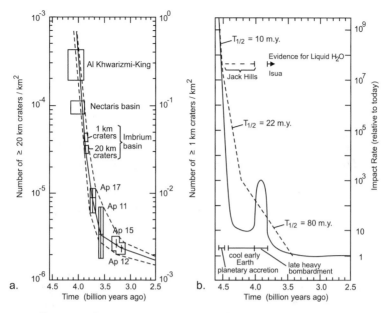

Fig. 3.6. Cratering frequencies as function of time from lunar studies. **a.** after Carr et al. (1984), **b.** modified after Valley et al. (2002) (with dashed curve from Hartmann et al. 2000)

years ago the formation of the terrestrial planets was completed, and that further development now occurred as a consequence of internal and atmospheric processes, as well as the evolution of the Sun. Note, that about 3.9 billion years ago a belated ocean-vaporizing impact might have occurred on Earth (Cohen et al. 2000), the origin and magnitude of which is still debated.

3.5 The Environment on the Early Earth

In the early evolution of the Earth, during the so-called Hadean era, which lasted from 4.56 until 4.0 billion years ago, an essential part was played by the heating effects of radioactive isotopes and the gravitational energy of the impacting planetesimals, both of which resulted in melting and the sedimentation of iron into the hot molten core. The lighter material remained in the Earth's mantle. Originally, most of the outer parts of the mantle consisted of hot liquid rock, a magma ocean devoid of any trace of organic material or life (Nisbet and Sleep 2001, Agee 2004). The primordial Earth was surrounded by a dense steam atmosphere derived from mantle evaporation and outgassed volatile material from planetesimals. By profusely radiating heat into space and by generating rain that poured down before being evaporated again, the atmosphere efficiently cooled the underlying planetary surface. A thin crust of solid rock eventually formed, which was repeatedly torn apart by volcanism and infalling planetesimals (see Fig. 3.7). After the surface temperature became low enough, presumably large oceans developed, which covered most of the Earth's surface probably as deep as today's oceans, which have a mean depth of 3000 m.

Fig. 3.7. Early Earth when the solid crust was forming (courtesy of HM Stationary Office, Institute of Geological Sciences)

But there must have also been large areas not covered by oceans. This is assured by a universal process that occurs both in stars and planets: convection. Driven by the great temperature difference between the Earth's surface and core, convection is a fluid motion by which hot material rises and cool material sinks and that at these early times was even more vigorous than today. Convection is discussed in more detail in Sect. 3.10. An astonishing property observed in both stellar and terrestrial convection is that the fluid motion does not only occur in a smooth orderly fashion (regular convection) but also in a very turbulent and uneven way, called plume convection (Sect. 3.10), that produces strong and localized upflows (hot spots) and downflows. Although it is thought that the Earth in the Hadean era, and during most of the subsequent Archean era from 4.0 until 2.5 billion years ago, had a much smaller landmass than today's continents, there is no doubt that by plume convection associated with hot-spot volcanism a large number of islands and mini-continents must have existed at these early times (Sect. 3.11).

There are three reasons why volcanic activity was extensive in the Hadean era. First, the solid surface in those times was rather thin and liable to frequent cracking, resulting in a much greater heat flow. Second, the Earth was still bombarded frequently by smaller comets and meteorites. Third, because

the Moon orbited at only about 1/10 of its present distance, it exerted very
strong tidal forces which facilitated crust cracking. In addition, large impacts
occurred occasionally, which where able to vaporize more than 30 m of water
from the oceans and which where up to 100 times more violent than the K/T
boundary event (Fig. 3.4).

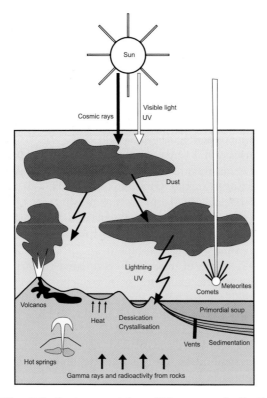

Fig. 3.8. Environmental conditions on early Earth

The outgassing of the Earth's mantle by volcanism and deep-sea vents,
combined with the accumulation of volatiles evaporated from impacting planetesimals, comets, and meteorites resulted in an atmosphere that consisted
mainly of the gases N_2, CO_2, CH_4, H_2O, CO, H_2, and NH_3. Large quantities
of dust were injected into this atmosphere producing an extremely hostile
environment, in which the total darkness at the terrestrial surface was only
broken by brightly glowing lava outbreaks and by constant lightning storms,
driven by the large amount of static electricity in the atmosphere (Figs. 3.7
and 3.8). Whereas today the blocking of sunlight by massive dust clouds
would result in strong cooling, in the Hadean era it had the opposite effect,
because of the greater heat flow from the molten interior. This very dusty
phase of the early Earth must have come to an end maybe already 4.4 bil-

lion years ago but certainly well before 3.8 billion years ago, by which time the appearance of photosynthesis indicates the presence of sunlight at the surface.

The precise composition of the Earth's atmosphere in the Hadean era, where life developed, is not known. Until fairly recently it was thought (e.g. Kasting 1993) that CH_4 and NH_3 may not have been present or were very rare in the atmosphere in these early times, because modern volcanoes emit mainly CO_2, N_2, and H_2O, as well as small amounts of H_2 and CO, while the production of CH_4 is extremely low. However, it has recently been recognized (Kasting and Brown 1998) that the CH_4/CO_2 ratio depends sensitively on the temperature and that at low temperatures the formation of CH_4 is strongly favored. Therefore, if CH_4 was primarily formed under much lower temperatures in hydrothermal vents and not in the hot magma of volcanoes, it would have been released at a much greater rate, particularly if one allows for the increased geothermal heat flow.

Moreover, hydrothermal vents could also have been abundant sources for NH_3, as has been experimentally demonstrated by mineral-catalyzed reduction of N_2 and nitrogen oxides (Brandes et al. 1998). In addition, the escape of hydrogen to space due to photolysis of water by UV light (Chap. 5) should have been reduced by dust that shielded the Earth in the earliest phases of the Hadean era. Finally, new hydrodynamic modeling by Tian et al. (2005) shows that hydrogen escape from the early Earth's atmosphere is decreased by two orders of magnitude compared to earlier estimates. All of this now points to the existence of a reducing (hydrogen-rich) atmosphere, which is required by laboratory experiments (Miller 1998) in order to produce the chemical compounds that are the building blocks for life (see Chap. 6).

Oceans, lagoons, and lakes constituted aggressive chemical laboratories, in which there was evaporation, desiccation, and an infusion of large numbers of different chemical substances via fumaroles and vents, as well as ionizing radiation from radioactivity and ultraviolet light from electrical discharges caused by dust and meteoritic infalls. This energetic chemical and physical environment can be reproduced in laboratories today and the results of these experiments suggest that the oceans rapidly filled with abiotically synthesized organic compounds (Fig 3.8), and were further enriched by organic material already formed in interstellar space and brought by meteorites and comets. But with no living organisms yet present to make use of these organic compounds, they accumulated, and their concentration increased steadily to create the so-called *primordial soup*. It was at this point that the stage was set for the appearance of life, which is discussed in Chap. 6. Because the formation of life as well as its development (discussed in Chap. 7) depends strongly on the properties of a terrestrial planet, particularly on the existence of oceans and whether land is present, it is necessary first to discuss in more detail the properties and internal structure of the Earth (for this see also the excellent books of Press et al. 2004 and Skinner et al. 2004).

3.6 Seismology and the Earth's Interior Structure

Similar to the distinct layering in other terrestrial planets, the interior of the Earth (see Fig. 3.9a), is divided into an outer *solid crust* (roughly 7–40 km thick), a *solid mantle* (extending to a depth of 2891 km), a *liquid outer core* (of 2259 km width, reaching down to 5150 km), and a *solid inner core* (with a radius of 1221 km) extending to the Earth's center at 6371 km.

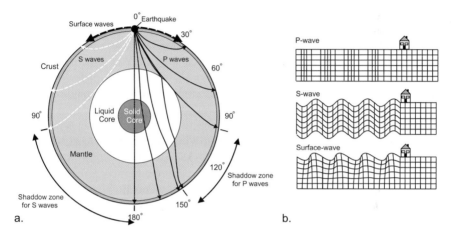

Fig. 3.9a. The interior of the Earth with the four main layers: crust, mantle, outer liquid core and inner solid core. An earthquake (*top*) generates three types of seismic waves that travel through the interior or along the surface. P-wave ray paths (*marked solid*) are shown in the right hemisphere while S-waves (*white-dashed*) are indicated in the left hemisphere. Surface waves (*black-dashed*) travel along the Earth's surface (modified after Skinner et al. 2004). **b.** Three types of seismic waves traveling from left to right on the terrestrial surface. P-waves are sound waves where parcels of matter get compressed and expanded. S-waves are shear waves where material gets displaced up and down or sideways. Surface waves travel only at and near the surface. Here a Raleigh surface wave is displayed

The Earth's interior structure has been unraveled using *seismic waves* generated by earthquakes. Similar to light rays from a light bulb, seismic waves can be thought of as traveling along individual ray paths from their source (see Fig. 3.9a). There are three types of seismic waves: *P-waves* (short for primary waves), *S-waves* (short for secondary waves) and *surface waves* of which there are two types: *R-waves* (short for *Rayleigh waves*) and *L-waves* (short for *Love waves*). Surface waves are the most destructive of the seismic waves.

The characteristic mode of propagation of the three types of seismic waves is shown in Fig. 3.9b. P-waves are sound waves that travel as a sequence of compressions and expansions along their ray path. Because during compression and expansion the rock particles move *along* the propagation direction these waves are so-called *longitudinal waves*. P-waves occur both in solids and liquids. In S-waves the rock packets move transverse to the propagation direction, for example up and down (Fig. 3.9b), or back and fourth in the horizontal direction. They are so-called *transverse waves* or *shear waves*. While present in solids they do not exist in liquids because liquids have no shear strength.

The distinctive feature of the third wave type, the surface waves, is that the biggest oscillation occurs at the surface and that with increasing depth the oscillation amplitude decreases rapidly to zero. Figure 3.9b shows a Rayleigh wave, where the rock parcels move in a rolling motion up and down as well as forward and backward in the propagation direction. Because these waves have both compression and shear, they are longitudinal-transverse waves that propagate purely horizontally. In Love waves the rock packets get displaced horizontally. They are transverse waves that also propagate purely horizontally.

After an earthquake the first signal detected by a seismometer somewhere else on the Earth's surface is the P-wave, which is the fastest seismic wave with a speed of about 5–6 km/s in crustal rocks. After that the S-wave arrives, which propagates with roughly half the speed of the P-wave, and finally the surface waves of which the Rayleigh wave is the slowest. Figure 3.9a shows the ray paths of individual seismic waves emanating from an earthquake at the top. It is seen that these waves move all the way through the Earth's interior and can be observed at the opposite side after about 20 min. The P-waves, shown solid in Fig. 3.9a, are seen to propagate from the surface into the mantle but curve back again towards the surface because the sound speed increases with depth and the wave propagation direction bends towards regions of lower sound speed. This behavior of the waves is called refraction.

The bottom of the crust that overlays the Earth's mantle is marked by the *Moho* (short for Mohorovičić) *discontinuity* where the P-wave speed jumps from 7.4 to 8.2 km/s. Here the temperature is about 1000 °C. This layer occurs about 5 km below ocean basins and 28–70 km under continents. Below the Moho the temperature increases further and at the bottom of the mantle reaches about 4000 °C, while the wave speed rises to 14 km/s. At this point, where the solid mantle borders on the liquid outer core, the wave speed jumps down to 8 km/s in the liquid, and, as seen in Fig. 3.9a, only P-waves can propagate deeper. Finally, another jump of the P-wave speed occurs when the solid core is reached. The inner core has temperatures of about 7000 °C.

At the *core–mantle boundary*, P-waves can be reflected back up or refracted down into the liquid. Since the wave speed in the liquid is less than in the mantle the angle of the refracted P-wave ray becomes smaller than the incident angle at this boundary (see Fig. 3.9a). This produces a zone of avoidance at the Earth's surface in a cone with an angle of 104–142° from the epicenter of the earthquake, called the *P-wave shadow zone*. A similar *S-wave shadow zone* is found between the angles 103° and 180° (see Fig. 3.9a). The arrival-time measurements of the reflected and refracted wave signals and the shadow zones allow us to find the exact location and depth of an earthquake and permits determination of the precise layering of the Earth's interior.

Figure 3.10 shows the locations and depths of large numbers of earthquakes that occurred over several decades. Surprisingly it can be noticed that the earthquakes do not occur at random over the Earth but concentrate along certain lines. These define the boundaries of 12 major (Antarctica, Africa, Eurasia, India, Australia, Arabia, Philippines, North America, South America, Pacific, Nazca, Cocos) and several minor *tectonic plates*. From the gray scale it is seen that although most earthquakes occur close to the surface, some take place at depths of up to 700 km. Precise locations using satellites and the global positioning system (GPS) show that the tectonic plates, also called *lithospheric plates*, are not stationary but typically move with respect to one another with speeds of as much as 1–14 cm per year. As discussed below this can lead to rifts between the plates where new crust is produced but also to collisions where sometimes one plate can be subducted below the other.

Measurements using Earth-orbiting satellites also allow very accurate mapping of the gravity from which the mass and density distribution within our planet can be determined. In addition, the reconstruction of the ray paths of seismic waves (see Fig. 3.9a) emanating from many thousands of earthquakes and ending in the hundreds of seismic stations all over the Earth's surface permits us to precisely determine the local wave speed at any point in the Earth's interior. This procedure, called *seismic tomography*, is similar to the nuclear magnetic resonance (NMR) tomography conducted on hospital patients to map the organs inside the body. Seismic tomography allows us to identify subducting tectonic plates as localized areas of slightly higher

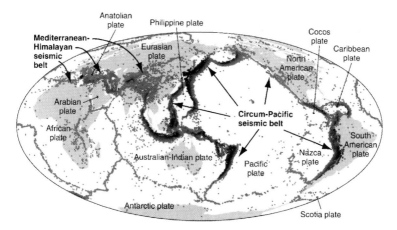

Fig. 3.10. Earthquake locations define tectonic plates that are found to move relative to one another. *Gray dots* mark earthquakes from 1–200 km depth, *black dots* from 200–700 km (courtesy of USGS)

wave speed and remarkably one can follow the plates all the way down (see Fig. 3.19a) to the bottom of the mantle (van der Hilst et al. 1997, Helffrich and Wood 2001). To fully understand the temperature and pressure distribution inside the Earth based on the behavior of the density and wave speed it is necessary to know the mineralogical and chemical composition of the rock material.

3.7 Volcanism and the Composition of Rocks

There are three types of rocks on Earth. *Sedimentary rocks* are solidified layered deposits of sand and mud (produced by weathering) or by carbonate shells (from dead animals) laid down in lakes, river mouths or in the sea. *Igneous rocks* are formed by the solidification of melted magma that originates from deep in the Earth's crust or mantle. *Metamorphic rocks* are transformations of originally sedimentary or igneous rocks under the influence of high pressure and temperature. Rocks are composed of *minerals* that are naturally occurring inorganic solid substances with a distinct crystal structure and specific chemical composition.

The composition of minerals is controlled by the abundance of the chemical elements found in the Earth's crust and mantle. As shown in the previous chapter (Table 2.1), the most abundant elements on Earth are O, Si, Al, Fe, Ca, Na, K, Mg of which two, O with 49% and Si with 26%, make up three quarters of all elements by weight. It is therefore not surprising that the most important constituent of minerals that make up igneous rocks from the Earth's crust and mantle is the chemical compound silica, SiO_2.

Table 3.1 shows the chemical composition, density and structure of minerals that are the constituents of igneous rocks. It is arranged by increasing density that also corresponds roughly to the depth where these minerals occur on Earth. It thus is not surprising that the lightest of these minerals, quartz, a special form of silica, is the most common mineral of the Earth's crust. Basic constituents of the minerals are tetrahedra of $SiO_4^{(-4)}$ that are

Table 3.1. Chemical composition, density and silicate structure of common minerals in igneous rocks of the Earth's crust and mantle arranged by density (modified after Press et al. 2004)

Mineral	Composition	Density (g/cm^3)	Silicate structure
Quartz	SiO_2	2.65	spatial structures
Orthoclase feldspar	$KAlSi_3O_8$	2.5–2.6	spatial structures
Plagioclase feldspar	$NaAlSi_3O_8$, $CaAl_2Si_2O_8$	2.6–2.8	spatial structures
Muscovite mica	$KAl_3Si_3O_{10}(OH)_2$	2.8–3.1	sheets
Biotite mica	$\{K, Mg, Fe, Al\}Si_3O_{10}(OH)_2$	3.0–3.1	sheets
Amphibole	$\{Mg, Fe, Ca, Na\}Si_8O_{22}(OH)_2$	3.1–3.3	double chains
Pyroxene group	$\{Mg, Fe, Ca, Al\}Si_2O_6$	3.1–3.6	single chains
Olivine group	$(Mg, Fe)_2SiO_4$	3.3–3.5	isolated tetrahedra

Fig. 3.11. Silicate minerals. Clockwise from *upper left*: feldspar, mica, pyroxene, quartz, olivine

arranged in different ways in spatial structures, sheets, chains and isolated tetrahedra, and where the four excess bonds indicated by the superscript couple with strategically placed ions of Al, Fe, Ca, Na, K and Mg in the crystal lattices. Note that Si has four, O, Ca and Mg have two, Na and K one, Fe two or three and Al three bonds. Figure 3.11 shows samples of principal silicate minerals.

In addition to the mineral content (where *felsic compositions* are rich in feldspar and quartz (silica), and *mafic compositions* have lots of magnesium and ferrum (iron)), igneous rocks are classified by their texture (see Fig. 3.12). While for instance *granite* has an easily visible grainy crystal structure, one needs a microscope to see the small crystals in *rhyolite* that has exactly the same composition as granite. Similarly, *diorite* and *andesite* as well as *gabbro* and *basalt* have similar mineral content but different textures (Fig. 3.12). The texture is a result of the depth (temperature) in which the rocks formed from the solidifying magma and the available time for the growth of the crystals.

Because molten rock, called *magma*, is less dense than the surrounding solid rock, it rises and upon reaching the surface erupts as lava in *volcanoes*. Since this rise is rapid, one has three major types of fine-grained magma, *basaltic magma* contains about 50% SiO_2, *andesitic magma* roughly 60% and *rhyolitic magma* about 70% (Fig. 3.12). Roughly 80% of the terrestrial volcanoes have basaltic magmas coming from deep to very deep in the mantle. As these magmas are very hot (1000–1200 °C), the lavas are fluid and move rapidly. This can lead to the formation of *shield volcanoes* that are flat and non-explosive since the lava has no problem running off a gentle slope. Examples of shield volcanoes with basaltic magmas are the Hawaiian volcanoes (see Fig. 3.13b).

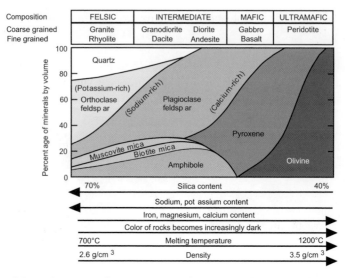

Composition	FELSIC	INTERMEDIATE		MAFIC	ULTRAMAFIC
Coarse grained	Granite	Granodiorite	Diorite	Gabbro	Peridotite
Fine grained	Rhyolite	Dacite	Andesite	Basalt	

Fig. 3.12. Mineral content of various types of igneous rocks. The rocks are classified by their composition and texture, the crystal size of the contributing minerals. Rocks formed deep in the Earth's crust or mantle where the cooling is slow develop large crystals and are coarse grained. Rapid cooling of magma at shallow depths leads to fine-grained rocks (modified after Press et al. 2004)

Fig. 3.13. Main types of volcanoes: **a.** stratovolcanoe Mount Fujiyama, Japan. **b.** shield volcano Mauna Kea, Hawaii (courtesy of Smithsonian Institution)

About 10% of the volcanoes erupt andesitic magmas. The magmas come from the upper mantle and the volcanoes are usually associated with island arcs or with continental margins where oceanic crust subducts and together with mantle material gets remelted. This is called *wet melting* because whenever water is added to hot rocks, the melting temperature is lowered and they melt more easily. Since the temperature of this magma is lower (800–1000 °C) it is more viscous and sometimes explodes as a shower of solid pyroclastic debris, called *tephra*. This material interlaced with layers of lava flows creates steep slopes producing *stratovolcanoes* of which Mt. Fujiyama and Mt. Vesuvius are examples (see Fig. 3.13a). Explosive release of dissolved gases in the magma can cause violent eruptions in stratovolcanoes.

Another 10% of the volcanoes have rhyolitic magma that is even less hot (650–800 °C) and has even higher viscosity. These magmas are the result of wet melting of continental crust. Rhyolitic volcanoes tend to erupt mostly tephra that accumulates in a distinctive cinder cone. They are primarily found on continents where one has rifting. Examples are the Yellowstone Caldera in Wyoming and many volcanoes on the North Island of New Zealand.

Next to the shield volcanoes and stratovolcanoes, a third type of volcanism occurs along elongated fractures usually at submarine spreading centers such as the Mid-Atlantic Ridge but also on land as on Iceland. Such fissure eruptions produce large amounts of basaltic magma that under the sea appears in the form of *pillow lava* (see Fig. 3.14a) that is produced by rapid cooling of erupting lava packets due to sea water. These layers of pillow basalt looking like an endless bed of sandbags form most of the oceanic crust. Pillow basalts represent the most common igneous rock on Earth.

Fig. 3.14. Volcanic-type eruptions in the deep sea. **a.** Pillow basalt. **b.** Hydrothermal vent with a so-called black smoker (courtesy of NOAA)

Finally, there is volcanism in the form of hot springs and geysers on land as well as at numerous *hydrothermal* (hot water) *vents* in the deep sea. Here, water percolating down through cracks in the underlying rock layers comes in contact with hot magma where it produces super-heated water, rises again and gets laden with minerals by leaching the surrounding rocks. Figure 3.14b shows such a mineral saturated flow from a hydrothermal vent called a *black smoker*. Due to high pressures in the deep sea these hydrothermal vent flows are still liquid despite temperatures of up to 400 °C.

The distribution of currently active volcanoes on Earth is seen in Fig. 3.15. Comparison with Fig. 3.10 shows that there is a close correlation of most volcanoes with nearby locations of earthquake centers and plate boundaries. An example of such a correlation is the huge continuous chain of volcanoes, called the *Ring of Fire*, that circles the Pacific from New Zealand, over the

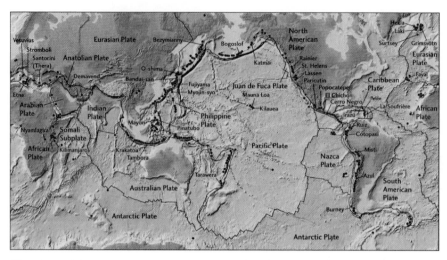

Fig. 3.15. Distribution of active volcanoes (*dots*) around the Earth (after Press et al. 2004)

Philippines, Japan, the Kurils, Aleutians, western North and South America. However, there are also a number of volcanoes that occur inside the plates such as the Hawaiian volcanoes.

3.8 The Earth's Core and Mantle

From satellite observations and seismic-wave measurements as discussed above one finds that the crust contains 0.4%, the mantle 67.2% and the core 32.4% of the Earth's total mass of 5.98×10^{24} kg. Additional information about the properties of the Earth's mantle and core comes from the study of material from past and present volcanoes and from laboratory mixtures of minerals confined to small cavities, so-called *diamond anvil presses*, where the crystalline state of the material under extreme temperatures and pressures can be investigated. Important knowledge is also obtained from *drilling projects* on continents and at the ocean floor that provide rock samples from depths of 2–5 km. Luckily this depth range can be extended by the discovery of *xenoliths*, unaltered pieces of rocks that sometimes are swept up from depths of up to 250 km in volcanic eruptions and provide direct insight into the properties at deeper layers of the upper mantle. But for most of the mantle and core only indirect methods can be employed, such as using the composition of meteorites.

It is remarkable that despite the Earth's core being most remote, its composition is more easily determined than that of the mantle. This is due to its high density and because iron meteorites are pretty good indicators of the conditions in the core of large planetesimals (Sect. 2.8). Of the six most abundant non-volatile rock-forming elements Si, Mg, Fe, S, Al and Ca, only Fe is frequent enough to explain the high density of the core. However, since the known density of pure Fe at core pressures is 10% higher than the observed value, there must be a 10% admixture of lighter elements in the core. Moreover, from the composition of iron meteorites, consisting of a Fe alloy with 5% Ni, a Ni fraction of this magnitude must also be assumed for the Earth's core. Because the density of Ni is similar to Fe this does not change the need for the admixture of light elements. After various arguments about this light element fraction the presently best estimate of the core composition by weight is thought to be 85% Fe, 5.2% Ni, 5.8% O, 1.9% S and 0.2% C but no Si (McDonough 2003).

But what materials make up the mantle? This can only be answered by considering the formation process of our planet. As discussed in Sect. 2.8, the Earth is formed by accreting hundreds of Ceres-, Moon- and even Mars-sized planetary embryos as well as a huge number of smaller planetesimals. The growth of the bulk of the Earth and other terrestrial planets took tens of millions of years and material from various heliocentric distances was involved. Because the Sun and the planetesimals formed out of the same solar nebula, the chemical element composition must be the same as that of the Sun, except that the planets did not accrete extremely volatile elements such as H, N, C, O and rare gases.

In Sect. 2.8 it was discussed that of the stone meteorites, the CI carbonaceous chondrites preserve best the type of material that once was involved in the formation of the terrestrial planets. Table 3.2 shows how the composition of the Earth's mantle is derived by a series of steps from the composition of CI meteorites (Hart and Zindler 1986). Starting with an amount of one Earth mass of material with a CI composition, one first removes the highly volatile elements (HVE) such as H_2O, S, C, N, etc.; they are supposed to have been lost in the accretion process. Then one takes away the Fe, Ni, O and S needed

Table 3.2. Relative composition by mass of the Earth's mantle and of CI-meteorites, modified after Palme and O'Neill (2005), White (2003), Hart and Zindler (1986). Also shown is the Primitive Mantle composition called the solar model. HVE indicates highly volatile elements such as H_2O, S, C, N, etc. The CI compositions add up to 100% for the entire meteorite, while the mantle compositions add up to 100% for the mantle separately

	CI-meteorite	Earth Mantle Solar model	Earth Mantle
	%	%	%
SiO_2	22.77	51.20	45.4
MgO	16.41	35.80	36.77
FeO	24.49	6.30	8.1
Al_2O_3	1.64	3.70	4.49
CaO	1.3	3.00	3.65
HVE	30.21		
NiO	1.39		

for the formation of the core. What remains is called the *Primitive Mantle* (PM) since it represents the mantle composition before the crust was formed during Earth's history. The composition of the PM (called the solar model) is shown in Table 3.2. Note that while the CI compositions add up to 100% for the entire body of the meteorite, the values for the mantle composition appear higher because after taking away the core composition they are made to add up to 100% separately for the mantle. By removing the known crust composition from the PM, one finally (Palme and O'Neill 2005, White 2003) arrives at the Earth's mantle composition shown in Table 3.2.

3.9 The Earth's Magnetic Field and Sea-Floor Spreading

The Mid-Atlantic Ridge is a narrow mostly underwater mountain range in the Atlantic Ocean that runs from 333 km south of the North Pole to the subantarctic Bouvet Island from where it turns into the Atlantic-Indian Ridge and other ridges (Fig. 3.18). Rising on average about 3000 m above the sea floor, its highest mountains are islands such as Jan Mayen, Iceland, the Azores, Ascension and Tristan da Cunha. Bisecting the length of the ridge is a trench that in places is more than 2000 m deep. In these trenches magma from the Earth's mantle erupts. One finds with ^{238}U–^{230}Th dating that the rocks at the ridge are very young and become progressively older away from it. Using magnetometers on aircraft and ships it has been discovered that parallel to the ridge extending on both sides to great distances there are stripes of different magnetizations (see Fig. 3.16a).

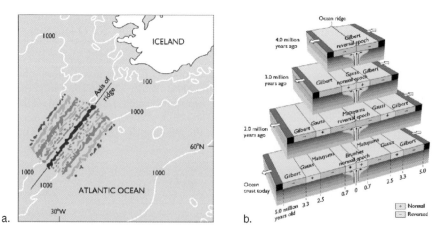

Fig. 3.16. Magnetic signatures parallel to the Mid-Atlantic Spreading Ridge. **a.** Stripes of similar magnetization south of Iceland. **b.** Generation of symmetric magnetic stripes in different magnetic epochs over the last 5 million years (courtesy of USGS)

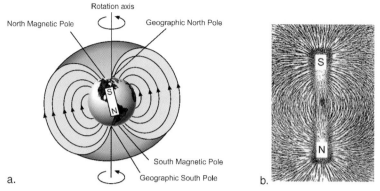

Fig. 3.17a. The Earth's magnetic field. **b.** The field of a bar magnet made visible using iron filings

These magnetic stripes are explained by the behavior of the Earth's magnetic field that is produced by convective motions in the liquid core of our vigorously rotating planet. The dipolar Earth magnetic field (Fig. 3.17a) is similar to that of a bar magnet placed at the center of the Earth. The field of a bar magnet (or compass needle) is shown in Fig. 3.17b where dusting with tiny specks of iron filings make the field lines visible. For Earth this magnet is inclined by about 9° to the Earth's rotation axis that runs through the geographic North and South Poles (Fig. 3.17a). But since the motions in the convecting liquid core are stochastic, the magnetic axis is not fixed and its exitpoints at the terrestrial surface, the northern one confusingly called *North Magnetic Pole* (Fig. 3.17a), have migrated by as much as 1000 km in the last 100 years. The confusion arises because the magnetic pole in the northern hemisphere is actually the south pole of the terrestrial bar magnet. This comes from the definition that the north-pointing part of a compass needle is a magnetic north pole and the magnetic field lines by definition go from the north pole of a magnet to its south pole (see Figs. 3.17a,b).

Two properties of the Earth's magnetic field are important for plate tectonics. First, the magnetic field lines at every location on the terrestrial surface point into a definite direction that can be determined by measuring their vertical and horizontal (from due north) angles, and second the stochastic convective motions in the liquid outer core occasionally generate complete reversals of the Earth's field. By this, a northward-directed field at a given location is changed into a southward-directed field and vice versa in a fairly irregular fashion in time spans that vary between a hundred thousand to many millions of years.

This is similar to the Sun, where convection in the outer envelope layers also generates a global magnetic field, although with a much more regular field-reversal rate of about 11 years, causing the *11-year sunspot cycle*. Since the erupting hot magma at the spreading ridges cools to become basaltic rocks containing magnetic minerals like *magnetite*, the instantaneous magnetic field direction at that moment gets frozen into the rocks. This produces stripes of similar magnetization at similar distances on both sides of the spreading ridge (see Fig. 3.16b).

Although the process of magnetic-field generation and reversal on Earth as well as on the Sun is not yet fully understood, the discovery of the magnetic stripes unequivocally proved the phenomena of *sea-floor spreading* and *plate tectonics*, the process of continental drift. Following the plate motions back in time by dating the rocks on the sea floor it is possible to draw maps (Fig. 3.18) of the plate motions over geological times and document the splitting of continents. From maps like this it can be seen that the North Atlantic opened in Jurassic times and the South Atlantic much later in the Cretaceous. This allows reconstruction of the distribution of continents over geological times and even predict their future position.

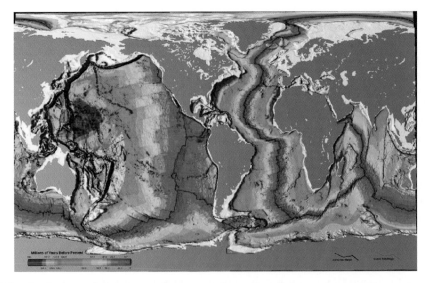

Fig. 3.18. Ages of emergence of new oceanic crust at the oceanic-spreading centers. Color code: dark blue Jurassic (180–145 million years ago), light blue and green Cretaceous (145–65 million years ago), yellow Paleocene to Oligocene (65–23 million years ago), and red Miocene till present (23–0 million years ago) (Müller et al. 1997)

3.10 Convection, Hot Spots and Plate Tectonics

The discovery of lithospheric plates, their piecewise rigid-plate motion, slab subduction and the distribution of andesitic volcanoes along the plate boundaries point to a powerful physical process going on inside the Earth's mantle: convection. *Convection* is driven by the great temperature difference between the Earth's surface and the core and lower mantle, where a large amount of heat from primordial impacts and from radioactive decay of U, Th and K is still present. The process of convection is well known from liquids heated on a domestic stove but also from stars where the convective medium is a gas. The driving force of convection is *buoyancy*, which in a surrounding denser medium drives lighter less-dense material up and lets cool more-dense material sink down. Hot material rises from the mantle interior and, after radiating the heat away at the surface, sinks back down again into the interior of the mantle.

Since the Earth's crust and mantle are polycrystalline solids one has solid-state convection, which is displayed in Fig. 3.19a. The creeping of a solid is made possible by the existence of so-called *Schottky holes*, *Frenkel defects* and dislocations in the crystals as well as by the grain boundaries between the crystals. The Earth's mantle convection, therefore, is considerably more complicated than the convection in fluids or gases.

Figure 3.19a shows that the convection moves the lithosphere that consists of quasi-rigid plates (white arrows). Each plate rotates en bloc with velocities of some cm/year around the Earth's center of mass, but with different rotation axes. This is possible because the *oceanic lithosphere* (brown) together with the *oceanic crust* (dark blue) slides on top of the *asthenosphere* in the outer mantle and is able to subduct at the oceanic trenches. The oceanic lithosphere is composed of a thin, roughly 7-km thick layer of oceanic crust with gabbroidic and basaltic composition and a roughly 60-km thick solid layer of ultramafic upper mantle material. That the lithosphere can slide on top of the asthenosphere is due to the reduced rigidity of this 100–300-km thick layer of solids, caused by temperatures that rise to values above 1300 °C where the rocks begin to melt. However, at the lower boundary of the asthenosphere the mantle rocks become rigid solids again due to the increasing pressure.

The *continental lithosphere* (shown in Fig. 3.19a, not to scale, brown) together with the *continental crust* (continents and island arcs, red) also slides on top of the asthenosphere but owing to its low density does not get subducted. The continental lithosphere consists of two layers: a solid, typically 40 km thick continental crust that is of felsic material and forms the continents as well as island arcs, and an attached more extensive layer of solid ultramafic upper-mantle material that is 50–300 km thick. Oceans are shown light blue. The bulk convection (black arrows) and the movements of the plates (white arrows) form an integrated system. New plates are created at the mid-ocean ridges. Ridges often but not always are at the place of an upwelling. Some subducting plates stay in the upper mantle that reaches to a depth of 660 km, others drop down all the way through the lower mantle to the core–mantle boundary.

Remarkably, it has recently become possible to numerically model the Earth's mantle convection including the complicated surface properties and explain, for example, why the 70-km thick Pacific plate with its numerous faults that extends from the Easter Island to as far as Japan (Fig. 3.20b)

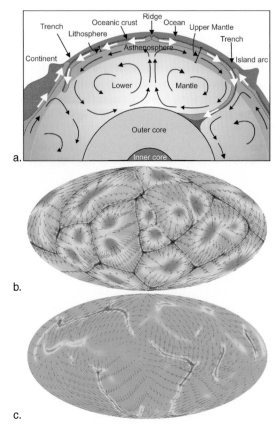

Fig. 3.19. Convection in the Earth's mantle. **a.** Cross-section of the solid-state mantle convection, details are explained in the text. **b.** The numerical convection simulation by Walzer et al. (2003) produces flows (*velocity arrows*) in irregular cells in which the temperature varies from 300–1300 K (*dark blue to red*). A layer at 135 km depth is displayed. The calculation using pure viscosity shows hot upwellings and thin cold sheet-like downwellings into a spherical mantle. **c.** A simulation with an improved treatment of viscosity that permits us to handle yield stresses is shown at the Earth's surface. Walzer et al. (2004) find a stable plate-tectonic behavior near the surface and simultaneously thin sheet-like downwellings at depth. Regions of high viscosity (very rigid) are shown *green* and of low viscosity (less rigid) *red*

behaves like a rigid body. This is achieved with time-dependent hydrodynamic viscous fluid simulations. Figure 3.19b shows a numerical convection calculation by Walzer et al. (2003) using a viscous fluid and Fig. 3.19c taking a viscous fluid that in addition permits modeling of *yield stresses* (Walzer et al. 2004). Yield stress is the level of stress that a material can tolerate before an irreversible plastic deformation sets in. The problem of the numerical simulation is to be able to treat the very thin sheet-like downwellings (Fig. 3.19a) and at the same time the piecewise rigid-plate motions near the surface. The calculation in Fig. 3.19b displays a layer at a depth of 135 km that shows hot upwellings in irregular cells and thin cold downwellings into a spherical mantle. Figure 3.19c presents a simulation where both a stable plate-tectonic behavior near the surface and at greater depth sheet-like downwellings are obtained.

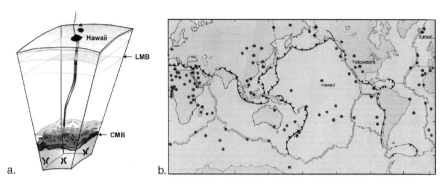

Fig. 3.20a. Hawaiian hot spot where due to plume convection one has chimney-like upflows all the way from the core–mantle boundary (CMB). LMB marks the lower-mantle boundary (Garnero 2004). **b.** Hot spots (*dots*) occur at plate boundaries, but also inside plates (Best 2003)

In addition to the regular *mantle convection* seen in Figs. 3.19a–c there is another type of convection called *plume convection* that occurs in narrow chimney-like flow channels (Fig. 3.20a). Plume convection is also known from the Sun where highly localized convective upflows and downflows are observed. In plume convection of the Earth's mantle that takes place beneath so-called *hot spots*, basaltic material flows up all the way from the core–mantle boundary. The rising high-temperature material melts when it arrives at the low-pressure regions near the surface and creates huge magma chambers from which large amounts of lava erupt in volcanoes. While many oceanic plates descend all the way to the core–mantle boundary there are computer simulations that suggest that some of the plates could be stalled at 660 km depth, the boundary of the lower mantle (Fig. 3.19a, left), until at irregular intervals they get flushed down to the core–mantle boundary, which might trigger the hot-spot events (Montelli et al. 2004).

On Earth there are about 40 hot spots some of which have produced large islands and island chains such as Iceland or Hawaii. As shown in Fig. 3.20b, these hot spots are located at plate boundaries but also occur inside plates, called *intraplate hot spots*. Hawaii, Samoa and the Society Islands are examples of intraplate hot spots in the Pacific, while in the Atlantic the hot spots Iceland, the Azores, Ascension, Tristan da Cunha are at plate boundaries, while the Canary Islands, Cape Verde Islands or St. Helena are intraplate hot spots together with all the hot spots on the African continent. Some of these hot spots produce shield volcanoes, while others create stratovolcanoes. From the 5-km deep ocean floor, for example, the Hawaiian shield volcano Mauna Loa, rises to 5 km above sea level, creating a mountain of 9710 m height.

Over the past 70 million years the continued movement of the Pacific plate over the Hawaiian hot spot has left a long trail of islands and seamounts with now mostly extinct volcanoes across the Pacific Ocean floor. This Hawaiian *Emperor Seamounts chain* extends for some 6000 km from Hawaii to the Aleutian Trench off Alaska with a sharp bend about 43 million years ago. It is not clear whether this bend is connected with the collision of India with the Asian continent or indicates that hot spots migrate. The volume of lava that erupted to form the Emperor Seamounts chain has been estimated to be at least 750 000 km^3 – more than enough to blanket the entire State of California with a layer of lava roughly 1.5 km thick. Other examples of the effect of plume convection over ancient hot spots are the *Deccan Traps*, an extensive mountain range in southwestern India, more than 2 km thick, covering an area of 500 000 km^2, that formed about 68 million years ago, or the ca. 1 000 000 km^2 sized flood basalt region north of Krasnoyarsk called *Siberian Traps*, which was the largest volcanic eruption in Phanerozoic times and occurred right at the time of the largest extinction event (Chap. 7) in Earth's history, 251 million years ago.

To permit motions of the solid lithospheric plates, three types of interactions occur at the plate boundaries (Fig. 3.21a): Transform faults, rifts and convergences. *Transform faults* are cracks in the solid lithosphere that allow the plates to slide past each other horizontally. A famous example is the San Andreas fault in California where the Pacific plate with Los Angeles slides with speeds of about 5 cm/year in a north-westerly direction along the North American plate. *Rifts* occur when plates move away from each other, which permits the formation of new crust. A well-known example is the Mid-Atlantic Spreading Ridge that primarily extends on the ocean floor but also on land as in Iceland.

There are three types of *convergent boundaries*, also called *thrust faults* where tectonic plates move against each other (Figs. 3.21b,c,d). *Oceanic–oceanic convergences* involve two oceanic lithospheric plates (Fig. 3.21b) where one gets subducted below the other. This produces an oceanic subduction trench and an island arc formed by volcanism due to partial melting in the plate that lies above the subducted plate. Examples of oceanic–oceanic convergences are the Japan and Mariana trenches associated with the Japanese and Philippine islands where the Pacific plate slides below the Eurasian and Philippine plates that have depths of up to 9000 and 11 000 m below sea level, respectively.

Oceanic–continental convergences such as on the west coast of South America consist of an oceanic lithosphere subducting below a continental lithosphere (Fig. 3.21c). Here associated with an oceanic trench, the Nacza plate is subducted below the South American plate and the andesitic vol-

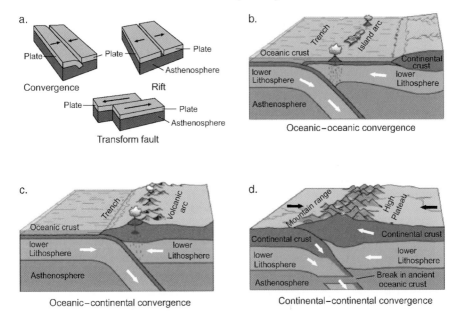

Fig. 3.21. Faults and convergent boundaries of lithospheric plates. The solid plates are composed of continental crust and lower lithosphere at continents as well as island arcs, and of oceanic crust and lower lithosphere below oceans. The plates slide on top of the asthenosphere. **a.** Types of faults: convergence or thrust fault, rift, and transform fault. **b.** Oceanic–oceanic convergence. **c.** Oceanic–continental convergence. **d.** Continental–continental convergence (modified, after USGS)

canic activity from the partial melting of the overlying plate wedge creates the mountain chain of the Andes.

In both types of convergences with oceanic plates so-called *wet melting* or *fluid-induced melting* occurs because trapped or chemically bound water carried by the subducting plates gets squeezed out and percolates up. This lowers the melting temperatures of the rocks in the overlying mantle wedge above the slab (Figs. 3.21b,c). It is thought that wet melting is essential for the initiation process of subduction of oceanic plates (Regenauer-Lieb et al. 2001). The melting produces differentiation of low and high density material (light rocks rise while heavy ones stay below) and to andesitic volcanoes that form island arcs or coastal mountain ranges on continents. This differentiation occurring throughout the Earth's geological history is the reason why the persistently growing continental crust is less dense.

The final type, the *continental–continental convergences* (Fig. 3.21d), created, for instance, the Himalayas where the Indian plate smashed into the Eurasian plate. Such continental–continental collisions result in great *mountain-building episodes*, also called *orogeny*, because the continental lithospheric plates cannot subduct below one another. Moreover, the lithosphere that originally existed between the colliding continents and the sediments from the continental shelfs get compressed and added to the mountain chains, while the already subducted parts of the compressed oceanic plates break off and descend into the mantle (Fig. 3.21d). Because of this compression, rock formations having undergone orogeny are frequently severely deformed and suffer *metamorphism* (see Sect. 3.7).

3.11 Mountain Building and the Evolution of Continents

Before discussing the reconstruction of the ancient terrestrial continents it has to be noted that there is a fixed amount of water on Earth. How it got there was discussed in Sect. 2.7. Figure 3.22 shows the height above and below sea level of various regions on Earth. It is seen that if the continents were flattened and evenly spread over the Earth then the terrestrial surface would be completely submerged under an ocean of 3 km depth.

Presently the land represents 30% of the Earth's surface with 70% covered by oceans. Taking the continental shelfs and slopes to be part of the continental crust one has a division of 40% continents and 60% oceans that have an average depth of 5 km. As will be discussed below, this ratio was different in the Hadean and Archean era because of the growth of continents. Moreover, as can be seen from Fig. 3.22, the erosion of the continental mountains leads to a great extent of low-lying continental regions. Flooding or exposing these regions because of slight sea-level changes, caused for instance by rises of the sea floor due to mantle convection or by glaciations that pile up ice on continents, the area ratio of land to sea can vary greatly. This severely affects the geographical reconstruction of ancient landmasses.

Figure 3.23 shows a reconstruction attempt of the motions of the continents over the last one billion years, called the *Paleomap Project* (Scotese

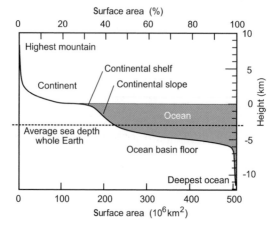

Fig. 3.22. Height above and below sea level of different surface regions on Earth (modified after Lunine 1999)

2002). This effort self-consistently combines the known indicators of former continent positions such as the fossil magnetic field directions (palaeomagnetism), the distribution of warm and cold-loving fossil plants and animals (palaeobiogeography), the distribution of arctic regions and deserts (palaeoclimatology) as well as what one knows about spreading centers from the ancient mountain-building episodes (orogeny) and about geologically inferred sea-level changes (tectonic and geological history).

For instance, it is seen in Fig. 3.23a that in the Early Jurassic, 195 million years ago, all continents were connected and formed a supercontinent *Pangea*. But the breakup of Pangea is progressing and the North Atlantic is about to open. In the Late Cretaceous, 94 million years ago (Fig. 3.23b), the South Atlantic opens as well and India – located still in the southern hemisphere – is seen to speed towards the north to eventually collide with Asia starting about 50 million years ago (Rowley and Currie 2006). Note that the continental shelfs are indicated light blue, which shows that most of Europe and parts of north Africa are submerged under a shallow ocean that is responsible for the Cretaceous rock formations all around the Mediterranean Sea.

Going back to the Cambrian, 514 million years ago (Fig. 3.23c), shows that the process of splitting up and rejoining of continental crust leads to

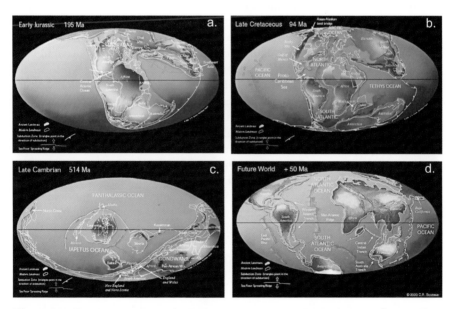

Fig. 3.23. Earth maps from the Paleomap project. **a.** Early Jurassic (195 million years ago), all continents are together forming the supercontinent Pangea. The North Atlantic is about to open. **b.** Late Cretaceous (94 million years ago). The South Atlantic opens and India is on its way to collide with Asia. **c.** Late Cambrian (514 million years ago), the continents look very different from today. **d.** Future world (in 50 million years), Africa has collided with Eurasia and obliterated the Mediterranean Sea (courtesy of Scotese 2002)

a very unfamiliar geography and land–sea distribution. The simulation of plate tectonics can also be carried into the future. In Fig. 3.23d the distribution of the continents is seen 50 million years in the future. The outstanding feature is that the collision of the African plate with the Eurasian continent will completely obliterate the Mediterranean Sea and form a huge continent-wide mountain range on the southern rim of Europe that surpasses even the Himalayas.

Rock falls, land slides, streams loaded with silt, great river deltas such as those of the Nile, Mississippi or Ganges all show the relentless erosion of mountains and land due to rain, ice and wind. Over many millions of years entire mountain ranges have been flattened and the resulting sediments were deposited on the continental margins. From the fact that one has a land area of only 30% as compared to 70% covered by oceans one might think that erosion should cause the combined continental area to decrease with time. The contrary is true because the total volume of the continents has increased over the last 4 billion years on average by about 2 km^3 per year.

Actually, all three types of plate convergences lead to the growth of continents. In continental–continental collisions widespread sediments between the two continents and even some oceanic crust get compressed, folded and uplifted to great heights during a process of mountain building. In oceanic–continental convergences the wet melting of some of the oceanic crust and of the continental mantle wedge above the subducting oceanic plate creates new low-density material via andesitic volcanoes. The oceanic–oceanic convergences create islands and island arcs that in a subsequent episode get accreted to the continents. Finally, one has the formation of flood basalts due to plume convection at hot spots. In summary, the convergences and the plume convection contribute to a permanently ongoing process of differentiation and separation of low-density material out of the high-density rocks of the mantle. This material due to its buoyancy resists subduction and increases the mass of the continents.

What do we know about the development of continents during Earth's history? We have already seen that when tracing the plate motions back in time to the Early Jurassic there was a supercontinent (Pangea) where all continents were joined together. Subsequently, that supercontinent broke apart and it is possible to follow the development of the fractions by dating of the oceanic crust (Fig. 3.18). The cause of this breakup apparently is the blockage of the heat flow by a vast land area above the convecting mantle. Such a blockage cannot go on for long because the convective upflows perpetually raise the temperature under the supercontinent and lift parts of it until thin weak regions of the supercontinent break apart. The splitting of Pangea created a northern part, *Laurasia*, with North America and Eurasia and a southern part, *Gondwana*. Subsequently in the Early Cretaceous, North America and Eurasia split, and in the Late Cretaceous, Gondwana broke up into South America, Africa, India, Australia and Antarctica.

After maximum splitting a reverse motion started. India has slammed into Eurasia, and the same is presently happening with Africa that in 50 million years will form a common continent with Eurasia. Going 250 million years into the future another supercontinent will be created. This behavior of splitting and reassembling of supercontinents driven by the mantle convection is called the *Wilson cycle*. Going back before Pangea two more Wilson cycles can be identified with supercontinents called *Rodinia*, about 1.1 billion years ago and *Columbia*, roughly 1.8 billion years ago with possibly another one about 2.6 billion years ago (Press et al. 2004). With the maximum of Pangea in the Triassic, about 240 million years ago, it is seen that the Wilson cycles range from 500–800 million years. Since there is no oceanic crust older than 180 million years (Fig. 3.18) this history had to be worked out by studying the continental crust, in particular by dating different geologically distinct continental regions called *terranes*, by investigating the episodes of orogenies that tell of continental–continental collisions and by decoding the magnetic, palaeobiogeographic and palaeoclimatological evidence.

Figure 3.24 shows the ages of different continental regions. The oldest continental nuclei are called *cratons*. Because they largely retained their identity throughout Earth's history they represent growth centers around which continental masses accreted. It is seen that cratons such as Laurentia (North America with Greenland), Southern Africa, Baltica, Western Australia, Eastern Siberia and South America perpetually continued to grow in size by accumulation of younger land masses. For a more recent account of the area of exposed and inferred Archean rocks with ages between 2.5–4.0 billion years see Valley (2005).

From these area distributions the growth of the continental crust over time can be derived as shown in Fig. 3.25. Such growth curves are uncertain as illustrated in the figure. It is important for the formation of life that continents and oceans were already present in the Hadean and Archean eras more than 3.8 billion years ago. This is assured on theoretical grounds because

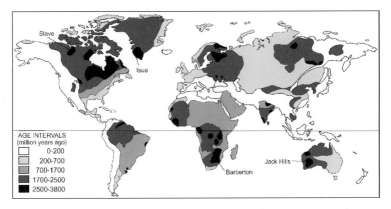

Fig. 3.24. Age distribution of the continental crust (modified after Lunine 1999)

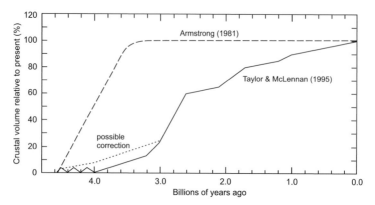

Fig. 3.25. Growth of the continental crust over time in percent of today's land area (modified after Deming 2002). While Armstrong (1981) favors an extremely rapid growth, Taylor and McLennan (1995) argue for a slower increase of the continents. Evidence from zircons discovered in the Jack Hills region of Western Australia proves the existence of considerable amounts of continental crust in the Hadean and early Archean eras and requires an upward revision of the growth curve (marked "possible correction"). The amount of revision is uncertain but could be bigger

at those times the mantle convection was more violent and one certainly had many mini-continents created by plume convection (Sect. 3.5). That mini-continents would have been present at those times can be seen from the examples of Hawaii and Iceland (Sect. 3.10) where hot-spot volcanism easily created mountains of up to 9 km height above the sea floor and would have penetrated the ocean surface even if in those days the Earth were covered by a 3-km deep global ocean.

In addition to theoretical reasons there is also solid empirical evidence for the existence of continents and oceans at these early times. As discussed in Sect. 1.5, from ^{206}Pb-^{207}Pb isotope dating it is now certain that continental crust already existed 4.4 billion years ago (Wilde et al. 2001). In addition, the existence of oceans as early as 4.2 billion years ago has been confirmed from $^{18}O/^{16}O$ isotope ratios (Valley 2005, Cavosie et al. 2005). On the basis of $^{176}Hf/^{177}Hf$ ratios there are even suggestions that continents and oceans might have existed as early as 4.5–4.4 billion years ago (Harrison et al. 2005). All these isotope ratios were measured in detrital zircons found in the Mount Narryer and Jack Hills regions of Western Australia.

There is no doubt therefore that oceans and continental shelfs existed, which are the requirements for the emergence and development of life (see Chap. 6) that took place near the end of the Hadean era. The presence of continents moreover ensured that when vertebrate animals conquered them 3.5 billion years later they developed appendages to lift their bodies up from the ground, which combined with the persistent evolution of intelligent behavior (as will be discussed in Chap. 7) became manipulators for tool use that represent the basis for our type of intelligence.

3.12 Plate Tectonics on Mars and Venus?

Is plate tectonics merely a specialty of the Earth or does it also happen on other terrestrial planets? Their history of formation indicates that these planets should be fairly similar to Earth in chemical composition because they were assembled from similar silicaceous, ferrous and ice planetesimals. This implies that they also have a well-determined internal differentiation in an iron core and siliceous mantle. In addition, theoretical simulations suggest that the mass of terrestrial planets should not vary greatly compared to Earth except for planets near the inner and outer boundary of the terrestrial planet zone. Here, tidal interactions from the central star and the Jupiter-type outer planets disturb the accretion and leads to the formation of low mass planets such as Mercury and Mars.

While Venus and Earth have similar mass (Table 2.1) and probably similar internal structure, the interior of Mars is different because of its much lower mass. On Earth the initial heavy bombardment by planetesimals liberated huge amounts of impact energy. This created a magma ocean that was roughly 1800 km deep (Agee 2004), bordering below on a solid silicate mantle and beneath that on the initially completely liquid Fe core. In this magma ocean metallic iron sedimented down to the solid mantle boundary. In addition, with temperatures as high as 3500 °C and pressures of 50 GPa at that boundary, FeO as constituent of olivine and pyroxene, the silicate components (Fig. 3.12) of the magma ocean, converted to metallic Fe and dissolved O. Subsequently, the metallic Fe after accumulating into large masses forced itself through the solid mantle into the core. In this way a lot of Fe got removed from the mantle. Finally, at a later stage, when the cooling of the Earth lowered the core temperature below the melting temperature of pure Fe, an almost pure Fe solid inner core developed together with a liquid outer core consisting mainly of Fe with dissolved O and the other lighter elements.

This history was different inside Mars (Agee 2004). Since that planet has a much lower mass (Table 2.1) the temperature at the bottom of its magma ocean (at a depth of 2000 km) did not exceed 2500 K with pressures of 25 GPa. Under these conditions FeO in the silicates did not get converted to metallic Fe and remained in the magma ocean. This can explain the high Fe content remaining in the outer layers including the surface that gives Mars its red color.

For Earth-like plate tectonics to function two important conditions must be met. There must be a large enough temperature difference between the core and surface to drive convection. This is assured by enough mass because more mass has higher heat capacity, produces greater amounts of radioactive heating in the core and leads to a greater core–surface temperature difference. Second, one must have enough water in the form of oceans such that hydrogenated rocks with their lower melting temperature can initiate the subduction process of oceanic plates (Regenauer-Lieb et al. 2001). In Mars the low mass and smaller temperature difference between core and surface appar-

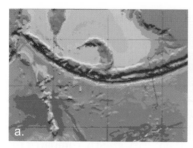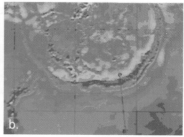

Fig. 3.26. a. Aleutian Trench with *color-coded* depths. *Violet* indicates greatest depth, *blue* to *green* represents shallower depths, *yellow* to *red* indicate highs. **b.** Artemis Corona, a trench-like feature on Venus, based on radar images and shown at the same vertical and horizontal scale (Sandwell and Schubert 1992)

ently stopped regular mantle convection long ago, possibly ending with final large-scale plume convection that created the huge shield volcanoes such as Olympus Mons. The decrease of volcanic activity due to the dying convection lowered the emission of greenhouse gases that lead to irreversible glaciation (Chap. 5) and the present unfavorable conditions for life on Mars.

With similar mass compared to Earth, Venus should have similar internal temperature differences and large-scale mantle convection. As in its early history Venus, like Earth, probably also had large amounts of surface water, one expects that it once had plate tectonics. However, the present absence of water on the planet, probably caused by a runaway greenhouse effect (Chap. 5), apparently stopped the plate motions by preventing subductions and continent building. Interestingly, there may be traces of ancient plate tectonics. Figure 3.26, on the same horizontal and depth scale, shows a comparison between the terrestrial *Aleutian Trench* and *Artemis Corona*, a trench-like feature on Venus.

However, crater-count statistics indicate that the Venusian surface is only about 300 million to 1 billion years old, which points to a catastrophic global resurfacing, possibly by large volumes of flood basalts from plume convection (van Thienen et al. 2005). This extensive plume convection could be the answer to the stopped plate tectonics caused by water loss by a runaway greenhouse effect that produced an interruption of the heat flow that normally emerged at the mid-ocean ridges of the planet. In 1979, the Pioneer-Venus spacecraft measured a high amount of sulfur in the upper atmosphere of the planet that then decreased over the next few years. This could be a signature of active volcanism since plume convection with its associated volcanism should still be going on. The presence of volcanism is supported by dramatic volcanic features discovered since 1990 using the Magellan spacecraft.

4 The Search for Extrasolar Planets

Only ten years ago, our knowledge of the existence of planets outside the solar system rested primarily on theoretical investigations. But since 1995, this has changed rapidly, with the discovery of more than 165 planets by October 2005. Unfortunately, so far, all of these planets, which orbit around main sequence stars, are of the Jupiter type and cannot be seats of life. In addition, four planets were discovered around pulsars, but here again the conditions for life are unfavorable. Nevertheless, great advances in instrumentation are expected to eventually result in the detection of Earth-like terrestrial planets. But how are such planets to be found?

4.1 The Recently Discovered Planets

Stars and planets exhibit a continuous range of sizes and a precise classification is a matter of definition. In astronomy, stars are defined as bodies which, at one time in their lives, show regular nuclear fusion of hydrogen to helium in their core. This definition clearly distinguishes stars from planets, where the core temperatures never reach high enough values for such fusion processes. But it also eliminates substellar objects such as *brown dwarfs*, a class of low-mass objects which are neither stars nor planets. During a star's formation, the temperature in the core of the protostar rises continuously, and the maximum temperature attained depends on the amount of accumulated mass (Chap. 1). Only for bodies with masses greater than 0.075 $M_\odot$ or 75 M_J does the temperature reach high enough values (10^7 K) to allow hydrogen to burn ($M_\odot$ is the mass of the Sun and M_J that of Jupiter). The surface temperature of these "minimal stars" is $T_{\text{eff}} = 2000$ K and they have a radius of about 70 000 km, which is roughly equal to that of Jupiter.

While planets have masses below 13 M_J, the mass range of brown dwarfs is between 13 M_J and 75 M_J. Two recent infrared sky surveys have found up to 100 brown dwarfs, from which the total galactic population can be estimated. Surprisingly, there are so many that they may outnumber stars by more than two to one. Brown dwarfs are distinguished from planets by the fact that their core temperatures reach 10^6 K and higher, which allows for short-lived nuclear burning of deuterium and lithium. While young brown dwarfs are similar to very-low-mass stars, older ones look very much like Jupiter. They

P. Ulmschneider, The Search for Extrasolar Planets. In: *Intelligent Life in the Universe*, P. Ulmschneider, Adv. Astrobiol. Biogeophys., pp. 73–85 (2006)
DOI 10.1007/11614371_4   Springer-Verlag Berlin Heidelberg 2006

Table 4.1. Some recently detected planets around main sequence stars. Shown are the distance from Earth and spectral type (Sp.T.) of the parent star, the planet's projected mass ($M \sin i$) in units of Jupiter and Earth masses as well as the semimajor axis (a), period, and eccentricity of its orbit (after Schneider 2005)

Star	Dist. (Ly)	Sp.T.	$M \sin i$ (M_J)	$M \sin i$ (M_E)	a (AU)	Period (d)	Ecc.
HD 75289 b	94	G0V	0.42	134	0.046	3.51	0.054
51 Peg b	48	G2IV	0.47	149	0.052	4.23	0.0
HD 187123 b	163	G5	0.52	165	0.042	3.10	0.03
HD 209458 b	153	G0V	0.69	219	0.045	3.52	0.07
v And b	44	F8V	0.69	219	0.059	4.62	0.01
c			1.89	670	0.83	242	0.28
d			3.75	1193	2.53	1284	0.27
55 Cnc b	44	G8V	0.78	249	0.12	14.7	0.02
c			0.22	69	0.24	43.9	0.44
d			3.92	1247	5.26	4517	0.33
e			0.045	14.3	0.038	2.81	0.17
HD 130322 b	98	K0V	1.08	343	0.088	10.7	0.048
ρ CrB b	54	G0V	1.04	331	0.22	39.9	0.04
HD 217107 b	121	G8IV	1.37	436	0.07	7.13	0.13
c			2.10	668	4.3	3150	0.55
HD 210277 b	72	G0	1.24	394	1.10	436	0.45
16 CygB b	70	G2.5V	1.69	537	1.67	799	0.67
Gliese 876 b	15	M4V	1.94	615	0.208	60.9	0.025
c			0.56	178	0.13	30.1	0.27
d			0.023	7.3	0.0208	1.94	0.0
47 Uma b	43	G0V	2.54	808	2.09	1089	0.061
c			0.76	242	3.73	2594	0.1
14 Her b	59	K0V	4.74	1507	2.8	1796	0.338
HD 195019 b	65	G3IV-V	3.43	1091	0.14	18.3	0.05
Gliese 86 b	36	K1V	4.01	1275	0.11	15.8	0.05
τ Boo b	49	F7V	4.13	1313	0.05	3.31	0.01
70 Vir b	72	G4V	7.44	2366	0.48	117	0.4
HD 114762 b	91	F9V	11.02	3504	0.3	83.9	0.34

have surface temperatures of about $T_{\text{eff}} = 900$ K, compared to Jupiter with 130 K, and their atmosphere contains large quantities of water vapor and methane. Likewise, their size is very similar to that of Jupiter. From physical appearance alone it is difficult to distinguish brown dwarfs from planets, but they can be differentiated by their spatial association. So far, planets have only been found as members of stellar systems, whereas brown dwarfs are predominantly free-floating objects, not associated with stars.

Table 4.1 presents a sample of the 165 planets (Schneider 2005) detected around main-sequence stars up to October 2005. Note that in the astronomical nomenclature additions such as lower case letters b, c, d, ... to the star names denote planetary companions, while upper case letters B, C, ... mark stellar companions. In this table, the distance from the Earth to the parent star is given in light years (1 Ly = 9.5×10^{17} cm), while the semimajor axis is given in astronomical units (1 AU = 1.5×10^{13} cm), which is the mean distance to our Sun. The masses are the projected masses ($M \sin i$; see below), which are close to the true masses, and for easy comparison they are given

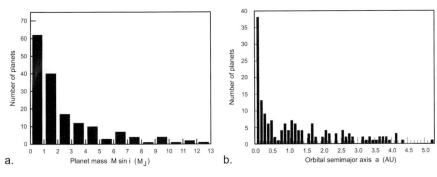

Fig. 4.1a. Mass distribution and **b.** orbital semimajor axis distribution of 165 planets with mass less than 13 M_J orbiting main-sequence stars, discovered up to October 2005. Data from Schneider (2005)

in Jupiter (M_J) and Earth (M_E) masses, where 1 M_J = 318 M_E = 2×10^{30} g. The planets found represent essentially a complete sample up to a distance of 90 Ly and are the result of a survey of more than 1000 stars. Using the data collected by Schneider (2005), Fig. 4.1 displays the mass and orbital semimajor axis distribution of the detected planets. It is shown that a third of the planets have masses of more than 2 M_J while the smallest detected planet Gliese 876 d has a mass of 0.023 M_J or 7.3 M_E (Table 4.1), so they all can be considered large or giant planets. Note that Beaulieu et al. (2006) just discovered an even smaller planet of 5.5 M_E using microlensing (see Sect. 4.5). The distance of Gliese 876 d with 0.0208 AU is the closest to a central star. Roughly half of the detected planets orbit within 0.5 AU, and of these the majority inside the orbit of Mercury (0.39 AU). This is puzzling, because the theory of jovian planet formation predicts that Jupiters form beyond the ice–formation boundary at 3 AU (discussed in Chap. 2).

That so far only large planets orbiting closely around their parent star are detected is very likely due to the peculiarities of the employed search method (discussed below) and one expects that the true distribution of planets will be different. Presently, no twin of the solar system has been found with jovian planets at distances comparable to or more than our Jupiter (5.2 AU). There is a system, 55 Cancri, with a jovian planet of 3.9 M_J at a similar distance (5.26 AU) but it has another giant planet of 0.78 M_J at 0.12 AU and two additional massive planets at 0.04 and 0.24 AU (see Fig. 4.1b and Table 4.1).

Note that in addition to the objects in Table 4.1, two jovian planets have been detected by microlensing (discussed below) and four planets around pulsars, two of which even have terrestrial-type masses (3.9 M_E and 4.3 M_E). However, as pulsars are the neutron-star remnants of a recent supernova explosion, these planets must have newly formed and are not likely to be seats of life.

4.2 Direct Search Methods for Planets

How does one find planets? The direct method of detecting planets is potentially the most powerful and will probably be the most productive in the
more distant future. With the direct method the reflected starlight or the
infrared radiation from a planet is observed. Presently, the problem with this
method is that at the great stellar distances the angle between the planet
and the star is so small that the weak emission from the planet is lost in the
blinding glare of the parent star. For example, if the Earth reflects 30% of
the visible light received from the Sun, the latter emits 2 billion times more
light than the Earth. This is different in the infrared, where at a wavelength
of 10 μm the Sun–Earth contrast is only 10 million. How can one resolve light
sources of such high contrast which are very close together?

While the diameter of our galaxy is about 100 000 Ly, with visible light one
typically sees only as far as about 3000 Ly in the galactic plane because of dust
obscuration. The angular distance between the Earth and the Sun as seen
from a point in the galaxy 3000 Ly away would be 1 milli arc sec, and the solar
radius would shrink to about 5 micro arc sec. In astronomical instruments,
a handy formula for the resolving power of a telescope in the optical spectrum
is $\alpha = 12/D$. Here D is the opening diameter of the main lens (in cm), and α
(in arc sec) is the angle that the telescope can resolve. For instance, to resolve
the radius of the Sun at a distance of 3000 Ly, one would need a telescope with
an opening of 24 km. This is much more than the largest existing telescopes
with an opening of 12 m, or even one of 100 m which is being planned at
the moment. But in future space projects, the construction of instruments of
this size, and even larger ones, is not insurmountable. It should be noted, as
discussed below, that for such instruments it is not necessary to have a full
parabolical mirror with a diameter of 24 km, but that it is sufficient to have
a few mirror segments in precisely controlled locations. Such instruments
could be set up on the Moon or on the asteroids, or built as free-floating
platforms in space.

Incidentally, the method of direct observation of planets would also permit analysis of their atmospheres. By finding the infrared absorption bands
of H_2O and particularly ozone O_3 in the 0.7–100 μm spectrum of terrestrial planets, one would be able to directly detect the existence of life (see
Fig. 8.13). Recently a high-mass companion has been imaged in the vicinity
of the K2V star AB Pictoris but this could be a brown dwarf. So far no single clear cut case of a planet near a main-sequence star has been found by
employing the direct method.

4.3 Indirect Search Methods

All the planets in Table 4.1 were found by the method of observing radial
velocity variations. When a planet orbits around its parent star, the two

bodies do not stay fixed, but instead orbit around a common *center of mass* (see Fig. 4.2a). One has the relation $M_S a_S = M_P a_P$, where M_S and M_P are the masses, a_S and a_P the orbital semimajor axes of the star and the planet, respectively. For the Sun with $M_S = 1$ $M_\odot$ and Jupiter with $M_P = 1$ M_J $= 1/1000$ $M_\odot$ as well as a semimajor axis of Jupiter of $a_P = 5.2$ AU, the semimajor axis of the Sun in its orbit around the common center of mass is $a_S = 0.0052$ AU $= 780\,000$ km, which is 10% larger than the solar radius. From large distances, therefore, it would be possible to deduce the existence of an unobserved Jupiter, because in the 12-year orbital period of Jupiter, the Sun's disk rotates around a point near its surface.

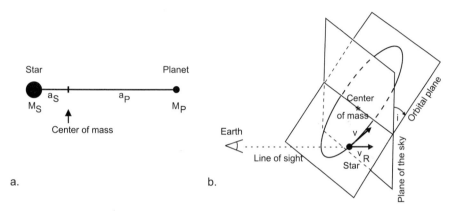

Fig. 4.2a. The center of mass of a star and its planet. **b.** The geometry of the orbit of a star around the center of mass

Although we cannot yet resolve stellar disks, except for those of a few very large supergiant stars, it is possible to detect the radial velocity variations of a star's orbit around the center of mass. This is achieved by carefully analyzing absorption lines in the stellar spectrum. Recent advances in the precision of these measurements now allow us to detect velocity variations of as little as 3 m/s. Figure 4.2b shows the geometry of such an orbit. The star orbits (solid) around the center of mass in a plane inclined at an angle i against the plane of the sky, the imagined celestial sphere perpendicular to the line of sight from the Earth. The radial velocity v_R (Fig. 4.2b) is measured by the Doppler shift of an absorption line of the stellar spectrum. Only the fraction $v_R = v \sin i$ of the true velocity v in the star's orbit is observed (see Fig. 4.2b).

Because of the inclination of the orbit, the mass of the planet from Kepler's law can only be determined as $M \sin i$, where $\sin i$ is a number between 0 and 1. This so-called *projected mass* is listed in Table 4.1. Only when $i = 90°$ and $\sin i = 1$, – that is, when we see the orbit edge on – is the projected mass equal to the true mass. However, since small angles i lead to low radial velocities and small Doppler shifts, it is clear that planets with such angles would

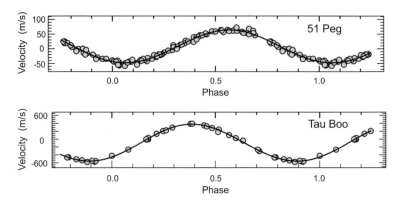

Fig. 4.3. Observed radial velocity variations due to planets (after Marcy and Butler 1998)

be difficult to detect. Therefore it is very likely that the unknown values of i for the detected planets in Table 4.1 are not much different from 90° and that the listed projected masses are close to the true masses. To illustrate the difficulty of detecting planets using the radial velocity method, it should be noted that in a recent monitoring campaign of 300 stars of spectral type F, G, K, and M (spectral types classify the stars according to their surface temperature, see Chap. 5) only nine stars were found to have planets with $M \sin i$ between 0.5–10 M_J.

Moreover, it takes a long time to measure tiny Doppler shifts over large orbital periods, such as the 12-year period of Jupiter. This is the reason why the detected planets mostly have small semimajor axes and short orbital periods, as the radial velocities are then large and one has a rapid variation with time. Figure 4.3, as an example, shows observed radial velocity variations for the stars 51 Peg and τ Boo. The Sun's radial velocity around the center of mass of the Sun–Jupiter system is 14 m/s and thus with the accuracy of Fig. 4.3 would not be detectable even if i were close to 90° because its variation would be equal to the width of the data points for 51 Peg.

The detection of planets of significantly smaller mass, such as terrestrial planets with typical masses of 1/300 M_J, would decrease the magnitude of the radial velocity variations by a factor of 300 compared to the values presently found for Jupiter-like planets. This clearly shows that for the detection of such planets velocity variations of 0.01 m/s should be measurable. As velocity observations with such amplitudes have been successfully made on the Sun, there is hope of being able to detect terrestrial planets using radial velocity variations in the future.

4.4 Circumstellar Disks

Circumstellar disks provide another indication of the existence of planets. Some of these disks are directly observed, while others are inferred from an excess of infrared radiation of their central star. Figure 4.4 shows the circumstellar disk of the A5V star β Pictoris, seen nearly edge on, which was first observed in 1984 and has an extent of about 1000 AU. A suspicious brightening of the star with a drop in intensity in the middle of this event was detected in 1995 and attributed to the passage of a giant planet in front of the star. Pronounced warps in the disk were discovered in 1998 and are also attributed to planets. However, no planet has been unambiguously detected so far. The same is valid for the disks of L1551 and BD+31°643 and for others around very young stellar objects (see Fig. 4.5) which were recently found with the Hubble Space Telescope in the Orion and Eagle nebulae.

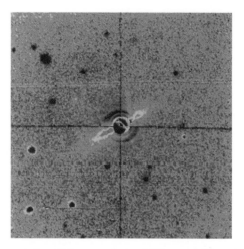

Fig. 4.4. The circumstellar disk around the A5V star β Pictoris. The *circle* in the center is a mask to block the light from the central star (courtesy of NASA)

Excess infrared emission, attributed to circumstellar disks, is detected both from young and old stars. Fifty percent of the young T-Tauri stars in the Taurus–Auriga as well as the Ophiuchus dark clouds, and 70–80% of the young stars in the dense star clusters of the Orion nebula have circumstellar disks (Sargent and Beckwith 1993), that are mostly accretion disks. But old stars (including the Sun) also have these disks. The solar dust disk is revealed by the *zodiacal light*, a diffuse glow, both in visible and infrared light, concentrated in the orbital plane of the Earth. Because after the T-Tau phase, the solar wind should have long ago swept any primordial dust out of the solar system (see Chap. 2), the excess infrared emissions indicate the presence of newly formed dust, generated by collisions of asteroids and by

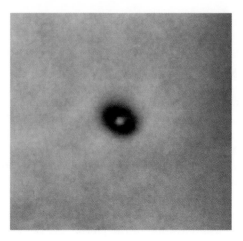

Fig. 4.5. A circumstellar accretion disk observed in the Orion nebula with the Hubble Space Telescope (courtesy of NASA)

evaporation from comets. The IRAS satellite discovered that 15% of A, F, G, and K main sequence stars have significantly (100 times) higher infrared excess emission than the Sun, which must be attributed to such debris disks. While these disks do not form planets, their presence nevertheless proves that planet-type bodies (asteroids and comets) had been produced during the original star formation process, which strengthens the theoretical picture that planetary systems are the natural byproduct of star formation.

4.5 New Search Strategies

In order to greatly increase the number of detected planets, a whole series of new imaginative instruments are being developed and put to work. One so-called *astrometric method* is to detect planets by watching out for the tiny wobbles in the star's position (as mentioned above) when it moves around the center of mass (Fig. 4.2). For this, the ESA mission *Gaia* (see ESA missions 2005) to be launched in 2011 is planned to increase the accuracy of the measurements of stellar positions a hundred times compared to ESA's very successful *Hipparcos* astrometry satellite, launched in 1989. Gaia is scheduled to orbit at the Lagrange point L2, 1.5 million km away from Earth in the opposite direction from the Sun. For stars within a distance of approximately 150 Ly, Gaia is expected to find every Jupiter-sized planet with an orbital period of 1.5–9 years. Estimates suggest that Gaia will detect between 10 000 and 50 000 planets beyond our solar system.

Scheduled to fly in the same year 2011 is the NASA mission *SIM Planet Quest* (see NASA missions 2005) that on the basis of ultra-precise astrometric observations can reveal even the much tinier wobbles of the target star due

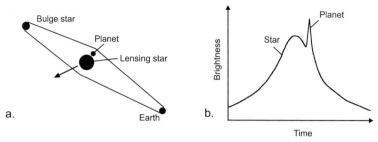

Fig. 4.6a,b. The geometry for gravitational microlensing and the light curve of a lensing star with a planet

to terrestrial planets. SIM stands for Space Interferometry Mission because its microarcsec angular resolution is achieved by an interferometer, where visible light from two telescopes mounted at the ends of a truss of 9 m length are brought to interference. SIM will survey roughly 200 stars within about 100 Ly for terrestrial planets. It will operate in an Earth-trailing orbit where the spacecraft will slowly drift away from the Earth at a rate of approximately 0.1 AU per year, reaching a maximum distance of about 95 million km after 5.5 years.

A different search method involves monitoring the intensity of stars in the densely populated galactic central bulge, in the hope of detecting *gravitational microlensing events* by stars midway to the bulge. Similar to that in other galaxies, the central bulge of our galaxy (see Fig. 1.3), is a large accumulation of stars near the galactic center, which protrudes considerably above the galactic plane, and thus can be observed relatively unobscured by dust. Bulge stars are employed in this search method because of their dense population, which allows us to simultaneously observe many stars in a small field of view.

The microlensing method detects the moment when a star and its planet move by chance through the line of sight from Earth to a background star (here a bulge star). Figure 4.6a shows the geometry of this event. The gravitational field of the parent star acts like a lens that focuses the light received from the bulge star. Upon moving into the light path, the lensing star causes the bulge star's light to brighten over the course of several days or weeks. A sudden jump in brightness above this bell-shaped light curve (Fig. 4.6b), lasting for a few hours, indicates the presence of a planet that also enters the light path.

With the distance to the galactic bulge of 24 000 Ly, the maximum efficiency of detecting a planet would be at a distance of about 12 000 Ly, for which the focusing power of a lensing star with a mass of 1 $M_\odot$ would allow planets as far away as 5 AU from its parent star to be detected. It is estimated that in this way hundreds of Jupiter-like and dozens of Earth-like planets might be detected. The disadvantage of the method, however, is that it is a one-shot observation in which little additional information other than the mass of the planet (derived from its diameter) and the distance to the parent star can be obtained. The first two planets detected by microlensing were OGLE235-MOA53 b and OGLE-05-071 b with masses of 2 and 2.7 M_J at distances of 2.9 and 3 AU, respectively (Udalski et al. 2005, Schneider 2005). Very recently, another planet named OGLE-2005-BLG-390Lb with a mass of 5.5 M_E was found by microlensing using bulge stars. Orbiting at a distance of 2.6 AU around an M dwarf star (Beaulieu et al. 2006), it has the smallest mass of all extrasolar planets except for pulsar planets.

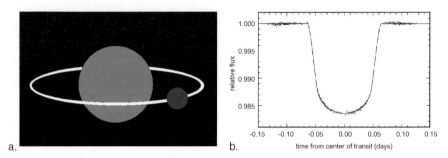

a.   b.

Fig. 4.7a. Occultation of a star by a planet. **b.** Light curve of the occultation of the solar-like G0V star HD 209458 by an orbiting jovian planet of 0.67 M_J at a distance of 0.045 AU (Schneider 2005)

Another photometric method is the high-sensitivity *monitoring of the stellar luminosity* of nearby stars; that is, looking for a brief decrease in the stellar intensity caused by a planet that transits in front of its parent star (Fig. 4.7a). As an example of this detection method Fig. 4.7b shows the light curve of the star HD 209458 (Table 4.1) that displays a photometric drop of 0.017 mag attributed to a planet. This planet had already been discovered earlier using the radial velocity method. To monitor photometric variations of a series of well-chosen stars and to search for planets, the spacecraft *COROT* (see CNES missions 2005) of the French Space Agency CNES, with a 30-cm telescope, will be launched in 2006 into a 896-km high polar orbit that every 6 months gets rotated to prevent the spacecraft from looking at the Sun. 6000 stars in selected star fields of $2° \times 2°$ size will be monitored five times per day for 150 days.

A more ambitious mission to find Earth-size planets by monitoring more than 100 000 stars in the constellation Cygnus brighter than 14th magnitude is the NASA project *Kepler* (see NASA missions 2005) to be launched in 2007. With a 0.95-m telescope and a $105° \times 150°$ star field it will circle the Sun in an Earth-trailing orbit. This mission could find 480 Earth-sized planets and 1400 larger ones. By excluding the very variable UV spectrum, the mission's photometers will be able to avoid spurious signals. That the detected photometric variation is due to a planet and not to sunspots or other stellar surface features is assured by the planetary transit times of 2–16 h and the very precise periods after which these events recur.

As already mentioned, direct methods should be the most rewarding in terms of information. There are two very promising approaches to a direct detection of planets, both try to dim the glare of the parent star compared to the planet. The invention of the coronagraph by B. Lyot in 1930 (see Fig. 4.8a) allowed us to view the tenuous corona outside a solar eclipse. Coronagraphs are now used also on spacecrafts such as SOHO to detect hundreds of comets in the corona (Sect. 2.7). Previously the corona was visible only during the infrequent and observationally difficult events when the Moon totally occults the solar photosphere.

In a coronagraph the image of the star, produced at the focus of the primary lens, is blocked by the first shade (Fig. 4.8a). At the focus of the secondary lens the primary lens is imaged and any stray light produced by imperfections in the primary lens or its holding is blocked by a second shade. The American *Terrestrial Planet Finder-Coronagraph* (TPF-C) mission (see NASA missions 2005) to be launched in 2014 will employ a coronagraph for planet search in visible light to detect terrestrial planets with masses greater than 0.5 M_E up to a distance of 30 Ly (see Fig. 4.9).

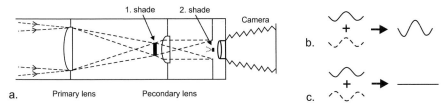

Fig. 4.8a. Light path in a coronagraph. **b.** Light waves with similar phase from two telescopes (*solid, dashed*) add. **c.** Light waves with opposite phase cancel. This is called constructive and destructive interference, respectively

Another very promising effect to block the light of the parent star, in addition to the *coronographic method*, is called *nulling*. Presently, nulling is extensively developed and tested at the ground in observatories such as Keck (on Mauna Kea, Hawaii), LBT (Large Binocular Telescope, on Mt. Graham, Arizona) and VLT (Very Large Telescope, on Mt. Paranal, Chile). Nulling works as follows. By adding the light from two telescopes without phase delay, that is by using precisely identical lengths of the light paths in the instruments, one gets constructive interference and the amplitude of the combined light wave is magnified (Fig. 4.8b). However, by slightly extending the light path in one of the telescopes, light waves of opposite phase can be produced that interact by destructive interference and cancel each other (Fig. 4.8c). Nulling thus allows us to block the unwanted light from the central star while the light of the planet is still visible.

Employing the nulling technology, the European *Darwin mission* (see ESA missions 2005) is scheduled to be launched in 2015. Darwin will consist of a flotilla of four free-flying spacecraft, three of them equipped with 3-m telescopes and a fourth one to serve as a communications hub. The success of Darwin and the later TPF-I projects will depend on the feasibility of controlling the length of the light paths in the cluster of spacecraft to distances of a few nm, when combining the light from the telescopes. This spacecraft cluster will be placed in an orbit at the Lagrange point L2, 1.5 million km from Earth in the direction away from the Sun.

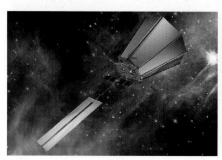 

Fig. 4.9. The Terrestrial Planet Finder TPF comprises two complementary observatories: a visible-light coronagraph TPF-C (*left*), to be launched around 2014, and a formation-flying infrared interferometer TPF-I, to be launched before 2020 (courtesy of NASA)

A follow-up mission to the TPF-C and similar to Darwin will be the *Terrestrial Planet Finder-Interferometer* (TPF-I) (see NASA missions 2005) that will employ the nulling technology in a cluster of free-flying spacecraft to be launched about 2020, and is expected to directly image terrestrial planets with masses greater than 0.5 M_E up to a distance of 300 Ly (see Fig. 4.9). With four 3.5-m telescopes, each on its own free-floating platform, 1000 m apart with a central spacecraft that houses the beam-combining apparatus these instruments will have the resolving power of a 1000-m telescope. In addition, the telescopes will operate in the infrared spectrum. Together with the fact (already mentioned above) that the infrared spectral range provides a much lower contrast between the central star and the planet, these nulling interferometers should be able to make planets directly visible and allow the discovery of a large number of planets in our galactic vicinity. In addition, by observing certain infrared spectral regions, these interferometers might detect molecular bands that indicate the presence of life (see Chap. 8).

For the years 2020 and later, much larger instruments, the LF (Life Finder) and the PI (Planet Imager) (see NASA missions 2005) are planned, based on the TPF concept, with baselines of 360–6000 km. These would look for life-revealing spectral signatures and even allow us to resolve the surfaces of nearby planets.

In addition to the many projects discussed above improvements are underway, both in space and on the ground, that although not aiming primarily for

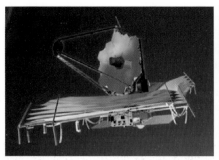 

Fig. 4.10. James Webb Space Telescope (JWST, *left*) and Atacama Large Millimeter Array (ALMA, *right*) (courtesy of NASA, ESO)

planet search will nevertheless have a great impact on the search for planets. A much more powerful space observatory, the *James Webb Space Telescope* (JWST) (see NASA missions 2005) will replace the ailing *Hubble Space Telescope* and operate mainly in the mid- and near-infrared (see Fig. 4.10).

On the ground, a particularly exciting project is the planned joint US–European Atacama Large Millimeter Array (ALMA) observatory (Fig. 4.10) (see ALMA 2005). This telescope would operate at millimeter wavelengths and consist of 64 antennas, each with a 12-m aperture and distributed over an area with a diameter of 10 km, located on a high plateau at an elevation of 5000 m in the Atacama desert at the Llano de Chajnantor, Chile. The observatory will use the atmospheric windows between 0.4 and 10 mm with a spatial resolution of 10 milliarcsec, which is 10 times better than that of the Hubble Space Telescope. ALMA will also be ideally suited to study the development of star-forming cloud cores and protoplanetary disks. It will allow us to observe the formation of planets and to search for protoplanets.

5 Planets Suitable for Life

When searching for extraterrestrial life, and particularly intelligent life, else-where in the solar system or in our galaxy, the obvious places to look are habitable Earth-like planets. This is because most living organisms are quite vulnerable to harsh conditions, and thus the presence of life will be most likely when very favorable conditions occur. Here organisms that survive un-der extreme conditions on Earth represent no contradiction, because they have adapted to their way of life by the fierce battle of survival on the basis of Darwin's theory (discussed in Chap. 7). But what are the conditions that are favorable for life?

Clearly, one of the most important requirements is energy. The fundamen-tal role of our Sun as a provider of energy, as a "source of life", has always been recognized by mankind. Too much sunlight, however, is life-threatening. In order to sustain life, therefore, a planet must orbit in the right distance range, the *habitable zone*, around its sun, and must have the right size. If, like Jupiter, it is too large, its surface will be entirely covered by a massive ocean of liquid hydrogen, where life is unable to form or survive. A planet also must not be too small – like our Moon, which, devoid of atmosphere and oceans, is likewise incapable of supporting life. In addition, the uninterrupted de-velopment of the present life forms needed billions of years of a relatively benign environment. The existence of extraterrestrial intelligent life there-fore requires a substantial list of favorable conditions, and the question is: Do such Earth-like planets with benign environments exist elsewhere, and how many of them can be expected in our galaxy?

5.1 Habitable Zones

As far away as we can see in our universe, out to the most distant galaxies and quasars, everywhere matter is composed of the same chemical elements as found on Earth. Thus extraterrestrial life, if it exists, must also be composed of these elements. However, one element among the roughly 110 is outstand-ing: carbon. This element is exceptional because it forms large numbers of different compounds. Presently, more than 10 million carbon compounds are known, the so-called *organic compounds*, compared to only about 200 000 *inorganic compounds* composed of all the other elements.

P. Ulmschneider, Planets Suitable for Life. In: *Intelligent Life in the Universe*, P. Ulm-schneider, Adv. Astrobiol. Biogeophys., pp. 87–113 (2006)
DOI 10.1007/11614371_5   Springer-Verlag Berlin Heidelberg 2006

In order to ensure the basic functions of life, even the most simple living organisms are made up of a large number of different building blocks of intricately designed form and complicated function. Only organic compounds have the huge variety, the different shapes and functions, required of such construction materials of life. It is thus no surprise that life on Earth is based on organic chemistry. Certainly, life could be imagined based on other types of chemistry; for instance, on silicon (see Chap. 9), which is used in the world of computers. However, it is most likely that on other planets the ready availability of a large variety of organic compounds will not be ignored and will invariably lead to organisms that employ a similar chemistry to that on Earth. Nature almost always attempts to take the easiest road to success. One must therefore assume that life on other worlds probably also started using organic chemistry.

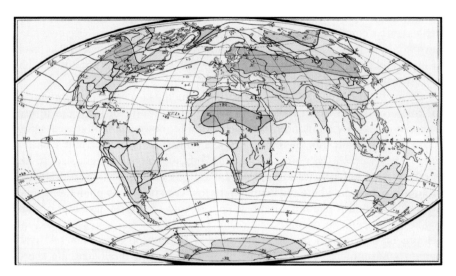

Fig. 5.1. Climatic zones on the Earth, with isotherms of the mean annual temperatures. The 30 °C isotherm encloses the central Sahara, marked *dark red*, the 0 °C and −10°C isotherms border Arctic and Antarctic regions, shown *light and dark blue*

5.1.1 The Solar Habitable Zone

The vast majority of biologically active organic compounds function best in a certain temperature range and under well defined conditions, like e.g. in solution of water. This temperature range can be seen from the variety of temperatures at which living organisms thrive. Figure 5.1 shows the habitable regions on Earth. In the very cold Arctic and Antarctic regions, where the mean annual isotherms fall below −10°C, little or no life exists. The same can be said for the extremely hot desert regions, where the mean annual isotherms rise above 30 °C. While the −10°C annual isotherms correspond roughly to lines at about 60 degree northern and southern latitude, the 30 °C isotherm roughly coincides with the central region of the Sahara close to the Equator. At a latitude of 60 degrees (see Fig. 5.2) the amount of solar energy received per cm^2 and sec, the so-called *solar radiative energy flux*, is only one half of that at the Equator (because $\cos 60° = 1/2$). Thus life needs a temperature range where the solar radiation flux is allowed to vary in a very narrow range of at most a factor of two.

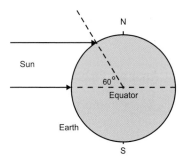

Fig. 5.2. The variation of the solar radiation flux with latitude on the Earth

Let us now conduct a thought-experiment. Assume that one could vary the distance of the Earth from the Sun (see Fig. 5.3). This distance is 149.5 million km, and is called one astronomical unit (AU). How close could one bring the Earth toward the Sun and still have life on it, and how far would one be able to move it away? Since the solar radiation flux varies with the square of the distance from the Sun and as the flux is allowed to increase or decrease by a factor of two, one could move the Earth inward as close as 0.7 AU (with $(0.7)^2 = 1/2$) or outward as far as 1.4 AU (with $(1.4)^2 = 2$). Therefore the distance from the Sun in which life is possible varies by a factor of two, and the solar radiation flux in this zone changes by a factor of four. These distances correspond roughly to the orbits of Venus and Mars (see Table 5.4). One calls the region where, in principle, life is possible the *ecosphere* or *habitable zone*. As the precise distances of the inner and outer boundaries of the solar habitable zone are determined from a number of time-dependent effects, it should be noted that the range of the solar habitable zone from 0.7 to 1.4 AU represents only a reasonable first estimate. More accurate values will be given later.

Moving the Earth to the inner boundary of that zone, the Sahara desert would extend essentially up to the poles and all land surfaces would become unbearably hot. On the outer boundary, the polar ice fields would extend all the way to the Equator and life would no longer be possible in a completely frozen world (Fig. 5.3). These two types of fate were essentially what happened to Venus and Mars.

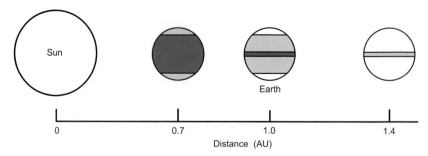

Fig. 5.3. Habitable regions (*light gray*) if Earth were at various distances from the Sun. Arctic regions are shown in *white* and desert regions in *dark gray*

5.1.2 Habitable Zones Around Other Stars

Now consider the habitable zones around other stars. Stars differ in size (there are giant and dwarf stars) and in surface temperature. Since the overwhelming majority of stars are dwarfs, or more precisely *main sequence stars* (see Chap. 1), and as the giant stars are late evolutionary stages of main sequence stars, we can exclude the few giants from our considerations. Table 5.1 displays the different types of main sequence stars.

For historical reasons, the stars are classified by the letters O to M (to be memorized by the famous phrase "Oh Be A Fine Girl (Guy) Kiss Me"). These letters denote *spectral types* and signify the variation of the star's surface temperature (or, more precisely, the *effective temperature*). An O-star (see Table 5.1) has an effective temperature of roughly 41 000 K and a G-star of about 5800 K, while an M-star of only about 3200 K. Since stars of a given spectral type have a range of temperatures, there is a finer subdivision; for example, G0 to G9. As a star also varies in size, an additional specification, the

Table 5.1. The spectral class, effective temperature T_{eff}, luminosity, lifetime, abundance, range of habitable zone, and tidal lock radius of main sequence stars (after Landolt-Börnstein 1982)

Spectral class	T_{eff} (K)	Stellar luminosity ($L_\odot$)	Stellar lifetime (y)	Stellar abundance (%)	Habitable zone (AU)	Tidal lock radius (AU)
O6V	41 000	4.2×10^5	10^6	4×10^{-5}	450–900	1.9
B5V	15 400	830	8×10^7	0.1	20–40	1.1
A5V	8 200	14	1×10^9	0.7	2.6–5.2	0.8
F5V	6 400	3.2	4×10^9	4	1.3 2.5	0.7
G5V	5 800	1	2×10^{10}	9	0.7 1.4	0.6
K5V	4 400	0.15	7×10^{10}	14	0.3–0.5	0.5
M5V	3 200	1.1×10^{-3}	3×10^{11}	72	0.07–0.15	0.4

luminosity class, is introduced, indicated by roman numerals. Dwarf stars – that is, main sequence stars (see Chap. 1) – have luminosity class V, giants class III, and supergiants class I. The stars mentioned in the table therefore have the spectral classes O5V to M5V. Our Sun, for instance, is a G2V-star.

To find the habitable zones for other stars, one uses the effective temperature to calculate the radiative flux and compares it with the solar radiative flux received on Earth. One then computes the distances at which the radiative flux of a given star is as large as that in the solar habitable zone. The results are shown in Table 5.1. It can be seen that the habitable zone of an O-star (450–900 AU) lies at a great distance because the star is very hot. For M-stars the habitable zone lies close, at 0.07–0.15 AU, as these stars are cool. Yet in each case, the distance from the star varies at most by a factor of two over the habitable zone.

5.2 Planetary Mass and the Evaporation of the Atmosphere

It is important for the existence of life that a planet has the right mass. As discussed in Chap. 2, planets can be divided into three broad classes: *Terrestrial planets* are Earth-size objects which occupy the inner regions of a planetary system. They have masses roughly like that of the Earth (1/10–5 M_E, where M_E (Table 2.1) is the mass of the Earth). The large Jupiter-like *jovian planets* consist mostly of hydrogen and have masses of 10–4000 M_E. Finally, the *Kuiper belt objects* are small planetary bodies and comet nuclei with masses less than 1/1000 M_E that orbit the Sun at large distances, beyond the belt of the jovian planets. Planet formation theories (see Fig. 2.5) provide estimates of the distances of the three types of planets around stars. These distance ranges for stars of different masses and spectral types are shown in Fig. 5.4. Terrestrial planets lie between the two dashed lines. The left line is the inner boundary for the formation of planets in a solar system, while the right line is the ice-formation boundary, which marks the beginning of the region where jovian planets form (Chap. 2). There is no well defined boundary between jovian planets and Kuiper belt objects further out. Note that the planets of the solar system (except Pluto) are indicated by dots.

The very massive jovian planets are completely covered by oceans of liquid molecular hydrogen (plus small amounts of helium). They are inhospitable to life, because any organic or inorganic compound would sink to the bottom of such an ocean, due to the very low specific weight of hydrogen. There, it would become entrapped in the region in which hydrogen becomes metallic.

Figure 5.4 also displays the habitable zones (HZ, gray) and it is seen that except for some M-stars they roughly coincide with the range of terrestrial planets. As Kuiper belt objects orbit far outside the habitable zones, they must be excluded from being thought of as life-supporting. This leaves

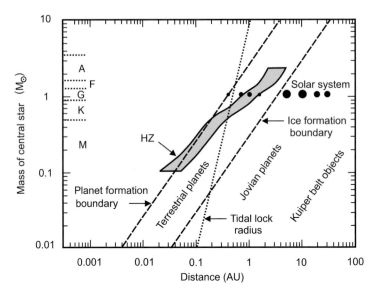

Fig. 5.4. Habitable zones (HZ) and the ranges of terrestrial planets for various stars (after Kasting et al. 1993)

only terrestrial planets as possible seats of life. However, not every terrestrial planet is suitable. As for the planets with a large mass, those with too little mass must also be excluded. This is because every life-bearing planet or moon must be able to retain an atmosphere. If the gravitational attraction is too small to hold an atmosphere, a planet in the habitable zone would lose its oceans by evaporation and eventually show only a solid surface similar to that of the Moon. This does not mean that, under unusual circumstances, an ocean might not be retained. In Chap. 8 we will discuss how far outside the habitable zone, Jupiter's moon Europa probably has an ocean under a surface layer of ice, in which primitive life might possibly exist.

To see why the planetary mass is important for holding an atmosphere, two characteristic speeds are discussed, the *escape speed* from a planet and the *thermal speed* of the gas. Consider some piece of matter, a meteorite, somewhere in space. When the meteorite comes close to a planet, it feels its gravitational pull and begins to fall toward it. The closer the meteorite approaches, the greater the planet's gravitational attraction becomes and the more the meteorite gets accelerated. Finally, just before it impacts on the surface of the planet, it reaches its maximum speed. This speed is called the escape speed, because a piece of matter from the planetary surface will have to be accelerated to exactly this speed in order to overcome the planet's gravitational attraction and escape from it. The escape speed does not depend on the size of the escaping object: it is the same for a gas molecule, a meteorite, or a space vehicle and only depends on the mass and radius of the attracting body. Table 5.2 shows the escape speed v_{esc} of some planets and the Moon.

Table 5.2. The mass, radius, and escape speed for various planets and the Moon

	Mass (g)	Radius (km)	$v_{\rm esc}$ (km/s)
Earth	6.0×10^{27}	6378	11.2
Mars	6.4×10^{26}	3395	5.0
Moon	7.4×10^{25}	1375	2.4
Ceres	1×10^{24}	450	0.54

To understand the other characteristic speed, assume for example that the Moon from an impact event temporarily obtains a tenuous atmosphere. The gas particles in this atmosphere, due to the intense radiation of the Sun, then attain temperatures as high as 390 K (117 °C) on the lunar day side, and move around rapidly, with a so-called *Maxwellian velocity distribution*.

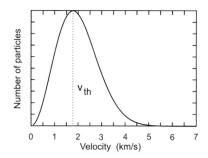

Fig. 5.5. The Maxwellian velocity distribution around the thermal speed $v_{\rm th}$ of a gas consisting of hydrogen molecules at a temperature of 390 K. Most particles have speeds in the range 1–3 km/s

Figure 5.5 shows the Maxwellian velocity distribution for hydrogen at a temperature of 390 K. It is seen that the H_2 molecules have different speeds and that, at this temperature, they move most frequently with a velocity of 1.8 km/s, called the thermal speed, $v_{\rm th}$. The reason why individual molecules move with different speeds is that they collide with each other, and due to this some get slowed and others accelerated. $v_{\rm th}$ depends on the temperature and also on the mass of the molecules. Other gas components of the atmosphere will have similar distributions. The lightest particles, H_2, have the highest velocity, while nitrogen molecules, N_2, are slower, with a thermal speed of 0.5 km/s. A tiny fraction of the molecules (in fact, 3×10^{-12} of those moving with $v_{\rm th}$) can even reach speeds of more than five times $v_{\rm th}$; that is 10 km/s for H_2 and 3 km/s for N_2.

As a considerable fraction of the H_2 and N_2 molecules have speeds greater than the lunar escape speed of 2.4 km/s (Fig. 5.5), they will be lost from the

Moon into space. However, collisions among the gas particles will reestablish the Maxwellian velocity distribution and thus rapidly moving particles will again be created in the remaining atmosphere. This will cause a continuous stream of particles to be lost into space, and it is clear that it will only be a matter of time until the entire atmosphere has evaporated. Therefore, if the mass of a planet is considerably less than that of the Earth (such as the Moon, with $1/81$ M_E; see Table 5.2), the escape speed will be so low that this body cannot retain an atmosphere, and oceans will evaporate as well.

5.3 The Lifetimes of the Stars

The right mass and orbiting in a habitable zone around its star are not the only requirements for a planet as a seat of intelligent life. As shown in Table 5.1, the stars do not only differ in temperature, but also have very different *lifetimes*. While every star eventually dies, some stars live much longer than others. The finite lifetime of the stars is a consequence of their energy loss by radiation. For main sequence stars, the energy loss from the emitted light is balanced by energy generated from nuclear fusion of hydrogen to helium in the stellar core. Since there is only a fixed amount of hydrogen in the core, stars can radiate only for a limited time. Once the hydrogen in the core is exhausted, the stars will develop into giants and subsequently degenerate into white dwarfs, neutron stars, or black holes (see Chap. 1). As these latter stages of stellar evolution are very hostile to life (Sect. 5.5), it is the main sequence hydrogen burning time of a star, called the lifetime in Table 5.1, which is important for the generation and support of life.

The massive O-stars expend huge amounts of energy and therefore have a very short lifetime of about a million years. G-stars, like our Sun, live for roughly 12 billion years. Since the Sun (together with the Earth and the other planets) was born 4.6 billion years ago, roughly half of its lifetime is already spent. The real masters of economic expenditure of energy, however, are the M-stars, which live for up to 300 billion years.

The development of intelligent life is very complex and time-consuming. Although life appeared on Earth about 4 billion years ago, intelligent life has only existed for 2.5 million years. As will be discussed in Chap. 7, there are good reasons why the development of higher forms of life and of intelligent life took so long. Here it suffices to conclude that in order to produce intelligent life, a star must live at least as long as it took to generate intelligent life on Earth; that is, about 5 billion years. If one insists on such a minimum lifetime for stars, it can be seen from Table 5.1 that because of their short lifetime, the O-, B-, A-, and F-stars are not suitable. This, however, should not cause too much concern, because these types of stars, as seen from the same table, are not very frequent among the stars of the universe.

5.4 Tidal Effects on Planets

There is another more important obstacle to the existence of life near other stars. In our thought-experiment above, the distance between the Earth and the Sun was changed. However, to move the Earth toward the Sun could be quite dangerous, since this would strongly increase the *tidal forces* exerted by the Sun on our planet. Tidal forces owe their existence to the fact that gravitational forces vary with distance. On the Earth it is the Moon that produces the largest tides. Figure 5.6 shows (exaggerated) the various forces exerted by the Moon on the Earth. Since gravity varies with the square of the distance, the gravitational force (dashed) is much larger on the side of the Earth closest to the Moon. The gravitational force is smaller at the Earth's center and even less on the far side of the Earth. The Moon and the Earth both orbit around their common center of mass, whereby centrifugal forces are created that prevent the two bodies from crashing into one another. These centrifugal forces (dotted) have the same magnitude everywhere on the Earth and balance the gravitational forces. When they are subtracted from the gravitational forces, they cancel at the Earth's center but produce residual forces called tidal forces at the Earth's near and far sides (Fig. 5.6, solid arrows), which pull oceans and landmasses in opposite directions relative to the Earth's center. As the response of the oceans to the tidal forces takes time to build up, the observed maximum of the tidal bulge at a given point lags behind the culmination direction of the Moon.

The deformation of land and the displacement of water in the oceans both use up energy, which slows the Earth's rotation (an effect called *tidal braking*) and increases the distance to the Moon. Because of tidal braking, the rotation rate of the Earth has decreased from about a 5-hour day, four billion years ago, to the present 24-hour day. This increase of the length of the terrestrial day has caused the Moon (due to the law of angular momentum conservation) to move away from a distance of about 22 000 km (Chap. 3) at its formation to one of 380 000 km today.

The Sun also exerts tides, which are weaker because of its greater distance, and the observed tides on the Earth are a combination of the strong lunar

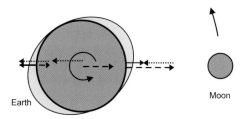

Fig. 5.6. Tides on the Earth generated by the Moon. Tidal forces are shown *solid*, gravitational forces *dashed*, and centrifugal forces *dotted*

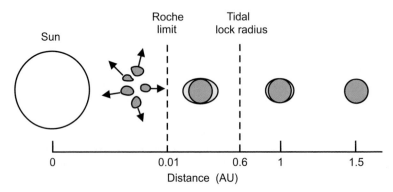

Fig. 5.7. Tidal effects as a function of distance

tides and the weaker solar tides. Now consider a planet without a moon (Fig. 5.7). As indicated in the figure, moving it toward the sun will increase the tides exerted on it. There is a certain distance, the *tidal lock radius*, where because of tidal braking the axial rotation of the planet has reduced so much that after times of a few billion years it becomes synchronized (locked) with its orbital rotation. In this state, an orbiting planet will always show the same side toward its sun. This is called tidal locking. An example of such a tidally locked rotation is the Moon, which always shows the same side toward the Earth. That the Moon's rotation is tidally locked and the Earth's rotation is not is due to the much stronger tides exerted by the Earth on the Moon. Other examples of tidal locking are the four Galilean moons of Jupiter. For a discussion about the possible presence of life under very special circumstances on such moons, see Chap. 8.

Since Venus is close to the tidal lock radius of the Sun, its rotation is almost locked (actually it is retrograde, the rotation axis being opposite to the orbital axis; see Table 5.4). One Venus day lasts 2800 hours; that is, 117 terrestrial days. Because of its highly eccentric orbit at 0.39 AU, well inside the tidal lock radius, Mercury has a locked rotation, with a 2/3 ratio of the rotational (59 days) and orbital (88 days) periods. Because it has only 1/20 the mass of the Earth, it holds only an extremely thin atmosphere. Its surface conditions are among the harshest in the solar system. During the Mercurian day the temperature rises to about 425 °C, and at night, drops to −180°C, which is among the coldest surface temperatures found in the Solar System.

This shows that on planets that orbit within the tidal lock radius, one expects extreme weather conditions. One side is persistently heated, and the other lies in long-lasting or permanent darkness. If the heat exchange due to atmospheric circulations or oceanic currents is not exceptionally efficient, this probably means that no liquid water can exist on a tidally locked planet. One thus must exclude such planets as carriers of life, because life cannot form or survive in an environment of purely solid and gaseous materials. Moving closer to the star, there is an even more dangerous boundary, the *Roche limit* (Fig. 5.7), where the tidal forces are so strong that they will break up the planet.

Considering the tidal lock radii for the different stars displayed in Table 5.1 and also shown dotted in Fig. 5.4, it can be seen that the habitable zone lies within the tidal lock radius for K- and M-stars. For these stars it means that a planet in the habitable zone is tidally locked and life cannot develop there. As can be seen from the frequency of different types of stars in Table 5.1, disregarding tidally locked planets now removes 90% of all stars as potential seats of extraterrestrial life. As a result, one finds that only G-stars, like our Sun, are suitable for life, because they are long-lived enough, and as their planets orbiting in the habitable zone would not suffer too much from tidal effects.

5.5 The Increase in Solar Luminosity and the Continuously Habitable Zone

So far, the requirements for life-supporting planets have been discussed in a static way, based on today's solar system. In reality, both the Sun and the planets show considerable development over time. While the Sun has increased its luminosity over 4.6 billion years, causing the habitable zone to move outward, the distance of the planets has changed very little. In order always to stay in the habitable zone, therefore, a life-carrying planet must orbit in a much narrower distance range, the continuously habitable zone.

Figure 5.8 shows the development of the luminosity during the evolution of a star like our Sun (1.0 $M_\odot$, spectral type G2V) from the main sequence phase to the red giant state. The luminosity L is the total energy loss of a star per second and in the figure is given in units of the present solar luminosity $L_\odot$. The time is measured from the start of the main sequence phase in which the star began to be powered by stable hydrogen fusion in its core. In this phase, mentioned in Chap. 1, the steady increase of luminosity is due to the fusion of four hydrogen nuclei into one helium nucleus, which decreases the number of particles in the stellar core. As a result, the gas pressure is lowered, leading to gravitational contraction and shrinking of the core, which raises its temperature. Since the fusion rates depend strongly on temperature, this causes an enhanced energy production and hence increased luminosity.

In addition, Fig. 5.8 displays the development of a slightly more massive star (1.1 $M_\odot$, spectral type F8V) and a less massive star (0.9 $M_\odot$, spectral type G5V), showing that the main sequence lifetime depends strongly on the stellar mass, as already mentioned previously. The dashed line indicates the present solar luminosity. It can be seen that, at the start of the main sequence phase 4.6 billion years ago, the Sun was only 65% as bright as today.

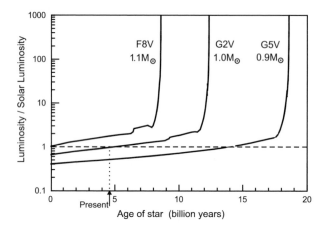

Fig. 5.8. The main-sequence evolution of the luminosity of solar type stars (after Bressan et al. 1993)

The distance of the habitable zone varies with the luminosity L as $\sqrt{L}$. Hence one finds that if today's habitable zone lies between 0.7 and 1.4 AU, it extended from 0.56 to 1.13 AU at the start of the main sequence phase. Therefore only those planets that orbit between 0.7 and 1.13 AU have always been in a habitable zone over the past 4.6 billion years. This narrower region, in which a planet where intelligent life develops must stay, is called the *continuously habitable zone* (CHZ). The quoted boundaries of the habitable zones can be determined more accurately when time-dependent instabilities are considered, it will be shown below that the true continuously habitable zone is even narrower (see Sect. 5.9). Seven billion years from now, the Sun will leave the main sequence phase and, starting with a rapid rise to 3000 times its present luminosity (Fig. 5.8), will reach the red giant stage (Chap. 1). In this stage, because of the scorching solar irradiation, the present forms of life will no longer be possible on Earth.

5.6 Instabilities of the Planetary Atmosphere

It is not only the evolution of the central star that restricts the number of habitable planets. There are also severe instabilities of the planetary atmosphere, caused by biological and chemical processes which, in the course of time, affect the habitability of planets. Generally there are two potential dangers for a planet: *irreversible glaciation* and the *runaway greenhouse effect*. Both processes must be avoided during planetary evolution if life is to survive to a stage at which intelligence develops. Another recently discovered danger is severe climatic changes due to the large *rotation axis variations* caused by gravitational interactions between the planets.

5.6.1 The Greenhouse Effect

After a terrestrial planet has formed, its atmosphere is generated mainly by impacting bodies (Chap. 2) and by outgassing of the upper mantle via volcanos and fumaroles, which produce gases such as nitrogen, carbon dioxide, methane, and water vapor (N_2, CO_2, CH_4, and H_2O). Because the latter three of these gases absorb infrared light, they are called greenhouse gases.

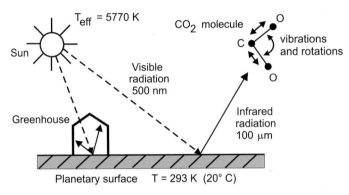

Fig. 5.9. The greenhouse effect, visible and infrared radiation

In Fig. 5.9, sunlight (dashed arrows), with a typical wavelength of 500 nm (the maximum of the spectral radiation curve for a G-star), falls onto the planetary surface and is absorbed. As a result, the Earth's surface is heated and radiates back in the infrared spectral region (Fig. 5.9, solid arrows). For a terrestrial surface temperature of 20 °C, this means light at a wavelength of 100 μm. However, infrared light easily gets absorbed by the atmosphere, where it leads to vibrations and rotations of the molecules of the greenhouse gases. Thus the incoming energy from the Sun gets trapped by the atmosphere and raises its temperature. This effect is called the *greenhouse effect* because it describes what happens in a greenhouse (Fig. 5.9), where visible light enters through the glass roof and walls, and the generated infrared light (heat) gets trapped inside, because glass is opaque for infrared radiation.

5.6.2 The Carbonate Silicate Cycle

Since the Earth possesses large amounts of water, atmospheric carbon dioxide can easily dissolve to form carbonic acid: $CO_2 + H_2O \rightarrow H_2CO_3$. This acid weathers silicate rocks (consisting mainly of $CaSiO_3$) and forms calcium carbonate (chalk $CaCO_3$) which precipitates as a solid. By this process, large amounts of CO_2 are taken out of the atmosphere and deposited in the form of calcium carbonate on the ocean floor. Without any means of replenishing the atmospheric CO_2, all carbon would be removed from the atmosphere in this way in about 400 million years. A main regeneration process is provided by plate tectonics, where the ocean floor together with the chalk gets subducted into the Earth's mantle where it is heated up and cracked, by which CO_2 is released and vented via volcanoes back into the atmosphere again. This circular process is called the *carbonate silicate cycle*. It rests primarily on the availability of liquid water. Together with the *biological cycle*, where CO_2 from the atmosphere is taken up in photosynthesis by biological organisms and subsequently returned by respiration and burning, the carbonate silicate cycle regulates the amount of carbon dioxide in the atmosphere and thus determines the magnitude of the greenhouse effect. The cycle strongly depends on temperature, since the efficiency of weathering is greatly reduced at low temperature. Long periods of cold climate will therefore cause a growth of the amount of atmospheric CO_2 due to reduced chalk formation, which in turn will enhance the greenhouse effect and raise the temperature. The carbonate silicate cycle thus acts as a thermostat.

5.6.3 The Runaway Greenhouse Effect

Consider a planet much closer to the Sun, where the surface temperature never reaches sufficiently low values (of $100\,°C$) to permit the formation of liquid water. On such a planet the weathering of stone and the subsequent removal of CO_2 from the atmosphere by carbonate formation never happens. On the contrary, the more CO_2 accumulates in the atmosphere by outgassing, the stronger is the greenhouse effect and the higher is the attained temperature. This effect, where the individual processes reinforce one another to raise the temperature, is called the runaway greenhouse effect. Venus very likely has become the victim of such an effect with the result that this planet today has a very dense CO_2 atmosphere and a surface temperature of $480\,°C$.

Although Venus has ultimately suffered a runaway greenhouse effect, it is presently thought that in the early history of the solar system, because of the much reduced solar luminosity, this planet not only had much lower surface temperatures but carried water in amounts comparable to Earth. That this water is all gone today is due to an effect called the *moist greenhouse effect*. When the surface temperature of the planet rises due to the growth of the stellar luminosity, the hot oceans produce large quantities of water vapor, by which the H_2O concentration is raised in the high atmosphere, the troposphere and stratosphere. In the stratosphere the H_2O molecule gets split up by the solar UV radiation and hydrogen escapes from the planet. This process is called *photolysis of water*, and through it large amounts of water can be lost such that eventually the oceans evaporate and a runaway greenhouse effect based on CO_2 sets in.

Computations by Kasting et al. (1993) show that these processes start when the global surface temperature reaches values of about $67\,°C$. From this temperature and today's solar radiation flux, the authors find a distance of 0.95 AU for the present inner boundary of the habitable zone, which is much larger than the value of 0.7 AU found above, when instabilities were not taken into account. This revision of the inner boundary of the habitable zone (and also of the CHZ) allows the Sun's luminosity to grow only by another 11% until the inner habitable zone reaches the orbit of the Earth and a moist greenhouse crisis will start. Therefore, from Fig. 5.8 it can be seen that mankind has only about another billion years before the Earth becomes uninhabitable.

5.6.4 Irreversible Glaciation

There is a second instability that threatens the habitability of planets. Table 5.3 shows the *albedo*, the reflectivity of different terrestrial materials. It can be seen that clouds and ice have a high, rocks a low, and oceans a very low albedo. Let us assume that by chance at a given site on the planetary surface the temperature becomes low such that ice precipitates. Since ice has a very high albedo, most of the incident solar radiation then gets reflected back into space (see Fig. 5.10), which reduces the heating of that planetary site. In addition, by emitting infrared radiation, the region loses energy and the surface temperature gets cooler, whereby more ice precipitates. This increases the size of the reflecting region, which lowers the heating even more and leads to still cooler temperatures, and so on.

In principle, such an instability process can progress until the whole planetary surface is solidly frozen. The ice sheets of the Earth's polar regions would constitute such a process, were it not for powerful atmospheric circulations and ocean currents from warmer planetary regions, as well as heat flow from the mantle, which strongly oppose this cooling instability. However, for planets where the temperatures of the hottest regions are barely above freezing, these global heat flows might become insufficient at some point in the planetary history, resulting in a complete freezing of the planet. This effect is called *irreversible glaciation* and is thought to have happened on Mars. Since the atmospheric cooling instability depends on the magnitude of the solar radiation flux, it determines the outer boundary of the habitable zone.

As a result of the carbonate silicate cycle discussed above, a complete glaciation could in principle be prevented or even lifted again by volcanism,

Table 5.3. The albedo of different materials

	Albedo (%)
Oceans	4
Rocks	15
Clouds	52
Ice	70

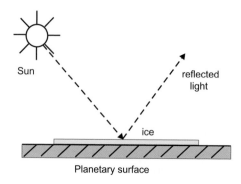

Fig. 5.10. The reflection of sunlight by ice prevents heating of the surface

which leads to an enhanced greenhouse effect due to the growth of the CO_2 concentration. But this warming reaches a limit when extensive CO_2 clouds appear, which due to their high albedo then strongly counteract the greenhouse effect. CO_2 cloud formation thus marks the true outer boundary of the habitable zone, which today would be at 1.67 AU (Williams 1998). However, such a life-saving greenhouse warming depends on the efficiency of plate tectonics; that is, on the heat convection from the planetary interior. Ultimately, this thermal convection is related to the amount of heat stored in a planet, which greatly depends on its mass.

Mars apparently had to suffer an irreversible glaciation because with only one tenth of the mass of the Earth it had a much lower heat content. The small mass of this outermost terrestrial planet is a peculiarity caused by the tidal effects of Jupiter on the Mars-forming planetesimals (see Chap. 2). Because of the low mass, Mars cooled much more rapidly than Earth, and volcanism died out billions of years ago, stopping the replenishment of the greenhouse gases. Since on Mars a global ice cover is not observed today, one might ask how the large amounts of ice expected from an irreversible glaciation vanished. This apparently was also related to the small mass of that planet, and to photolysis of water by UV light. The hydrogen from the split water molecule easily escaped into space. In addition, the small mass led to a much slower sedimentation rate of iron to the planet's core at the time of its formation by which more iron remained at the surface and in the outer mantle of Mars. Oxygen, the other splitting product from the photolysis, was used to oxidize iron to Fe_2O_3 (rust), causing the red color of Mars. By this process large quantities of ice have probably been removed until the surface soil shielded the buried ice from further destruction by UV light. It is very fortunate for our understanding of the ancient history of Mars (Boynton 2002) that the missions Mars Odyssey and Mars Express (see Mars missions 2005) could detect subsurface permafrost ice layers with gamma-ray spectrometers and radar observations (see Chap. 8).

5.7 Axis Variations of the Planets

Another important atmospheric instability arises from variations of the inclination of the planetary rotation axis. For the Earth that axis, also called the *spin axis* presently has an inclination of 23.5° against the axis of the Earth's orbital motion around the Sun (see Fig. 5.11). Table 5.4 displays the inclination angles of the orbital axes of the planets of the solar system against the Earth's orbital axis. Except for Pluto and Mercury the planetary orbits have a relatively small inclination against the Earth's orbit. The table also shows the spin axis inclinations. Here one finds that except for Uranus, all of the spin axes of the planets are more or less aligned with their orbital axes. The deviation of Uranus is generally explained by a giant collision at its formation, while the retrograde rotation of Venus is attributed to tidal effects due to its proximity to the Sun.

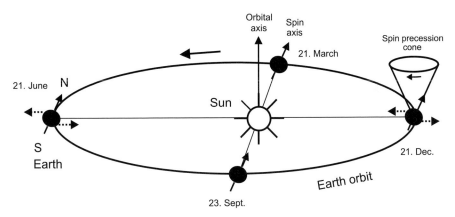

Fig. 5.11. The Earth's orbit around the Sun, the orbital and spin axes, and the precession cone

The planet's spin axis inclination is responsible for the seasons. For the Earth, this is seen in Fig. 5.11. In the Northern Hemisphere during summer, the spin axis is inclined toward the Sun (most strongly on June 21), giving this hemisphere more light than the Southern Hemisphere, while in the winter it points away from the Sun (with a maximum on December 21). Because of the rotation around its spin axis, the Earth does not have a perfectly spherical shape, but is slightly extended at the Equator. The solar gravitational forces work on this so-called equatorial bulge by exerting a torque (shown as dashed arrows in Fig. 5.11). The torque in turn creates an effect called *precession* – or, more precisely *spin precession* – which changes the Earth's spin axis and moves it around a cone aligned with the orbital axis (see Fig. 5.11).

While the inclination of the Earth's spin angle remains constant during the spin precession caused by the Sun, this angle changes due to the gravita-

Table 5.4. The distance r, the orbital and rotation periods (in years, days, hours), and the orbit and spin axis inclinations of the planets (after Allen 1973)

Planet	r (AU)	r (10^6 km)	Orbital period	Rotation period	Orbit axis inclination	Spin axis inclination
Mercury	0.387	57.9	88.0 d	59 d	7.0°	28°
Venus	0.723	108.2	224.7 d	244 d	3.4°	3°
Earth	1.000	149.6	365.3 d	23.9 h	–	23.5°
Mars	1.524	227.9	687.0 d	24.6 h	1.9°	24.0°
Jupiter	5.203	778.3	11.86 y	9.8 h	1.3°	3.1°
Saturn	9.539	1427	29.46 y	10.2 h	2.5°	26.7°
Uranus	19.18	2870	84.01 y	10.8 h	0.8°	97.9°
Neptune	30.06	4497	164.8 y	15.8 h	1.8°	28.8°
Pluto	39.44	5900	247.7 y	9.4 d	17.2°	

tional forces exerted by the Moon and the planets. In addition, the Earth's orbital axis changes (called *orbit precession*) due to the gravitational forces exerted on the Earth by the other planets (mainly Jupiter). One might think that all these various forces acting on the Earth would cause only minor, slowly varying changes of the spin axis inclination. However, this is not so due to an effect called *spin – orbit resonance*. The power of such a resonance is best understood from our childhood experience on a playground swing. We all remember that when repeatedly applying small accelerations to a swing at the right moment, large and even dangerous excursions can be generated. Therefore, when the spin and orbit precession rates become similar (due to tidal braking, then such a spin-orbit resonance takes place and large variations of the spin axis inclination will occur.

If and when such resonances happen for a planet can be found by detailed computations. Indeed, recent numerical simulations have shown that the inclination of the spin axis of planets can vary by large amounts (e.g. Laskar and Robutel 1993, Williams 1998). These authors find that inclination angles

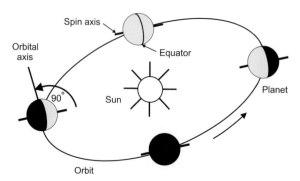

Fig. 5.12. The Earth orbiting the Sun with a spin axis inclination of 90°

of up to 60° can occur for Mars and up to 85° for an Earth without the Moon. However, the Moon has efficiently stabilized the Earth's inclination angle.

Since the presence of a massive moon is probably rare for most terrestrial planets, the important question is: Will large inclination angles on a planet without a moon lead to a drastic destabilization of the planetary climate? The worst case would be if the planet had an axis inclination of 90°, where the spin axis would lie precisely in the orbital plane. In this situation, displayed in Fig. 5.12 for an orbit similar to the Earth with a year length of 365 days, there would be periods of darkness lasting for about 160 days at the poles. In addition, one would find darkness episodes of 90 days in regions of 45° latitude, decreasing to half-day periods at the Equator. Quite different from our Earth, however, this means that the total amount of sunlight received at the polar regions would be much greater than that at the equatorial regions. This is because the Sun would spend much less time directly over the Equator than over the poles.

Detailed atmosphere calculations by Williams (1998), who modeled the heat flow over the latitudes, show that one might have equatorial glaciations and very warm poles, with summer ocean temperatures above 47 °C and even higher temperatures over land. Yet these models are not yet able to take into account the extensive cloud formation from the production of water vapor, as well as the generated ocean currents (such as the Gulf Stream), which depend on the instantaneous location of the continents being moved around by plate tectonics. Depending also on the land to ocean surface ratio, these effects would significantly reduce the temperature extremes on the planet. Williams finds that due to the carbonate silicate cycle, planets in the outer regions of the habitable zone with a denser CO_2 atmosphere should be less affected by inclination axis variations. As mentioned above he determined that the outer limit of today's habitable zone is at 1.67 AU which would translate into a distance of 1.35 AU when the Sun started on the main sequence. He concludes that although climate variations due to orbital and spin axis changes are important, they do not appear to be fatal for the development of life.

5.8 Biogenic Effects on Planetary Atmospheres

Up to now, various physical processes that affect the habitability of planets have been considered. There are also important biogenic effects which, because life is present, affect the later habitability of a planet. Although it is difficult to imagine that the survival of life, once it has appeared, will be threatened by its own actions, it might well be that early biogenic effects may later slow or even prevent the formation of very advanced life. As will be discussed in Chap. 7, photosynthetic bacteria generate oxygen as a waste product, which in the first 2 billion years on the Earth was essentially used up to oxidize the iron in the sea water, but later enriched the atmosphere and destroyed the greenhouse gas methane.

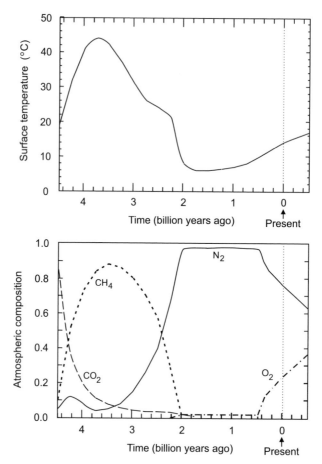

Fig. 5.13. The evolution of the mean terrestrial surface temperature and atmospheric composition with time (after Hart 1978)

Figure 5.13 shows the time evolution of the terrestrial atmosphere over 5 billion years, from a simulation by Hart (1978). The atmosphere was computed on the basis of a simplified model of an average Earth with a mean surface temperature. However, this model was already quite sophisticated, taking into account the Sun's time-dependent increase of luminosity, the outgassing by volcanism, the greenhouse effect, the carbonate silicate cycle, and the reactions between nine atmospheric gases, as well as the photolysis of water with the subsequent escape of hydrogen. From the composition of the atmosphere, it can be seen by comparing the upper and lower panels of Fig. 5.13 that the removal of methane (CH_4) by oxidation about 2 billion years ago caused a strong reduction of the greenhouse effect, which lowered the mean surface temperatures to values of only 6 °C. These low temperatures were also caused by the much lower solar luminosity at those times and it appears that habitability on the Earth went through a bottleneck 2 billion years ago. It can be seen that subsequently the Earth's mean surface temperature increased with time, and will continue to do so in the future. This is primarily a consequence of the increase of the solar luminosity.

5.9 Proterozoic Glaciations and Snowball Earth

That the habitability of our planet was severely endangered by the significant rise in oxygen levels and reduction in methane due to oxygen-producing bacteria combined with the low luminosity of the Sun in the Proterozoic era (2500–542 million years ago) is suggested by geological evidence for severe global glaciations in this epoch. Glaciations can be detected in the geological record by the presence of *tillites*, packed conglomerates of pebbles, sand and mud, by *glacial striations*, scratches on rocks dragged by moving ice, or by *dropstones*, rocks transported by icebergs and dropped into finely laminated sediment.

In Fig. 5.14 the temperature variations on Earth derived from $^{13}C/^{12}C$ isotope ratios show that the first evidence of ice-ages on Earth occurs in late Archean and early Proterozoic times approximately 2.9 and 2.3 billion years ago, called *Pongola* and *Huronian glaciations* respectively (Kopp et al. 2005). Both glaciations are attributed to effects caused by methane. While the Pongola glaciation is supposed to occur due to the high methane abundance generated by methanogenic bacteria that produced an organic haze such as on Saturn's moon Titan (Trainer et al. 2004), which due to the reflected solar radiation generated an anti-greenhouse effect on the terrestrial surface, the Huronian glaciation is attributed to the destruction of the methane greenhouse gases from the oxygen produced by the newly developed cyanobacteria. These ice ages are followed by three late Proterozoic glaciations named Sturtian (ca. 725–710 My ago), Marinoan (ca. 635–600 My ago) and Gaskiers (580 My ago) (Narbonne 2005). Finally, as can be seen in Fig. 5.14, there were three additional ice-ages at Phanerozoic times, the Ordovician-Silurian (460–430 My ago), Permo-Carboniferous (320–250 My ago) and our recent Pleistocene (1.8 My–12 000 y ago) glaciations that were less severe.

From these ice ages it appears that the late Proterozoic ones were the most severe bottlenecks for the evolution of life and at their end led to an extraordinary explosion of life in the Ediacaran period 630–542 million years ago (discussed below in Sect. 7.8). Plate-tectonic reconstructions (Sect. 3.9) show that these glaciations apparently happened at the times when most continents were connected and clustered around the equator in the supercontinent Rodinia. From palaeomagnetic measurements in deposits from North America, Asia, Australia and Africa it was found that glaciations occurred at low geographic latitudes of less than $10°$ from the equator. Although glaciers can exist in the tropics, this is unusual because even at the last Pleistocene glacial maximum tropical glaciers in the Andes did not descend to levels lower than 4000 m above sea level.

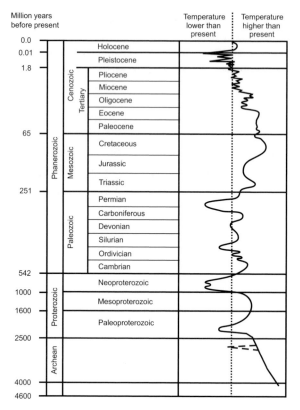

Fig. 5.14. Temperatures on Earth through the geological eras, modified after Clefs CEA 49 (2004) with evidence for the Pongola ice age (*dashed*) 2.9 billion years ago

This led to the picture of a *Snowball Earth*, where not only the continents but also the oceans were frozen and covered by km-thick ice layers (Hoffman and Schrag 2002). New evidence for the severity and duration of these glaciations was recently found by the discovery of thin extended layers of the element iridium that are thought to have formed by accumulation of interplanetary dust on the ice for millions of years and that, upon melting of the ice cover at the end of the glaciation, dropped to the ocean floor (Bodiselitsch et al. 2005). On the other hand there is fossil evidence during the same glacial times discovered in California (Corsetti et al. 2003) and Brazil (Olcott et al. 2005) of abundant marine life of prokaryotic and eukaryotic photosynthetic bacteria that could only survive, living in a stratified ocean with thin or absent sea-ice.

Another problem with a completely frozen surface (called *hard Snowball Earth hypothesis*) is how such a severe state can be terminated. For this it is assumed that the reduced weathering of continental rocks allowed the carbon dioxide produced by volcanism to accumulate in the atmosphere and lead to an increased greenhouse warming. However, recent computer simulations show that for the termination of a hard Snowball Earth an unrealistically high amount of CO_2 (more than 550 times the present value) would be required (Pierrehumbert 2004). This shows that very likely the Snowball Earth glaciations were less severe and that large areas must have existed where the conditions for microbial life were relatively benign.

5.10 The Requirements for Continuous Habitability

It was argued above that for uninterrupted evolution over 4.6 billion years toward intelligent life, an Earth-like habitable planet must orbit in a continuously habitable zone (CHZ) which, due to the rising solar luminosity and the instabilities of the planetary atmosphere, was found to extend between 0.95 and 1.35 AU. In this range the inner boundary of the CHZ zone is the present distance from the Sun, where the moist greenhouse effect would take place, and the outer boundary the distance where, 4 billion years ago, CO_2 clouds would have formed and with their high albedo would have prevented further greenhouse warming, leading to an irreversible glaciation. Here the outer boundary of the CHZ is too optimistic, because it does not take into account the tidal effects of Jupiter on the mass of the outer terrestrial planets. As discussed above for the case of Mars, planets at the outer range of the habitable zone are expected to have a smaller mass and to cool more rapidly. Due to a decreased efficiency of plate tectonics and a reduced greenhouse effect, they should thus have suffered an irreversible glaciation more easily.

The realistic distance of the outer boundary of the CHZ must thus be reduced. For this reason the values of Kasting et al. (1993), 0.95 to 1.15 AU, appear more appropriate. However, considering the above-discussed severe Proterozoic glaciations caused by biogenic effects, his outer limit appears to be still too optimistic and should be reduced further. This fits with computer simulations by Hart (1978) who found an outer boundary of the CHZ of 1.01 AU. I therefore assume that the CHZ extends over a distance of only 0.06 AU from 0.95 to 1.01 AU. The question now is how many planets orbiting in such a narrow CHZ around the right kind of central star can be expected?

5.11 The Drake Formula

In our search for extraterrestrial intelligent life, looking for Earth-like planets is only the first step. The next step (discussed in Chaps. 6 and 7) will be to consider the probability of the formation of life – and particularly of intelligent life – on such planets. It was the American radio astronomer Frank Drake, who in 1961 first suggested that the number of intelligent societies in our galaxy can be inferred, starting with an estimated frequency of Earth-like planets. He proposed a formula in which the total likelihood is expressed as a product of the individual probabilities for the various necessary conditions. This formula, henceforth called the *Drake formula*, roughly estimates the number of extraterrestrial intelligent civilizations that are expected to communicate by using radio waves.

The great advances in radio technology during the Second World War led in the 1950s to the construction of large radio telescopes, both in England (Jodrell Bank, near Manchester) and the United States (National Radio Astronomy Observatory, NRAO, at Green Bank, West Virginia). The prime aim of these telescopes was to look at a large variety of astronomical objects (galactic and extragalactic sources, interstellar gas, and the Sun) by using the "radio window" through the Earth's atmosphere. It is worth noting that aside of the narrow optical window, the radio window in those days was the only other means of observing the universe from the Earth's surface. Today, through the use of satellites, this limitation imposed by the atmosphere no longer exists.

As early as 1959, Giuseppe Cocconi and Philip Morrison, as well as a number of Russian physicists, considered the possibility of interstellar radio transmission, and whether extraterrestrial intelligent beings could use it to communicate with themselves and us (see Chap. 10). To search for such radio messages from nearby stars, project *Ozma* was initiated by Drake in 1960, and a year later at an important conference at Green Bank he presented his Drake formula, which gives the total number N of communicating extraterrestrial civilizations as the following product:

$$N = N_S f_P n_E f_L f_I f_C L / L_S \ .$$

Here N_S is the number of stars in the galaxy suitable for life, f_P is the fraction of such stars which have planets, n_E is the number of Earth-like planets which orbit in the habitable zone for each star, f_L is the fraction of planets on which life develops, f_I is the fraction of planets on which life evolves into intelligent life, f_C is the fraction of civilizations which engage in radio communications, L is the lifetime of communicating societies, and L_S is the time-span over which Earth-like planets and their parent stars have formed in the galaxy. Note that for an existing communicating society the fraction L/L_S represents the likelihood that it can be detected.

While the complete Drake formula will be discussed in Chap. 10, we will presently consider only that part of the formula which estimates the number of habitable planets. For this purpose, it is useful to split the formula into two parts containing astronomical and biological factors. The first group with three factors, $N_{HP} = N_S f_P n_E$, represents the number of habitable planets in the galaxy, while the second group, $f_{IC} = f_L f_I f_C L / L_S$ specifies the fraction of habitable planets that develop intelligent communicating life. This allows the Drake formula to be written more simply as $N = N_{HP} f_{IC}$.

It is difficult to provide actual numbers for these groups. While recent progress in our astronomical knowledge about the observation and formation of planets should allow us to make improved estimates for N_{HP}, the second group f_{IC} is much more speculative, as it involves the formation of life and the creation of technological intelligence. Note that f_{IC} requires knowledge about the factor L, the lifetime of a communicating intelligent civilization, which cannot even be reliably answered for our own society, as it involves the prediction of the future development of mankind (see Chaps. 9 and 10). The much argued question among biologists and geologists, about how life formed, developed, and led to intelligent life, will be discussed in Chaps. 6 and 7.

Before estimating the number of habitable planets, it should be pointed out that in the literature the Drake formula is often written in another form by introducing a formation rate R of stars suitable for life:

$$N = Rf_P n_E f_L f_I f_C L .$$

This second formula was actually the one that Drake first wrote on the blackboard at the Green Bank conference. It can be reduced to the first form by substituting $R = N_S/L_S$.

5.12 The Number of Habitable Planets

Let us now attempt to estimate the number N_{HP} of habitable planets in our galaxy. Table 5.5 shows estimates by various authors since 1963. The values N_S listed for Sagan (1963) as well as Rood and Trefil (1981) were not provided by these authors but have been derived for this table by using their rates $R = 10$ and $R = 0.05$, respectively, and multiplying by the time $L_S = 1 \times 10^{10}$ years. As that value of L_S will also be used in the complete Drake formula in Chap. 10, the final estimate of the number of communicating societies N given by these authors will not be changed. In addition, it should be noted that Rood and Trefil (1981) as well as Goldsmith and Owen (1993) describe their estimated numbers as intermediate between optimistic and pessimistic values.

Table 5.5. Estimates of the number of habitable planets in the galaxy, made by different authors

Author	N_S	f_P	n_E	N_{HP}
Cameron (1963)	4×10^{10}	1	0.3	1×10^{10}
Sagan (1963)	1×10^{11}	1	1	1×10^{11}
Rood & Trefil (1981)	6×10^8	0.1	0.05	2×10^6
Goldsmith & Owen (1993)	9×10^{10}	0.025	10	2×10^{10}
The present author (2002)	1.4×10^9	1	0.003	4×10^6

It is the wide variation of the numbers N_{HP} of habitable planets in Table 5.5 which suggested repeating such an estimate by using more recent data. Taking the mass of the galaxy as $6.4 \times 10^{10} M_\odot$ from Méra et al. (1998)

and dividing by an average mass per star of 0.4 $M_\odot$ (derived from the mass distribution of Table 5.1), we find a total number of 1.6×10^{11} stars in our galaxy. As discussed above, only a fraction of these stars are suitable for harboring seats of intelligent life. In view of the long formation time of intelligent civilizations, the lifetime of the parent star of a habitable planet must exceed 5 billion years. This excludes O-, B-, A-, and F-stars. In addition, the habitable zone must be outside the tidal lock radius, which excludes K- and M-stars, thus leaving only G-stars, which represent a fraction of 10% of all stars. In addition, the low metal abundance of population II stars is assumed to lead to terrestrial planets which are too small to hold an adequate atmosphere. We therefore consider only the 60% of stars which belong to the metal-rich population I (Méra et al. 1998).

Stellar systems with multiple stars are excluded, as their persistent tidal interactions will probably disrupt planet formation. Therefore we multiply by 0.3, because only 30% of the stars are single. This would give $N_S = 2.9 \times 10^9$ stars suitable for life. However, in a system of habitable planets one probably always has giant jovian planets that together with the central star will exert severe tidal perturbations at the inner and outer boundary of the terrestrial planet zone. Planets in these regions will consequently be too small for suitable seats of life. Judging from the solar system, where from the 5 terrestrial planets (counting the asteroids as a failed one) only Venus and Earth are suitable, we therefore multiply by an additional factor of 0.5 and finally obtain

$$N_S = 1.6 \times 10^{11} \times 0.1 \times 0.6 \times 0.3 \times 0.5 = 1.4 \times 10^9 \ .$$

Assuming that the accretion disks from which our estimated stars formed are similar enough to that of the solar system and will create the full range of terrestrial planets, jovian planets, and Kuiper belt objects, it appears reasonable to take

$$f_P = 1.0 \ .$$

As is the case for the solar system typical simulations of planet formation (Chap. 2) find about four terrestrial planets within the Jupiter distance of 5.2 AU. Furthermore, an Earth-like planet must orbit in the narrow continuously habitable zone of width 0.06 AU (Sect. 5.9). This is to ensure favorable conditions for life and to avoid life-threatening time-dependent developments of the planetary atmosphere associated with the brightening of the parent star during its main sequence evolution. In addition, depending on the mass of the accretion disk, migration of jovian planets will take place and can destroy terrestrial planets. As discussed in Chap. 2, only about 6% of the planetary systems are presently thought to possess terrestrial planets that are undisturbed by migrations. From this, we conclude that

$$n_E = \left(\frac{0.06 \text{ AU}}{5.2 \text{ AU}} \right) \times 4 \times 0.06 = 0.003 \ .$$

With these values, as shown in Table 5.5, the total number of Earth-like habitable planets in our galaxy is then roughly given by

$$N_{HP} = 4 \times 10^6 .$$

This estimate indicates that only one in every 40 000 stars has an Earth-like planet in a continuously habitable zone. The reason for this low estimate compared to the values of most authors in Table 5.5 is that for a star suitable for life only single G-stars of population I have now been selected, which lowers the value N_S. Rood and Trefil (1981) have an even smaller value of N_S, because they assume an unusually low number of G-stars compared to the observed 10% fraction of G-stars in the solar neighborhood. That estimate would be even lower if their value of $f_P = 0.1$, attributed to counting only single stars, were incorporated into the value of N_S. The second main reason why my value of N_{HP} is so low arises from the simulations of jovian planet migrations. These tentative results indicate that there are only very few planetary systems where terrestrial planets are not disturbed by migrating Jupiters. The low number of undisturbed terrestrial planetary systems decreases the value of n_E significantly.

The estimates in Table 5.5 could still be too optimistic if the existence of a giant moon were necessary for the development of higher forms of life on Earth. This is because the occurrence of a large moon of a terrestrial planet is thought to be very rare. However, in our estimate we have assumed that the stabilization of the Earth's rotation axis by the Moon was not of prime importance, and that an Earth-like planet without a Moon would not have prevented the evolution of life. This is suggested by the above-mentioned studies of the terrestrial climate under different axis inclination angles by Williams (1998).

On the other hand, as suggested by the small fraction of detected stellar systems with closely orbiting jovian planets, it could well be that the importance of migration of jovian planets is severely overestimated and that my low values of n_E and N_{HP} must be considerably increased. Only additional observations and more refined theoretical simulations of planet formation will give us an answer here. The above estimate, therefore, seems to be a reasonable compromise. Note finally that from our estimate of habitable planets, and with a total number of 10^{11} galaxies, we obtain a staggering number of 4×10^{17} Earth-like planets in the universe.

Part II

Life

6 Life and its Origin on Earth

Suppose that somewhere in the universe there is another Earth-like planet. How likely is it that life will develop there? Two approaches can be taken to answer this question. The first is to set up detailed search programs for extraterrestrial life, both inside and outside the solar system. Although unsuccessful so far, this procedure, as will be seen in Chap. 8, is potentially very powerful and has a high chance of detecting extraterrestrial life in the near future. The second approach is to study how life formed and evolved on Earth, based on the assumption that on other Earth-like planets life does appear for similar reasons. The present chapter therefore outlines the basic chemical tools and processes employed by life, and summarizes the biology of cells, because these are the basic units of living organisms. After considering the likely environment on the early Earth, the important question is then addressed: How did life form? Let us begin first by asking: What is life?

6.1 What is Life?

This question has always puzzled man and seems very difficult to answer even today. However, with the recent DNA sequencing of a wide variety of organisms, and particularly of the most primitive life forms successfully completed, the goal of understanding what constitutes life is within reach. As the complete information about the mycoplasmas (see below) is contained in a few hundred genes, the full unraveling of their function will give us the precise definition of life.

In antiquity, Aristotle in his lecture notes *De Anima* argued that life and the soul are the same. Aristotle taught that everything that lives has a soul and that at death the soul leaves the body. He distinguished three levels of life: vegetative life (plants), sensitive life (animals), and conscious life (man). The large qualitative differences between these three levels of life that impressed Aristotle are seen even today as fundamental stages of sophistication of biological organisms, ultimately made possible by the laws of nature. Since we are interested in the future development of life, we may well ponder the question of whether nature holds more of these fundamental stages in store both for mankind and for the extraterrestrials, stages of greater

P. Ulmschneider, Life and its Origin on Earth. In: *Intelligent Life in the Universe*, P. Ulmschneider, Adv. Astrobiol. Biogeophys., pp. 117–147 (2006)
DOI 10.1007/11614371_6   Springer-Verlag Berlin Heidelberg 2006

sophistication which, since the birth of our universe, are waiting to be realized (this question will be discussed in more detail in Chap. 10).

In textbooks of biology, life is usually described phenomenologically. An organism lives when it shows the following basic properties: metabolism (application of chemical processes), growth (directed development), energy utilization, individuality (preservation of information of its own identity), procreation, and mutation (change of the hereditary information). In addition, a characteristic of living organisms is that they are in principle able to maintain these properties in a completely abiotic environment. Viruses, for example, do not satisfy these criteria, because they do not show metabolism and they need biological organisms for survival and reproduction. Therefore they cannot be considered as "alive".

6.2 The Special Role of Organic Chemistry

Life on Earth is intimately connected with organic chemistry. This is very likely a consequence of the unique compound-forming role of carbon among all other elements, resulting in a huge number of different substances. We know today that for every inorganic compound there are at least 50 organic compounds (Chap. 5). The reason for this special role of carbon is that it has four directed atomic bonds, which allow the building of extended spatial structures, as opposed to the three bonds of nitrogen and the two bonds of oxygen, which produce planar and linear structures, respectively. Elaborate spatial structures are also not possible with undirected ionic bonds such as those of Na or Cl. In addition, carbon often has the strongest binding energy and only this atom forms aromatic rings. Finally, CO_2 is a gas that can easily interact in biological processes while, for example, SiO_2 (quartz) or Al_2O_3 (corundum) are hard solids that do not interact. For the uniqueness of carbon as the universal building block for life and the low probability for life based on silicon see also Schulze-Makuch and Irwin (2004).

While the most frequent elements (except Si and Al) of the Earth's mantle usually play some role in living organisms, the prominent role of carbon is surprising because, although being the fourth most abundant element in the cosmos, it is not very abundant in the outer crust of terrestrial planets (see Table 1.2). The reason for the intimate relation between organic chemistry and life therefore seems to be the ability of carbon to provide the huge variety of specialized compounds and building blocks for the very complex structures of biological organisms.

6.3 The Elements of Biochemistry

Since life on Earth is based on organic chemistry and employs biochemical processes, it is necessary, if we want to understand what life is, that the elements of biochemistry are briefly discussed. In biological systems there are four major classes of organic compounds: *proteins*, *carbohydrates*, *lipids*, and *nucleic acids*. In addition to the catalytic substances that help to synthesize carbohydrates and lipids, there exist *three characteristic mechanisms* to replicate the master archive DNA, to transcribe the information from DNA onto the blueprint RNA, and finally to translate it into proteins. In addition, there is a *language* (genetic code) in which DNA and RNA are written, and an *energy carrier* (ATP) that powers all this construction activity.

6.3.1 Proteins, Carbohydrates, Lipids, and Nucleic Acids

In biological systems, more than 170 *amino acids* have been identified. Of these, there are 20 (listed in Table 6.1) that are coded by DNA and function as building blocks of proteins. They are all of the geometrically distinct l-chiral (left-handed) type, except for glycine. Amino acids are both acidic and basic; that is, they have a COOH carboxy end that easily loses an H^+ ion and an NH_2 amino end that readily takes up an H^+. The 20 coded amino acids consist of the five elements C, O, N, H, and S.

Table 6.1. The 20 amino acids coded by DNA (after Hart et al. 1995)

Alanine	Glutamic acid	Leucine	Serine
Arginine	Glutamine	Lysine	Threonine
Asparagine	Glycine	Methionine	Tryptophan
Aspartic acid	Histidine	Phenylalanine	Tyrosine
Cysteine	Isoleucine	Proline	Valine

Proteins consist of chains of amino acids linked together by so-called peptide bonds. These amino acids are selected from the set of 20 amino acids coded by DNA. While water makes up 70% of a cell's weight, proteins account for more than 50% of the cell's remaining weight (Alberts et al. 1994). One currently knows more than 20 000 different proteins in man (HPRD 2005). They have many functions: structure proteins, enzymes, hormones, transport proteins, protection proteins, tractiles, toxines, and so on.

The sequence of amino acids determines the biological function of proteins. Proteins consist of several tens to several thousand but typically around 300 amino acids, and many of them occur in the form of several subunits which are held together by bonds. These bonds between different subunits and parts of the same subunit give proteins a distinct three-dimensional spatial structure, that is essential for their specific biochemical function. The most important bond in proteins is the *peptide bond*, where the COO^- end of one amino acid is joined to the H_3N^+ end of the next amino acid, forming the sequence OC–NH, whereby a water molecule, H_2O, is liberated. The macromolecules resulting from this fusion are also called *polypeptides*.

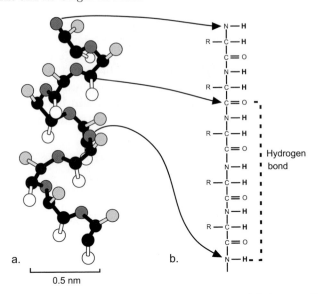

a.   b.

0.5 nm

Fig. 6.1. The structure of a protein. **a.** Ball and stick model showing C (*black*), N (*dark gray*) and O (*white*) atoms. The H-atoms are not shown. R (*light gray*) are the amino acid rests. **b.** The structural formula. *Arrows* show corresponding atoms (after Green et al. 1993)

Proteins have a hierarchy of structure. The primary structure is the amino acid sequence determined by the DNA (Fig. 6.1b). The secondary one is either the winding up of this chain to form an α-helix type spiral (see the CCN backbone in Fig. 6.1a, marked solid), or a back and forth folding of the chain onto itself, called a β-sheet. Both polypeptides are stabilized by hydrogen bonds. The tertiary structure is the three-dimensional shape into which these macromolecules fold due to disulfide bonds and the hydrophobic interaction. In a fourth type of structure, several protein subunits combine into a complete three-dimensional protein. For example, hemoglobin consists of four subunits.

Carbohydrates and *lipids* are two other basic building blocks of living organisms. Among the carbohydrates, cellulose serves as a structural material in plants. Starch and glycogen are the most important storage forms of energy in plants and animals, while sugars are used in energy transport. Sugars also constitute essential components of the nucleic acids. Lipids are familiar as fats and oils. Composed of glycerol and fatty acids, they are insoluble in water but soluble in organic solvents. Lipids serve many functions, of which their capacity for efficient energy storage is best known. One class, the *phospholipids*, constitute the structural basis of the cell membranes, both of the outer cell membrane and the inner membranes separating the organelles from the cytosol (cell plasma) in eukaryotic cells. Figure 6.2 shows a section of the cell membrane, which is built employing a phospholipid bilayer. In this bi-

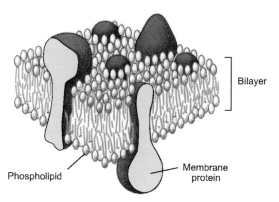

Fig. 6.2. A section of the cell membrane, showing the bilayer consisting of phospholipids (after Hart et al. 1995)

layer the lipid molecules are arranged in such a way that their hydrophobic (water-averting) tails are turned into the layer, while the hydrophilic (water-preferring) heads face the surfaces. The bilayer is about 5 nm thick. Figure 6.2 also shows that the membrane is permeated by various so-called membrane proteins, which serve as transport gates and carry out various control functions.

The fourth type of basic building blocks are lengthy molecules: the *nucleic acids* DNA and RNA, which consist of sequences of nucleotides. While DNA contains the master archive and building plan of the organism, RNA stores the blueprints for the individual construction processes. RNA is the abbreviation for ribonucleic acid, and DNA that for deoxyribonucleic acid. Nucleic acids are built up from large numbers of nucleotides, which themselves consist of three building blocks (see Fig. 6.3): a phosphoric acid (indicated by a ring), a sugar (pentagon), and a base (rectangle). In RNA the sugar is always d-ribose, and in DNA, deoxy-d-ribose. RNA and DNA each have only four types of base. In RNA the bases are cytosine, uracil, adenine, and guanine, abbreviated to C, U, A, and G, respectively; in DNA they are C, A, G, and thymine, abbreviated to T (Fig. 6.3). The bases C, U, and T are pyrimidines, while A and G are purines. The purines contain a double ring and thus need more space than the pyrimidines. The nucleotides are formed by joining the phosphoric acid to the 5′ carbon atom and the base to the 1′ carbon atom of the sugar. The five nucleotides, composed of three building blocks each, are called CMP, AMP, GMP, UMP, and TMP, where MP stands for monophosphate.

To create an RNA or DNA molecule, an arbitrary number of these nucleotides is joined together to build up a chain. For RNA, the nucleotides containing C, U, A, and G, and for DNA, those with C, T, A, and G, are used. As shown in Fig. 6.3, the joining occurs by connecting the phosphoric

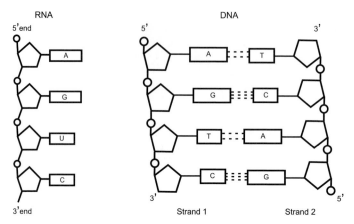

Fig. 6.3. A single-stranded RNA chain and a double-stranded DNA chain (after Green et al. 1993)

acid, attached at the 5′ carbon atom, with the 3′ carbon atom of the sugar of the next nucleotide, producing so-called phosphodiester bonds. Figure 6.3 shows how long RNA chains with 3′ and 5′ ends are formed.

But while RNA remains single-stranded, DNA forms a ladder in which two single DNA strands are tightly coupled together (Fig. 6.3). This is done in such a way that opposite to base A of one strand lies base T of the other strand. Likewise, G always lies opposite C. One always has a space-saving pyrimidine opposite a more voluminous purine, fixed by multiple hydrogen bonds (dashed). Similar to proteins, the double-stranded DNA is wound up in an α-helix type spiral ladder, where the A–T and G–C base pairs constitute the rungs. The many hydrogen bonds, the similar size of the base pairs, and the fact that the information is now stored more safely on two strands make DNA much more stable and suitable for a memory molecule than RNA.

In cells of advanced life forms, the DNA molecule becomes so long (about a mm in cells with a diameter of only a few μm), that it has to be wound up tightly. DNA, which is only 2 nm wide, is first wound around bead-like bodies, the so-called *nucleosomes*, with diameters of 10 nm (Fig. 6.4, top), and then into larger star-like structures with diameters of about 30 nm. Further looping and supercoiling winds DNA into narrow spirals with a diameter of 700 nm (Fig. 6.4). These spirals of DNA are called *chromosomes*. The chromosomes of prokaryotic cells are a single large ring, while eukaryotic cells possess sets of chromosomes. The difference between prokaryotic and eukaryotic cells is discussed in Sect. 6.4. For example, humans have 46 chromosomes, consisting of 22 pairs, and men in addition have an X- plus a Y-chromosome, while women in addition have an X-chromosome pair. Figure 6.5 shows the set of human chromosomes. The banding is achieved by staining. The length marked by arrows indicates 50 million base pairs.

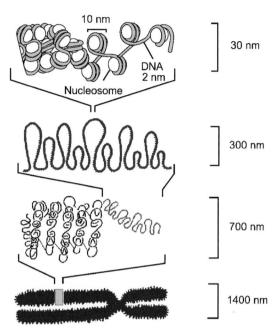

Fig. 6.4. The supercoiling of DNA and the formation of chromosomes (modified after Alberts et al. 1994)

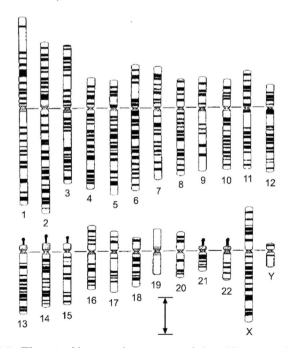

Fig. 6.5. The set of human chromosomes (after Alberts et al. 1994)

6.3.2 The Genetic Code

The genetic code specifies the language in which the building plan of the organism is written on DNA. A slightly different version of this genetic code (displayed in Table 6.2) represents the language in which the individual construction blueprints on RNA are written. The genetic code is a triplet code; that is, every triplet sequence of letters on RNA corresponds to one of the 20 coded amino acids of Table 6.1. It is read in the following way: the first letter is selected on the left-hand side in Table 6.2, the second letter on the top, and the third letter on the right-hand side. Each triplet is called a *codon*. There are three stop codons (UAA, UAG, UGA), while AUG (= methionine) is the start codon. The start and stop codons signal where synthetization processes must begin and end, respectively (see below).

Table 6.2. Genetic code (after Alberts et al. 1994)

First Letter		Second Letter			
	U	**C**	**A**	**G**	Third Letter
U	UUU UUC = Phe UUA UUG = Leu	UCU UCC UCA UCG = Ser	UAU UAC = Tyr UAA = Stop UAG = Stop	UGU UGC = Cys UGA = Stop UGG = Trp	U C A G
C	CUU CUC CUA CUG = Leu	CCU CCC CCA CCG = Pro	CAU CAC = His CAA CAG = Gln	CGU CGC CGA CGG = Arg	U C A G
A	AUU AUC = Ile AUA AUG = Met	ACU ACC ACA ACG = Thr	AAU AAC = Asn AAA AAG = Lys	AGU AGC = Ser AGA AGG = Arg	U C A G
G	GUU GUC GUA GUG = Val	GCU GCC GCA GCG = Ala	GAU GAC = Asp GAA GAG = Glu	GGU GGC GGA GGG = Gly	U C A G

6.3.3 ATP, the Energy Currency of the Biochemical World

Before we discuss the mechanisms which synthesize DNA, RNA, and the proteins, we must talk about the essential role of ATP (adenosine triphosphate), which provides the energy for carrying out these constructions. What the dollar or euro is to our human society, that ATP is to the biochemical world: the universal energy source to enable change. By adding one and two phosphate groups to the AMP discussed above, the molecules ADP and ATP are made. When, for example, in a cell a compound X–Y needs to be synthesized from the two components X–OH and H–Y, then one component must first be energized. This is done when the high-energy ATP transfers a phosphate group to X–OH. By this reaction, ATP is degraded to the low-energy ADP. The two components now combine and release a phosphoric acid. Finally, ATP is regenerated again from ADP by taking up the phosphoric acid and releasing a water molecule. However, this regeneration reaction occurs only when energy is added. In plants the energy is provided by sunlight, and in animals by the combustion of food.

6.3.4 Synthesizing RNA, DNA, and Proteins

As the final topic of this brief review, we now discuss the different conversion and synthetization processes that produce DNA, RNA, and proteins. The duplication of DNA is called *replication*. The procedure of manufacturing RNA from DNA is called *transcription*, and that of proteins from RNA is named *translation*.

When cells are duplicated, it is also necessary to duplicate the DNA. In this replication process DNA unwinds and the two strands separate. Then each strand is duplicated in such a manner that the synthesis always runs in the direction from the 5′ end toward the 3′ end. In this duplication process there are several proofreading procedures due to which mutated or destroyed pieces of DNA are repaired. These correction procedures and the fact that the information is stored on both strands generate a very high accuracy of duplication, which is necessary to ensure that in the human body, for example, 10^{16} cell divisions can be carried out without major problems in the course of a lifetime.

The manufacture of proteins is a two-step process. From the main archive DNA, the hereditary information is first transcribed to RNA. This single-stranded RNA, called *messenger* RNA (mRNA) then directs the synthesis of proteins. As mentioned above, in prokaryotic cells, DNA is present in the form of a ring and the information is stored in a continuous sequence, start codon to stop codon, start codon to stop codon, and so on. In eukaryotic cells the information on the DNA is not stored continuously. It is divided into *exons* – that is, pieces which carry information – and *introns*, pieces that carry no information (see Fig. 6.6). As a first step in eukaryotic cells, the DNA information is transcribed as a whole onto RNA. This RNA, which exists only in the cell nucleus, is called heterogeneous nuclear RNA (hnRNA). From hnRNA, the introns are then removed and a completed mRNA, ready for protein synthesis, leaves the nucleus through the pores in the nuclear envelope.

The protein synthesis works with the information supplied on the mRNA. It needs accessory molecules: *transfer* RNAs (tRNA) and the *ribosome*. The latter consists of two subunits between which the mRNA is read. The translation process starts with the 20 types of amino acids of Table 6.1 and

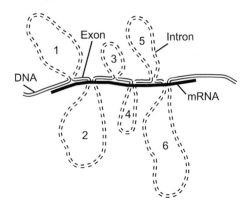

Fig. 6.6. The relationship between DNA and mRNA. The six intron loops of DNA will not be incorporated in mRNA (after Alberts et al. 1994)

the tRNAs present in the cell plasma. Each tRNA has a so-called anti-codon section, consisting of a complementary triplet to the 64 triplets of the genetic code table (Table 6.2), and is loaded with its specific amino acid. The ribosome (together with an initiator tRNA that has a comple-mentary anticodon UAC and is loaded with the amino acid methionine) searches on the mRNA for the start codon AUG and attaches itself there (Fig. 6.7). Now the ribosome moves to the next codon and accepts a tRNA with the next correct anticodon. The unloaded amino acid of this tRNA is then attached to the previous amino acid to form a protein chain. The depleted tRNA reloads its specific amino acid, and the ribosome moves to the next codon, where the translation procedure continues until a stop codon is reached. Here the ribosome falls off by splitting into its two sub-units and the released newly synthesized protein folds into its correct spa-

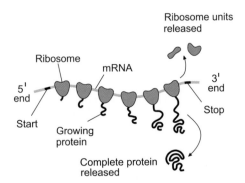

Fig. 6.7. Many ribosomes (*dark gray*) read the same gene, attaching themselves at the start codon, and falling off by splitting into their subunits at the stop codon (after Alberts et al. 1994)

tial shape. As soon as the start codon is free, a new ribosome attaches itself, such that at any one time many ribosomes read the same gene (see Fig. 6.7). The translation always proceeds from the $5'$ to the $3'$ end of the mRNA. While in prokaryotes translation is initiated independently at every AUG of the sequence, in eukaryotes usually only the first AUG is read.

Carbohydrates, lipids, proteins, and nucleic acids, together with the characteristic synthetization processes and the genetic code, constitute the basic biochemical building elements of all life forms on Earth. As the type and frequency of the building blocks are controlled by the master plan on the DNA, one can call this entire building industry the *DNA machinery*. It will be mentioned below that in the early history of the Earth, when life started, there were other types of machinery.

6.4 Cells and Organelles

From the discussion of the building materials, we now turn to the assembled product, the *cell*. All living organisms are made of cells. By enclosing itself in a cell membrane, a basic unit of life defines its identity by separating the outside world from its own self. From the absence or presence of a nucleus, one classifies the cells as either *prokaryotic cells* or *eukaryotic cells*, respectively. Figure 6.8 shows some examples of cells, magnified to the same scale. Prokaryotic cells have diameters between 0.1 and 10 μm. Simple prokaryotes such as mycoplasmas possess only a scanty internal structure, while more advanced prokaryotes have a cell wall and a sizeable number of organelles.

Eukaryotic cells possess a nucleus, and with diameters between 10 and 100 μm are much larger than prokaryotic cells. They show a very complicated internal structure, with a large number of different organelles (see Fig. 6.8). Organelles are organ-like subunits in the cells in which specialized processes take place. Typical organelles of eukaryotic cells, as well as their form and function, are listed in Table 6.3. The most conspicuous organelle is the *nucleus* (Fig. 6.8), where the chromosomes reside and where the transcription to RNA is carried out. It is enclosed by the nuclear envelope, which allows communication with the other cell parts through small openings, the nuclear pores.

Arrays of protein filaments, microtubules, and actin fibers, called the *cytoskeleton*, give the cell its shape and support, and allow internal movements. Whip-like *flagella* provide the cell with the means to move through the external fluid medium. *Lysosomes* contain enzymes which carry out intracellular digestion. *Mitochondria* are the power plants of the cell. By burning food using oxygen, they produce ATP. *Peroxisomes* are involved in the lipid metabolism; in the *endoplasmic reticulum* lipid synthesis and protein formation are carried out; while in the *Golgi apparatus* the manu-

Prokaryotic cells   Eukaryotic cell

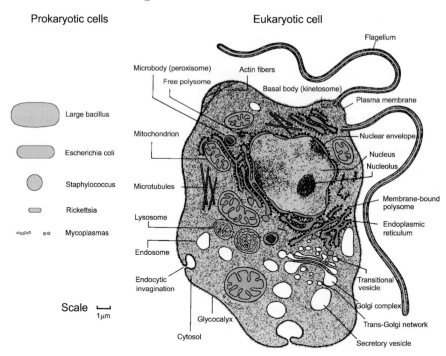

Fig. 6.8. Prokaryotic and eukaryotic cells on the same scale (modified after de Duve 1991)

Table 6.3. Organelles of eukaryotic cells, their form and function

Organelle	Form	Function
Nucleus	Enclosed by envelope	Archive, DNA, RNA synthesis
Cytoskeleton	Tubes, fibers	Internal transport, support
Flagellum	Tubes, fibers	Cell movement
Lysosome	Enclosed by membrane	Digestion
Mitochondrion	Enclosed by membranes	Burns food using oxygen, produces ATP
Peroxisome	Enclosed by membrane	Fat metabolism
Endoplasmic reticulum	Enclosed by membrane	Lipid and protein synthesis
Golgi apparatus	Enclosed by membrane	Storage and modification of lipids and proteins
Plastids (only plants)	Enclosed by membranes	Photosynthesis

factured proteins and lipids are modified, sorted, and packaged for delivery to other organelles. All these organelles are enclosed by membranes. This and their similar size hints that a long time ago their ancestors had been independent prokaryotic cells, that managed to survive inside the eukaryotic cell by becoming *endosymbionts*. Endosymbionts are organisms which live in a symbiotic relationship (a relationship with mutual benefits) with their host.

6.5 Sequencing and the Classification of Organisms

The discovery that the nucleotide sequences of the DNA can be used to classify all life forms has revolutionized our understanding of the relationships between the most abundant organisms, the bacteria, in particular. In former times, bacteria were all thought to be minor variants of single-cell life forms. Now, by sequencing, a staggering difference between individual bacterial species has come to light. But sequencing has also become a tool to assess the relationship between higher animals and plants, which has allowed us to check the results of biological classification.

6.5.1 Classification by Sequencing

To see how sequencing leads to a powerful method of classification, consider the following example. Suppose that we have a series of four organisms, O1, O2, O3, and O4, with a line of descent O1 → O2 → O3 → O4, where each differs from the previous one by a single mutation in a segment of its DNA chain (Fig. 6.9). Now assume that the knowledge about how these four organisms descended from each other has been lost. How can one reconstruct the correct line of descent? In the true line of descent one has a total of three mutations (indicated using lower-case letters). For another assumed line of descent, for example, O1 → O3 → O4 → O2, one would have a total of five mutations, because from O1 → O3 one has two, from O3 → O4 one, and from O4 → O2 two. In the same way, for the sequence O1 → O4 → O2 → O3 one needs a total of six mutations. Since successful mutations (where the mutated organism survives) are rare, it is necessary to arrange the four sequences in such a way that a *minimum number of mutations* occurs. This brings us to O1 → O2 → O3 → O4 or to O4 → O3 → O2 → O1.

O1: GGC ATC TCC GAA GAA TGT

O2: GGC AcC TCC GAA GAA TGT

O3: GGC AcC TCC GgA GAA TGT

O4: GGC AcC TCC GgA aAA TGT

Fig. 6.9. DNA base sequences of four related organisms. Mutated bases are shown with lower case letters

6.5.2 The Molecular Clock

By arranging the organisms with respect to the closeness of sequenced DNA (or RNA) segments, an evolutionary relationship can be constructed, based on the principle that the number of mutations should be a minimum. This

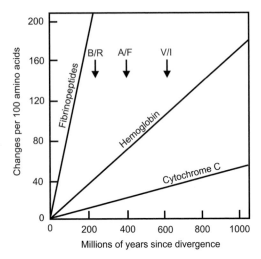

Fig. 6.10. The number of mutations per 100 amino acids as a function of time for three indicated cell compounds. V/I marks the time of the vertebrate/invertebrate, A/F the amphibia/fish, and B/R the bird/reptile divisions (after Alberts et al. 1994)

relationship can be made even more interesting if a time-scale is associated with it. Clearly, when two DNA sequences differ by many mutations one expects only a distant relationship, while for sequences that vary by few mutations one assumes that the organisms must be closely related. Distant relationships are far removed in time from a common ancestor, while close relationships hint to a common ancestor not too far away in time.

The use of the number of mutations as a measure of time is called the *molecular clock*. In the concept of a molecular clock one assumes that mutations occur at a constant rate with time. One can test this assumption by comparing sequences of different organisms. For the cell compound cytochrome C, taken from birds, reptiles, amphibia, fish, and invertebrates for instance, Fig. 6.10 shows that in 200 million years one has 10 mutations (per sequence of 100 amino acids), in 400 million years 20, and in 800 million years 40 mutations. That a molecular clock runs at a constant rate can also be checked by comparing different molecular clocks. Figure 6.10 shows that in the same 200 million years one has 40 mutations in hemoglobin and 200 mutations in fibrinopeptide, which shows that these clocks run more quickly. The linear growth with time of all three clocks, running at different speeds, shown in Fig. 6.10 demonstrates the stability and reliability of molecular clocks.

6.5.3 The Evolutionary Tree of Bacteria

By sequencing ribosomal RNA of bacteria and other organisms, Woese (1987) derived a tree of descent, called *phylogenetic tree*, shown in Fig. 6.11. The

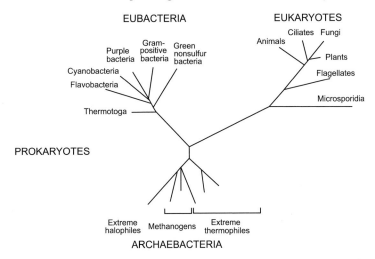

Fig. 6.11. The phylogenetic tree of living organisms (after Woese 1987) Woese, C.R.

length of the lines in that figure corresponds to the number of mutations and thus indicates time differences. This length also indicates the evolutionary distances between different bacteria, plants, and animals. The fact that three relatively closely related groups of organisms are connected over large distances in the figure allowed Woese to classify the world of living organisms into three main branches *archaebacteria*, *eubacteria*, and *eukaryotes*.

Note that it is from the eukaryote line that the higher life forms – the plants, fungi, animals, and man – developed. The surprising result was that plants and animals were much more closely related than the various bacteria among themselves. Note also that the prokaryotes (eubacteria and archaebacteria) are all unicellular organisms, and that this is also the case for the microsporidia, flagellates, and ciliates from the eukaryotes, called *protists*. Animals, plants, and almost all fungi are multicellular organisms.

6.5.4 The Timetable of the Evolution of Life

Of great importance for the origin of life is the fact that all organisms discovered on Earth up to the present day have a similar internal biochemistry based on DNA, RNA, the genetic code, and proteins built from the 20 amino acids of Table 6.1. Since this complicated DNA machinery could not have developed in two life lines independently, one concludes that the three branches of bacteria must have had a common ancestor, called the *Last Universal Common Ancestor* (LUCA), which must already have had the complete DNA machinery, and which very likely was similar to our simplest prokaryotic bacteria. From this cell, all life forms that presently inhabit the Earth descended. It is significant that this cell lived as early as about 3.9 billion years ago (see Fig. 6.12).

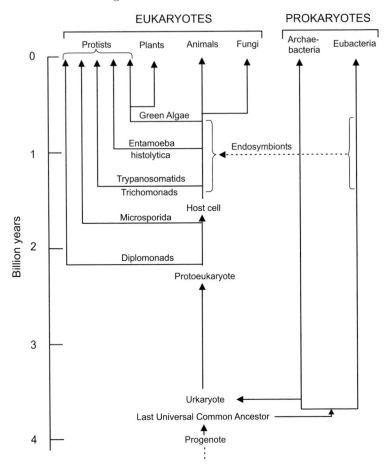

Fig. 6.12. Timetable of the phylogeny of living organisms. Life started with the Progenote that evolved in the RNA-world to the Last Universal Common Ancestor (LUCA), the ancestor of all present life forms in our DNA-world. Subsequently life branched into the domains Archeabacteria and Eubacteria, and later Eukaryotes (modified after de Duve 1991)

It is thought that the ancestral cell line split into the two prokaryotic bacterial branches, the archaebacteria and eubacteria. At about 3.7 billion years ago, the ancestor of the eukaryotic cell line, the so-called *Urkaryote*, branched away from the archaebacteria. Figure 6.12 shows the subsequent evolution of the eukaryotic cell line to a *Protoeukaryote*, and a host cell for endosymbionts. This will be discussed in more detail in Chap. 7.

It is very significant that LUCA (as will be discussed below) was already so complicated that it is not possible that this cell was the first living organism on Earth. The complicated DNA machinery, with its transcription to RNA, very likely had a predecessor with a much simpler, pure RNA machinery.

This cell belonged to the so-called *RNA-world* in which the genetic code must have been much simpler than today. Even these organisms, which had RNA as a memory molecule, may not have been the first living creatures on Earth, but could have been preceded by even simpler organisms that formed in the often hot (by frequent volcanism and very numerous hydrothermal vents), energetic, and chemically aggressive environment of the early Earth. The oldest ancestor of all life on Earth, which must have developed from prebiotic organic chemistry, is called the *Progenote* (see Fig. 6.12): it lived about 4.0 billion years ago, not long after the time at which the giant ocean-evaporating impact events ceased to occur.

Aside of sequencing, an insight into the history and evolution of life on Earth can be obtained by laboratory experiments and by computer simulations. One can experimentally and computationally recreate the conditions on the early Earth and try out ideas about the abiotic chemical evolution of organic molecules and complexes.

An impressive computational effort to simulate an entire cell, called the *E-Cell project* (Tomita 2001, 2005), has been under way since 1996. A simple virtual cell model with 127 genes is patterned after Mycoplasma genitalium. It takes up glucose from a culture medium, generates ATP by catabolizing glucose to lactate, and exports lactate out of the cell. Enzymes and substrates are spontaneously synthesized and degraded over time to simulate life. Protein synthesis is implemented by modeling the molecules necessary for transcription and translation, namely RNA polymerase, ribosomal subunits, rRNAs, and tRNAs. The cell also takes up glycerol and fatty acid for membrane structure using a phospholipid biosynthesis pathway. The model cell is self-supporting, but not capable of proliferation because the cell does not have pathways for DNA replication or the cell cycle.

After completion of this simple model more ambitious studies and simulations of more advanced cells such as S. cerevisiae (Gavin et al. 2002), E. coli, myocardial cells, neural cells and plant cells (Tomita 2005) are under way. It would be very desirable to have computer simulations also for earlier life forms but unfortunately, at the present time we do not yet have a sufficient understanding of the steps and the fossil signatures of the chemical evolution that led to the formation of the Progenote, the RNA-world organisms and the Last Universal Common Ancestor.

6.5.5 Sequencing and the Complete Genome

Today, an intense effort has succeeded in unraveling the complete genome of man and of a whole series of other organisms (see GOLD-EBI 2005). By Nov. 2005, we already had 305 completely sequenced genomes of microbial and advanced organisms. These consist of 24 archaebacteria, 242 eubacteria, and 39 unicellular and multicellular eukaryotes. For the human, chimpanzee, mouse, rat, dog, chicken, mosquito, fruit fly, rice genomes, for instance, drafts or already highly reliable versions of the complete sequences have been published.

At the moment the sequencing of 753 prokaryotes and 531 eukaryotes is underway; for example, gorilla, orang-utan, rhesus monkey, horse, cat, sheep, pig, cow, wheat, corn, cotton, soya and many insects.

Table 6.4. The genome sizes in base pairs of some fully and partly sequenced prokaryotic and eukaryotic organisms. The column "Genes" denotes nonidentical start sequences

Organism	DNA Bp.	Exon %	Exon Bp.	Genes
Prokaryotes				
Mycoplasma genitalium	5.8×10^5	100	5.8×10^5	468
Pelagibacter ubique	1.3×10^6	100	1.3×10^6	1354
Heliobacter pylori	1.7×10^6	100	1.7×10^6	1590
Bacillus subtilis	4.2×10^6	100	4.2×10^6	4099
Escherichia coli	4.6×10^6	100	4.6×10^6	4289
Eukaryotes				
Saccharomyces cerevisiae (yeast)	1.2×10^7	50	6×10^6	6294
Caenorhabditis elegans (nematode worm)	9.7×10^7	25	2×10^7	19 099
Drosophila melanogaster (fruit fly)	1.4×10^9	8?	1×10^7?	14 100
Arabidopsis thaliana (flowering plant)	1.2×10^8	20?	2.5×10^7?	25 498
Fritillaria (flowering plant)	1.2×10^{11}	0.02?	3×10^7?	30 000?
Protopterus (lungfish)	1.4×10^{11}	0.02?	3×10^7?	30 000?
Mouse	3.5×10^9	1?	3×10^7?	30 000?
Man	3.4×10^9	1?	3×10^7?	30 000?

Table 6.4 shows the size in DNA base pairs of some fully and partly sequenced genomes of organisms. The first five entries are for prokaryotes. Column "Exon %" lists the percentage of the DNA that represents exons. Note that higher life forms usually have a large percentage of introns. The column "Genes" refers to the number of nonidentical start sequences found. Question marks indicate estimates. For the estimates in the column "Exon base pairs" it was assumed that a gene has an average length of roughly 1000 base pairs. While unicellular organisms contain a few hundred to a few thousand genes, worms, insects, fish, and flowering plants have in the order of 14 000–30 000 genes.

Man has roughly 30 000 genes. Presently, the exact number of human genes is not yet known and could be as low as 20 000–25 000 (IHGSC 2004, Bertone et al. 2004). However, it has recently been discovered that in addition to these genes there are so-called noncoding RNAs (ncRNA), short interfering RNAs (siRNA) and other switches that provide instructions about how (much, little or not) genes are expressed. The number of these additional instructions are estimated to be as large as the number of genes (Claverie

2005). Clearly, to compare the genomes of humans with those of the chimpanzee, gorilla and other primates to find out what makes human intelligence unique, it is necessary not only to make an exact comparison of genes but also of these additional instructions.

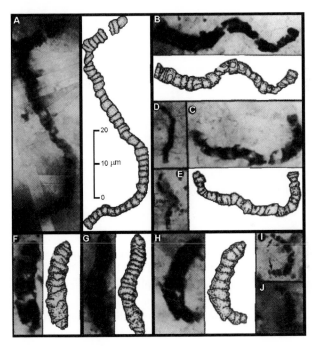

Fig. 6.13. Fossilized alga-like prokaryotic cells from the 3.5-billion-year-old Warrawoona Formation of Western Australia (after Schopf 1992)

6.6 Geological Traces of Life

What is the geological evidence for life? As discussed in Chap. 3, there are only a few sites on Earth where we have sedimentary rocks from the early history and where fossil traces of life can be found. The oldest indications for life are obtained from the unusual $^{13}C/^{12}C$ carbon isotope ratios found in carbon globules in metamorphic rock in the 3.8-billion-year-old Isua Formation and on the nearby Akilia island in southwestern Greenland (Schidlowski 1988, Mojzsis et al. 1996, Rosing 1999, Westall 2004), which points to photosynthetic bacteria. The ratio $^{13}C/^{12}C$ is 2.5% lower in biological organisms, since they preferentially take up the lighter ^{12}C isotope in photosynthesis.

Dauphas et al. (2004) found that 3.83-billion-year-old metamorphosed rocks on Akilia and other southwest Greenland locations are strongly enriched

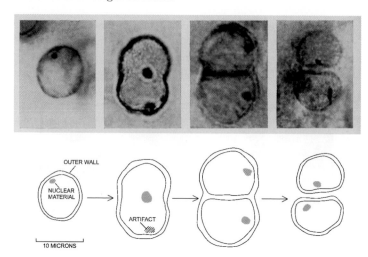

Fig. 6.14. Fossilized eukaryotic cells in various stages of cell division from the 1-billion-year-old Bitter Springs Formation of Western Australia (after Barghoorn 1971)

in heavy Fe isotopes. They pointed out that this suggests that the original rocks were *Banded Iron Formations* (BIFs) where Fe_2O_3 (with Fe^{3+}), formed from oxidation of Fe^{2+} by anoxygenic photosynthetic bacteria, fell out and sedimented in layers on the ocean floor constituting the oldest sediments on Earth. This view and the age of the Akilia rocks have been questioned (Moorbath 2005, Whitehouse et al. 2005). However, this does not concern the Isua Formation where uranium-rich Archean sea-floor sediments from Greenland indicate oxygen generation by photosynthetic bacteria around 3.7–3.8 billion years ago (Rosing and Frei 2004).

Banded fossils, called *stromatolites*, which were produced by successive growth layers of photosynthetic bacteria, have been discovered in Western Australia, where similar colonies in Shark Bay still exist today. These and other fossils shown in Fig. 6.13 date to about 3.5 billion years ago. Additional fossils of that age are reported by Schopf et al. (2002). At later times, the fossil record becomes richer. In the 2.1-billion-year-old Gunflint Formation of Ontario, Canada and particularly in the 1-billion-year-old Bitter Springs Formation of Western Australia (see Fig. 6.14), fossil eukaryotic cells with a clearly visible nucleus can be seen and the full mitotic cell cycle (discussed in Chap. 7) is beautifully recorded.

6.7 The Stage for the Appearance of Life

How life began on Earth is intensely debated at the present time and there are several theories about its origin. Nevertheless, great progress has been made on this question by approaching the problem from very different angles. Geological and astronomical studies of planet formation and the development of the early Earth provide a better understanding of the stage that was set for the appearance of life. DNA sequencing and analyses of the peculiarities of the genetic code point to simpler ancestral codes and restrict the time frame. There are even attempts recently, by using experimental and computational methods, to recreate the Last Universal Common Ancestor, that is, to produce an organism that can survive with a genome reduced to an absolute minimum consistent with the DNA machinery. Finally, great advances have been made in the understanding of the basic chemistry involved in the processes of cell function, prebiotic evolution, and abiotic production of the necessary organic compounds.

6.7.1 The Origin of the Genetic Code

It is thought that the genetic code must have had simpler codes as predecessors, which would explain the peculiarities (for example, that leucine has six codons, while tryptophan has only one) of the genetic code table. Eigen and Winkler-Oswatitsch (1982) suggested that these older tables already used a *triplet code*, and that in the oldest predecessor of the genetic code table only the middle letter was important, while the first letter was a G and the third a pyrimidine C or U. With this so-called one-dimensional (1D) code, proteins with a maximum of four different amino acids can be formed. In a subsequent hypothetical 2D code, the first letter also became important. It is thought that this table developed from the previous one by changing the purine G into A. Here a maximum of eight different amino acids could be encoded. The 2D code was subsequently extended to a 3D code when the third letter became important. Instead of a pyrimidine, now also a purine G or A was acceptable. This modification allowed the coding of 12 different amino acids out of a maximum possible number of 16. Finally, by allowing the full set of four nucleotides, G, A, C, and U, in all three letters, the existing genetic code was developed.

In this hypothetical evolution of the ancestral code tables, two facts remain difficult to explain. Why was the oldest code already a triplet code, and why do all these genetic codes accept only l-chiral amino acids? It was discussed above that the amino acids coded by DNA are all of the l-chiral (left-handed) type, while d-chiral (right-handed) amino acids are also known in organic chemistry. Indeed, the analysis of meteoritic material shows, that l- and d-chiral amino acids both occur in abiotic environments, and with roughly the same abundance.

Although these questions cannot currently be answered, there is an interesting suggestion by Mellersh (1993), which gives a type of answer that we are looking for to explain both facts in a simple way. He points out that an RNA strand deposited on the surface of a solid substrate experiences electrostatic repulsion at certain regular points along the chain (see Fig. 6.15). This leads

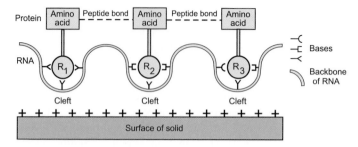

Fig. 6.15. The possible formation of the triplet genetic code and its fixation to specific amino acids (after Mellersh 1993). R_1 to R_3 are the side chains of the amino acids

to a wave-like folding of the RNA, with clefts in which three RNA bases are exposed. Only a specific amino acid fits precisely into each particular cleft, and by geometrical reason this just happens to be the l-chiral type.

6.7.2 The Urey–Miller Experiments

Another approach to shed light on the formation of life is laboratory simulations of the chemical conditions on the early Earth. The most famous of these experiments on primeval chemical evolution were those by Urey and Miller. In 1953, Harold Urey and Stanley Miller took a mixture of gases (CH_4, NH_3, H_2, and H_2O) presumed to be present in the atmosphere of the early Earth, but ultimately provided by rare excentric planetesimals (Chap. 2). In a circular flow process (Fig. 6.16), the mixture was heated and sparked with a 60 000 V electrical discharge, which also bathed the gas in its associated UV radiation. Electrical activity and UV light are assumed to have been present in primeval times due to frequent lightning, caused by static electricity in the extensive dust clouds, thrown into the atmosphere from volcanos and infalling meteorites (Chap. 3).

After running the experiment for a week, Miller found an impressive number of organic compounds, which were synthesized using the carbon provided from CH_4. Table 6.5, from a review by Miller (1998), shows that among these compounds were several biologically important amino acids (marked by *).

In later years, these experiments were criticized because it was pointed out that modern volcanos emit mostly CO_2 and, moreover, due to photolysis of H_2O by solar UV light (Chap. 5), the produced free O_2 might have oxidized CH_4, with the result that carbon in the early atmosphere should have occurred mostly in the form of CO_2. Under such conditions, the yield of amino acids would be much lower. However, as discussed in Chap. 3, based on recent investigations of the formation of CH_4 and NH_3 in hydrothermal vents, as well as if one takes into account the large amounts of dust in the Hadean era which reduced the influx of solar UV and therefore the photolysis

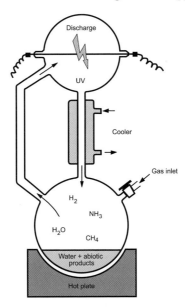

Fig. 6.16. The apparatus for the Urey–Miller experiment

of water, it now appears that the assumption of a reducing atmosphere originally made by Urey and Miller is highly likely. Later simulations by Miller and colleagues (Miller and Orgel 1974, Robertson and Miller 1995) showed that it is even possible to form purines (adenine, guanine) and pyrimidines (cytosine, uracil) with a yield of 30–50% in evaporating lagoons or in desiccating pools near beaches. These compounds had earlier been believed, to be very difficult to produce.

6.7.3 The Search for the Last Universal Common Ancestor

Another experimental method to shed light on the early stages of life is the quest for the structure of the Last Universal Common Ancestor (LUCA) of all living organisms, which already had the full DNA machinery (Fig. 6.12). There are two ways to approach the problem of how the Last Common Ancestor looked like. It is clear that it was a prokaryotic cell and there is the possibility that it looked not too different from the most primitive prokaryotic cells, the mycoplasmas.

As the DNA sequence of *M. genitalium* (with 468 genes, see Table 6.4) and other microbes with small genomes have been completely sequenced, it is possible to compare them in detail. In a computational approach Mushegian and Koonin (1996) counted the number of genes that are equal or similar in M. genitalium and Haemophilus influenzae, the only two fully sequenced microbes at that time. The 241 so-called orthologous genes that were found are presumably basic and absolutely necessary for survival. To this number,

Table 6.5. Prebiotic organic compounds from Miller's experiments (after Miller 1998)

Compound	% of CH$_4$ carbon
Formic acid	4.0
Glycine*	2.1
Glycolic acid	1.9
Alanine*	1.7
Lactic acid	1.6
β-Alanine	0.76
Propionic acid	0.66
Acetic acid	0.51
Iminodiacetic acid	0.37
α-Aminobutyric acid	0.34
α-Hydroxybutyric acid	0.34
Succinic acid	0.27
Others	0.62

the authors added 15 genes that were different but perform the same essential functions in the two organisms. By this procedure a minimum set of 256 genes was obtained, which might constitute the minimum set for survival. As it is very difficult to identify the latter type non-orthologous genes the gene-comparison approach very likely gives an underestimate of the minimum gene set.

In an experimental approach, Hutchinson et al. (1999) disabled specific genes of *M. genitalium* and investigated whether this organism still survived in a nutritious laboratory environment. They found that possibly only as few as 265 genes are needed for the organism to survive, a number not too different from that found by the first method. Since the minimum gene set found by the experimental approach depends on the experimental details and the content of the nutritious laboratory environment it very likely gives an overestimate of the number of essential genes.

More recent work on genome-wide inactivation of genes reported by Koonin (2003) confirmed the above picture of 265–380 essential genes for M. genitalium and M. pneumoniae and 271 essential ones for Bacillus subtilis despite the fact that this cell has a total of 4099 genes (Table 6.4). For these microbes, which all belong to the group of (Gram-positive) cells with a single membrane, it is found that the essential genes are involved in cellular functions in roughly the following proportions: translation (34%), RNA metabolism (5%), replication/repair (10%), cellular functions (18%), metabolism (28%) and miscellaneous (5%). Microbes that belong to the group of (Gram-negative) cells with a double membrane apparently need more proteins for their cellular functions, H. influenza with 1714 genes has 670 essential genes, while E. coli with 4275 genes has 708. Eukaroyotic organisms with their complicated internal structure need even more essential genes, S. cerevisiae (Table 6.4) has 1124 and C. elegans, 1080.

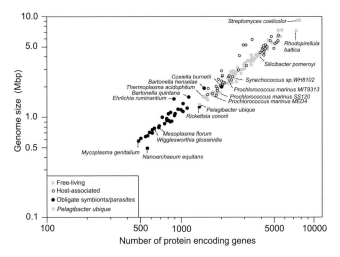

Fig. 6.17. Number of predicted protein-encoding genes for 244 complete genomes of Eubacteria and Archaebacteria. Free-living cells (*gray dots*), parasites (*solid dots*) and host-associated cells (*rings*), after Giovannoni et al. (2005)

This computational and experimental research shows that it might be possible to develop a *minimal organism*, which in a suitable laboratory environment is able to survive, and has the full DNA machinery (replication, transcription, and translation) with roughly 200–300 genes. It is not clear whether there is a possibility of different types of such minimal organisms.

An effort to create a living cell using standard chemical compounds is underway in several laboratories around the globe. By assembling a cell with an artificially created genome of 300 000 base pairs length and a minimum of 300 genes Craig Venter's group plans to create a living organism from scratch (Westphal 2003). At Los Alamos a group around Steen Rasmussen and Norman Packard (Holmes 2005) attempts to construct a simple cell, the so-called *Los Alamos Bug*, by mixing four ingedients: fatty acids that in water self-assemble into a container, peptide nucleic acids (PNA) for memory, a fatty acid and PNA producing a simple metabolism that uses the precursors of these molecules as nutritional input, and light to power the metabolism. If this artificial cell could be made to be self-sustaining it would allow Darwinian evolution and realize four basic properties of life: containment, heredity, metabolism and evolution.

The important question for the formation and evolution of life is, however, whether with these methods a minimum gene model for the Last Universal Common Ancestor (LUCA) can be developed that existed on Earth around 4 billion years ago. There are two problems. The first concerns the environment in which LUCA lived. For the minimal organisms discussed above very special conditions characterize their life style. Figure 6.17 shows that the microbes with the fewest genes (correlated with the smallest genomes) are parasites or host-associated organisms that find very special environmental conditions, which might not be appropriate for LUCA. Note that the free-living microbe with the smallest set of genes, *Pelagibacter ubique* (Table 6.4), needs at least 1354 genes (Giovannoni et al. 2005).

Another problem arises from the well-known fact that prokaryotic cells exchange genes and that therefore a considerable *horizontal gene transfer* (HGT) could have happened between organisms at those early times, which implies that the typical vertical gene transfer expressed in terms of evolutionary trees such as those in Figs. 6.11 and 6.12 would need to be modified. By assuming frequent horizontal gene transfers the whole concept of an evolutionary tree came under attack (Doolittle 1999) because under those conditions the history of life cannot properly be represented as a tree. Reconstructing LUCA on the basis of genome sequencing of living microbes appeared to be futile according to these arguments. Luckily, it has recently been found by reconstructing the microbial phylogenetic network of a large number of sequenced organisms (Kunin et al. 2005) that horizontal gene transfer is not very important and that the vertical inheritance of genes represents the bulk of gene transfer in the tree of life. This gives hope that LUCA with about 300 (Koonin (2003) assumes about 600) essential genes can be determined in the near future.

LUCA is essential for understanding the earliest phases of all present life on Earth but also because it will show what type of cell the previous RNA-world will have needed to develop. Clearly, the organisms of the RNA-world would have had even fewer genes, since many accessory molecules involved in the DNA replication and the DNA to RNA transcription would not have been needed. Here much fewer than 200 genes (maybe as few as 100 or less) might have been sufficient. It is very likely, therefore, that in the near future we will be able to create organisms that can survive without DNA; that is, pure RNA-world organisms. All these considerations suggest that the amount of coded information necessary for the earliest life forms – and particularly for the first one, the Progenote – is greatly reduced.

6.7.4 Summary: The Boundary Conditions

To put the question of how life formed on early Earth into perspective, we now summarize the boundary conditions that set the stage for the appearance of life and thus limit the possible answers. For a recent review of these conditions see also Westall (2004).

The Time window: We know today that life must have appeared very rapidly. From the oldest meteorite (the Efremovka chondrite), the formation of the Earth started 4.567 billion years ago (see Chap. 1). The last ocean-vaporizing impacts by planetesimals occurred 4.2 billion years ago (see Chap. 3). The oldest traces of photosynthesis (and thus of life) are found in the 3.8 billion year old Isua Formation of southwestern Greenland (see Sect. 6.6). The oldest microscopic fossils date from the Warrawoona Formation of Western

Australia and are 3.5 billion years old. At the same time one also finds bacterial mats called stromatolites. As photosynthesis indicates an already quite advanced stage of life, presumably with a DNA machinery, the Last Universal Common Ancestor must be considerably older than 3.8 billion years and can be assumed to have existed around 3.9 billion years ago (see Fig. 6.12), which leaves about 300 million years for the prebiotic stage, the Progenote, and the RNA-world. This time-span might be even shorter if the cataclysmic event 3.9 billion years ago (see Chap. 3) caused the evaporation of the Earth's oceans.

The Environment: The time between the vaporizing impacts and the Last Universal Common Ancestor is characterized by strong volcanism and intense emission by hydrothermal vents, due to the thin crust and the much greater heat flow. The result of this was a very dusty, often hot, and chemically aggressive environment (see Fig. 3.8): desiccating lakes, evaporating lagoons, hot springs, and oceanic fumaroles with hot acidic effluence, as well as energetic UV radiation from lightning caused by the large amount of static electricity generated by the dust clouds. A gigantic and efficient "chemistry laboratory Earth", with lots of energy available, was able to synthesize organic compounds under a huge variety of different conditions.

The microbial and genetic evidence: The connecting point (Fig. 6.11) of the three branches of bacteria (eubacteria, archaebacteria, eukaryotes) is populated by extreme temperature-loving (thermophilic) and acid-loving (halophilic) bacteria. The molecular clock points to a common ancestral cell well before 3.8 billion years. Computer and laboratory experiments point to a minimal organism with only about 300 genes. In previous stages, the genetic code was probably a 1D code and memory very likely based on RNA. This indicates that the RNA-world organisms probably had of the order of 100 genes, and the Progenote at the beginning of that era had even fewer. A strong accumulation of organic material in the ocean and in desiccating lagoons, called the *primordial soup* might have taken place before the time of the Progenote, as there were no biological organisms that would use up the organic compounds for food. This gave the prebiotic organic chemistry ample building material to work with.

6.8 Abiotic Chemical Evolution and the Theories of How Life Formed

The detailed way in which life came about is presently not known. However, a general picture of its formation process has emerged in the past few years. The fact that we find progressively simpler creatures the further we go back in time is a clear sign that life originated on Earth, because any life from a developed extraterrestrial source would have arrived on Earth already in an advanced state. In the literature one sometimes finds probability arguments about the likelihood of primitive life being spontaneously assembled from many components. These arguments typically lead to extremely low probabilities. As was pointed out by de Duve (1991), such arguments are misleading, since nature does not proceed that way. Similarly to when we assemble airplanes from millions of components, nature always builds up large units from smaller subunits. Thus, it is almost certain that life came about by a sequence of small steps, each with a plausible probability.

The present view is that life could have formed at two places, at the shores of the oceans or at deep-sea hydrothermal vents. For these environments one discusses three theories by which life may have formed: "metabolism came first", "genes came first", and the "iron–sulfur world". Relatively undisputed in these scenarios (as discussed above and in Chap. 3) is the fact that from the time when the last ocean-evaporating impact occurred, the chemical laboratory Earth has abiotically produced an ample supply of thiols, organic acids, amino acids, and hydroxy acids.

If the early atmosphere were CO_2 dominated then these abiotic sythetization processes would be difficult (e.g. Kasting 1993). However, there is new evidence that the early Earth had a reducing atmosphere because of a larger fraction of CH_4 relative to CO_2 produced by the much more numerous hydrothermal vents at those times (Chap. 3). The existence of a reducing atmosphere has recently been strengthened with new hydrodynamic modeling by Tian et al. (2005) who show that hydrogen escape from early Earth's atmosphere is reduced by two orders of magnitude compared to earlier estimates and that one now must assume that volcanic outgassing produces a hydrogen-rich atmosphere with a mixing rate of more than 30% hydrogen. Under these conditions abiotic synthesis of organic compounds is much easier.

But how does the development proceed from the formation of these simple compounds to the complex biochemistry of the RNA-world, where there were cell-like units enclosed by semipermeable membranes, memory molecules to define identity, and metabolic mechanisms to maintain and replicate these organisms? In addition, because it is recognized that the metabolic chemical processes would be much too slow if they were not catalyzed by enzymes, how did the necessary enzymes come about? As a definitive answer to these questions cannot be given at the present time (for a recent review see Botta 2004) a possible answer is now outlined that appears to me particularly convincing. This answer is of the "metabolism came first" type.

De Duve (1991, 1998) reserves the term "protein" for polypeptides made from an RNA-based machinery using the 20 coded amino acids of Table 6.1 and the term "metabolism" to denote the chemical reactions that are mediated by protein enzymes. He also takes "multimeres" to mean peptide-like substances of a more heterogeneous composition than proteins, and "protometabolism" to denote the chemical reactions that generated the RNA-world and sustained it before the invention of protein enzymes. In his view, that can be named "protometabolism came first", de Duve assumes that chemical evolution proceeded in the following way. The basic building blocks found local conditions that favored their condensation into thioesters. The thioesters spontaneously assembled into a variety of multimers. Some multimers, short-chained phospholipids, formed the first protocellular semipermeable membranes, which encased reaction centers. Other primitive multimers included a number of enzyme-like catalysts which initiated protometabolism. Thioesters and iron–sulfur compounds harnessed energy from UV light. Phosphoric acid reacted with the thioesters, creating inorganic pyrophosphate, a predecessor of ATP. This energy source allowed a full protometabolism of the multimere population and led to a *Thioester-world*.

The next step after de Duve was that ATP took over the role of pyrophosphate as the main energy source. The buildup of the first nucleoside triphosphates, and through these of the first polynucleides as memory molecules, led to the RNA-world working at first with a protometabolism. This produced the first living organisms enclosed in lipid membranes, the Progenotes. The invention of protein synthesis then led to a fully developed RNA-world. Finally, a genetic takeover of RNA by DNA brought about the DNA-world, by creating the Last Universal Common Ancestor.

Woese (1998) has pointed out that in this scenario, Progenotes should not be pictured as well-defined organisms because of their rudimentary and inaccurate translation mechanisms, and as cell division occurred in a very simple manner, in which a set of small linear chromosomes, each present in multiple copies, became sometimes divided into quite unequal parts. Therefore, Progenotes should rather be viewed as a whole population, as a communal ancestor, where genetic information could also be transferred by lateral gene transfer; that is, by cells cross-feeding one another.

The "genetics came first" view (Orgel 1998) is based on the important discovery of ribozymes, RNA enzymes that can catalyze the formation of other RNA molecules. Note that Johnston et al. (2001) have found ribozymes that can make complementary copies of RNA molecules up to 14 nucleotides long, regardless of their sequence, with impressive accuracy. The other basis for this view is the realization that the highly complex and specific reactions required by metabolism must have come about by Darwinian evolution (Chap. 7), which in order to work properly requires that the genetic material of the parent must be quite faithfully reproduced. In addition, Orgel (1998) points out that oxidation–reduction, methylation, and oligosaccharide synthesis supported by nucleotide-containing coenzymes, probably became part of the chemistry of the RNA-world long before the invention of protein synthesis.

However, the main problem with the "genetics came first" theory is how to produce the first RNA molecules without the benefit of ribozymes, because the spontaneous chemical synthesis of even a short structurally and chirally homogeneous piece of RNA is implausible in the extreme (Schwartz 1998). Therefore, some other form of organized chemistry must have preceded the RNA-world. As mentioned by Orgel (1998), the only potential informational systems, other than nucleic acids, which have been discovered, are closely related to nucleic acids. Pyranosyl-RNA (pRNA) and peptide nucleic acids

(PNA) can form long chains and could have suffered a later genetic takeover by RNA. Yet these possibilities do not appear to be overly promising (Orgel 2000).

A very different possibility, suggested by Cairns-Smith (1982), is that the first form of life was a self-replicating clay, the inorganic genome of which was subsequently taken over by RNA. Indeed, experimental studies (Ferris 1998) have shown that it may have been possible to make the transition from the prebiotic world to the RNA-world by a montmorillonite clay-catalyzed formation of RNA. Smith et al. (1999), Hazen (2001), and others have suggested that common minerals develop microscopic pits by weathering, which could have housed the first self-replicating biomolecules that later developed a biological cap and emerged into a nutrient-rich primordial soup.

While the "metabolism came first" and "genes came first" scenarios are both thought to have taken place at the terrestrial surface, the third theory of the origin of life, the "iron–sulfur world" of Wächtershäuser (1988, 1998), places the origin of life near deep-sea hydrothermal vents. In this view, life began on the surface of solids, in particular pyrite (FeS_2), and a primordial soup theory that abiotically creates vesicles and RNA should be rejected. Negatively charged organic compounds get electrostatically attached to the positively charged pyrite surfaces, where they react with each other and form cellular aggregates. The beauty of this scenario is that it automatically incorporates three main advantages: the FeS_2 surface keeps all the constituents together, it controls the metabolic reactions, and it provides energy.

The "iron–sulfur world" model has recently been extended by Martin and Russell (2003) who propose that the first cellular life could have formed in so-called *black smokers* at hydrothermal deep-sea vents. These vents emit black flows of superheated water that are rich in sulfides of Fe, Mn and Cu and create porous tower-shaped deposits with microscale caverns that are coated by thin metal-sulfide walls. These micro-caverns solve several critical points of the Wächtershäuser model, they provide a means of concentrating newly synthesized molecules, the steep temperature gradients inside a black smoker allow for establishing optimum zones for various reactions at different locations, and the flow of hydrothermal water through the structure provides a constant source of energy and building blocks from freshly precipitated metal sulfides.

Here, a single structure would facilitate the exchange between all developmental stages, and the synthesis of lipids as a means of closing the cells against the environment is not necessary until basically all cellular functions are developed. The last evolutionary step would be the synthesis of a lipid membrane that finally allows the organisms to leave the micro-cavern. In this model the Last Universal Common Ancestor would occur inside a black smoker, rather than as a free-living form. Although the black-smoker scenario is supported by the discovery of early microbial life in deep-sea volcanic rocks that are as old as 3.2 billion years (Rasmussen 2000), the questions remain as to whether hydrothermal life preceded photosynthetic life and how in the "iron–sulfur world" the RNA memory molecules are created.

In summary, we presently do not know in detail how life first appeared, whether it was one of the above scenarios or some combination of them, although it is generally felt that there must have been a straightforward chemical route to life. However, it is clear that once life started, intense competition (Darwin's theory) led to the survival of the most efficient organisms. How life subsequently evolved on Earth is discussed in Chap. 7.

7 Evolution

When we are looking for intelligent life outside the Earth, there is a fundamental question: Assuming that life has formed on an extraterrestrial planet, will it also develop toward intelligence? As this is hotly debated, we will now describe the development of life on Earth in more detail in order to show that there are good reasons why evolution should culminate in intelligent beings.

From the time at which the Earth became hospitable enough for organic chemistry to function, the formation of the first cells took only a few hundred million years, while the development of higher life forms, such as multicellular organisms, required an additional 3 billion years. Only recently did the highly evolved and very complicated nature of eukaryotic cells become fully appreciated, together with the realization that the enormous timespan, during which the Earth was populated by single cells only was not a period of stagnation, but actually witnessed a surprising pace of persistent development. Based on the creation of highly specialized organs, made possible by multicellularity and centralized control from the cell's nucleus, the development of intelligent life took another 800 million years. This evolution is driven by two fundamental processes: mutation and natural selection, as described by *Darwin's theory*. While the first of these processes is a pure chance event, the second is directional, since there is usually a very good reason why an organism survives in a given environment.

7.1 Darwin's Theory

Every living organism on Earth fights for its existence (food, light, territory, and shelter) and for its successful reproduction. This effort is called the *struggle for survival*. In this battle, only the most successful organisms survive, a fact that is termed *natural selection*. An additional fact of life is that in the process of reproduction there are *mutations*, caused by changes of the DNA. Mutations are unavoidable and occur at random. They are caused both by the environment (chemicals, radiation, and energetic particles) and by internal processes (faulty DNA replication). Darwin's theory (or principle) states that mutated organisms compete in the struggle for survival, with the consequence that, by natural selection, new, more efficient life forms appear. As

P. Ulmschneider, Evolution. In: *Intelligent Life in the Universe*, P. Ulmschneider, Adv. Astrobiol. Biogeophys., pp. 149–199 (2006)
DOI 10.1007/11614371_7   Springer-Verlag Berlin Heidelberg 2006

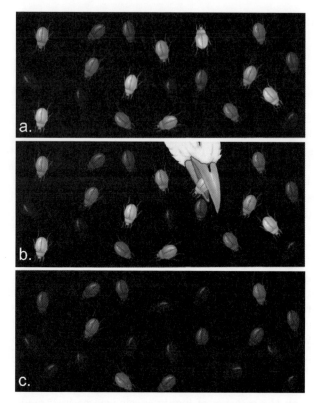

Fig. 7.1a–c. Progressive change of a group of bugs that live in a dark environment due to natural selection. Because of the preferential predation on the more easily seen white bugs the group evolves towards a darker population (after Bennett et al. 2003)

the less capable individuals get outcompeted by the more efficient organisms a group evolves towards a population with more efficient members.

The effect of natural selection on a population of bugs living in a dark environment is illustrated in Fig. 7.1. By random mutation individual bugs have different genes for coloring; the available genes of the bug population is called the *gene pool*. As white bugs are more conspicuous in front of a dark background they are preferentially caught by predators and their genes are eliminated. Since the dark bugs have more success to pass on their genes to the next generation the gene pool becomes that of a darker population. Although the predation depends on chance meetings of predator and prey the preference of selection is due to the laws of nature, in this case color contrast and visibility.

This example illustrates that Darwin's theory describes a powerful and basic physical process that lies at the core of the biological evolution. Note that the selective advantage of different organisms (e.g. to be dark in the above example) does not need to be large. Because evolution has lots of time available, even rather small advantages can eventually lead to the dominance of the favored species. Because any efficient process outperforms an inefficient one, Darwin's theory not only applies to the evolution of biological organisms, but was already at work in the prebiotic world, where it influenced chemical evolution.

There presently is a heated debate among evolutionary biologists, physicists, and chemists about the long-range effects of Darwin's theory (for a more detailed discussion, see Sect. 7.17). Does the directional aspect of the "survival of the fittest" predict the eventual emergence of intelligence? Or is the directional quality of Darwin's theory only valid over a short range? The answer to this question can probably only be found experimentally: by careful studies of the natural (physical, chemical, and environmental) reasons why the successful organisms survived, and by simulations. Since we are looking at terrestrial evolution in order to learn about the behavior of extraterrestrial life, it is these natural reasons governing evolutionary developments that are the most interesting, because they will also determine the path of evolution on other planets. This chapter therefore concentrates particularly on the question of *why* biological evolution on the Earth happened in the historically documented way.

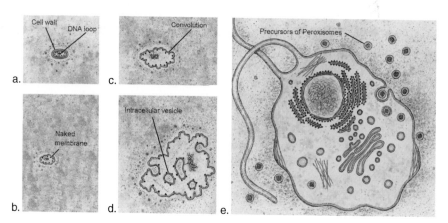

Fig. 7.2. The evolution of the eukaryotic cell (drawn to the same scale) (after de Duve 1996)

7.2 The Development of Eukaryotes and Endosymbiosis

We have seen in Chap. 6 that there are three major branches of prokaryotic cells that separated very early on from a common ancestor, the Last Universal Common Ancestor. Sequencing indicates that the eubacteria and archaebacteria separated first, and that subsequently the ancestor of the eukaryotes, the Urkaryote, split from the archaebacteria (Fig. 6.12). This Urkaryote was in almost all respects still a typical prokaryote, and like these, to provide rigidity and protection, had a cell wall, on the inside of which a single ring-shaped chromosome was attached (Fig. 7.2a). There was a long process of evolution from this simple prokaryotic cell to modern eukaryotic cells with their organelles, sets of chromosomes in a nucleus, and the processes of mitosis and meiosis, during which the cell volume grew by a factor of 10 000. According to de Duve (1996), this development happened in the following stages.

Prokaryotic cells feed by shedding digestive enzymes into their surroundings and subsequently taking up the processed food through the surface membrane (Fig. 7.2a). The first step in the evolution of eukaryotes was probably that the cell lost its wall and was enclosed only by a soft deformable membrane, by which the feeding process was made easier (Fig. 7.2b). By extensive folding of this membrane, shown in Fig. 7.2c, the cell subsequently increased its surface area. Since the amount of matter that can be absorbed increases with the surface area, the cell could take up more food. In addition, in these folds the digestive enzymes became less diluted and thus ensured better processing. The more efficient food handling allowed the cell to grow even larger. Eventually (see Fig. 7.2d), the cell learned to pinch off the inward folds of the membrane to create vesicles (vacuoles) into which food (bacteria) could be swallowed wholesale and treated with undiluted enzymes. Thus eukaryotic cells could take up food both from the outside membrane and from the inside vacuole, and became very efficient hunters.

About 2 billion years ago, an eubacterial predecessor of the organelles, which had been captured in a vesicle, succeeded in avoiding digestion and remain as a guest in the eukaryotic cell. By making itself useful for its host, a symbiotic relationship started (Fig. 7.2e). This was the first of many *endosymbionts*, which from that time onward were able to gain access to the eukaryotic cell and became their organelles. The first organelles were very likely the fibers and microtubules that gave the cell rigidity, and the flagellae, the whip-like projections that propel them in the liquid surroundings (see also Fig. 6.8). The peroxisomes came next, and afterwards the mitochondria. Finally, the cyanobacteria arrived, which brought the plastids that carry out photosynthesis in eukaryotic plant cells (see Fig. 6.12). The use of endosymbionts greatly improved the efficiency and power of the eukaryotic cell.

During this time, between the stages shown in Figs. 7.2d and 7.2e, but probably starting as early as 2.5 billion years ago, another fundamental step occurred in the development of the eukaryotic cell. The originally single chromosome ring was split up into sets of chromosomes, packaged into a nucleus (from which the eukaryotic cells got their name), and became encapsulated in the nuclear envelope (Fig. 6.8). The formation of the nucleus together with the invention of sexuality constitute a major selective advantage of the eukaryotic cell.

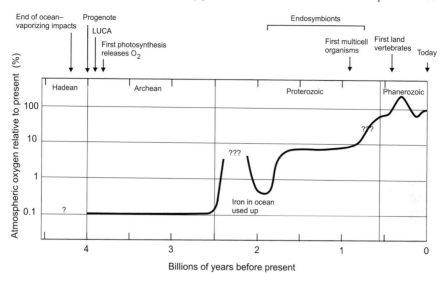

Fig. 7.3. Time evolution of the oxygen concentration in the Earth's atmosphere relative to today. The fraction before 2.5 billion years ago was probably even lower. At 2.45–2.1 billion years ago the so-called Great Oxidation Event occurred to levels that are difficult to constrain (modified after Canfield 2005)

7.3 Oxygen as an Environmental Catastrophe

Already, the prokaryotic cyanobacteria produced oxygen as a waste product of photosynthesis. Via sunlight, these bacteria used carbon dioxide and water to synthesize, for example, formaldehyde:

$$CO_2 + H_2O \rightarrow CH_2O + O_2 .$$

The O_2 liberated in this process subsequently oxidized the abundant iron dissolved in the oceans and produced Fe_2O_3, which fell out and deposited the Banded Iron Formations, which can be as old as 3.7 billion years (Mojzsis et al. 1996). However, at about 2.3 billion years ago, essentially all of the Fe had been used up, and free O_2 started to enrich the surface waters of the oceans and the atmosphere (Fig. 7.3, see also Bekker et al. 2004), although at first in tiny amounts as compared with today. As O_2 oxidizes the organic compounds in the cell, it can act as a toxic substance. Yet it also represents a source of chemical energy which can be exploited. While photosynthetic organisms synthesize ATP with the help of sunlight, in other organisms ATP is produced using the energy released in respiration by aerobic oxidation of food molecules.

Before the appearance of free oxygen, all forms of life had been adapted to an oxygen-free environment and were extremely sensitive to oxygen, like the anaerobic cells of today. To survive poisoning in an "oxygen catastrophe", the anaerobic ancestors of the eukaryotes either found refuge in oxygen-free locations, such as the deep seas, or enlisted help from nearby aerobic bacteria that were able to use up the oxygen. Note that the deep-sea remained anoxic until about 1.8 billion years ago (Rouxel et al. 2005). A most ingenious way to fight oxygen poisoning became to employ endosymbionts inside the cell (Fig. 7.2e). Eventually, after peroxisomes served this function, mitochondria appeared. They not only were capable removers of O_2, but in addition turned out to be very efficient producers of ATP. As already discussed in Chap. 5, another problem with free oxygen was that it had severe effects on the atmosphere, because by oxidizing greenhouse gases such as methane it reduced the greenhouse effect, which led to very low terrestrial temperatures.

Note in Fig. 7.3 that in Phanerozoic times (542 million years ago to the present) there were considerable variations of the abundance of atmospheric oxygen from 10% to as much as 35% during the Permo-Carboniferous compared to today's 21% (Graham et al. 1995) and, for instance, doubled from the Triassic–Jurassic boundary until today (Falkowski et al. 2005).

7.4 The Cell Nucleus and Mitosis

By concentrating the transcription and replication into the nucleus, from which specialized instructions can be sent out into the cell in a controlled manner, a much better regulatory power over the biochemical processes and the organelles was achieved. Expressing only specific genes became particularly important for the development of organs in multicellular organisms. Moreover, a nucleus permits the existence and removal of introns, and prevents unwanted protein synthesis. The function of introns is not yet fully understood, but they are possibly involved in the variation of gene expression and serve as "scratch gene material" which, by mutation, can be employed to modify or create genes. New genes can be created, for instance, by exon shuffling, where an exon from another gene is anomalously inserted into an intron. Putting the genetic material into the nucleus had also the important advantage that it allowed the lengthy DNA to be subdivided into sets of chromosomes. This powerful invention not only sped up the replication process but permitted sexuality, by which genes from different parents could be combined and reshuffled. However, to realize these advantages, very complicated reproduction processes of the nucleus, called mitosis and meiosis, had to be developed.

Mitosis is the division of a parental nucleus in an eukaryotic cell which produces two daughter nuclei with the same complete set of chromosomes. It consists of five phases (see Fig. 7.4). The *interphase* is the synthesis and growth phase, in which the cell spends most of its time. Just before cell division, the DNA of each chromosome is replicated. Two identical chromosomes for each parent chromosome are formed, held together by a centromere. Outside the nuclear envelope there are the two centrioles (Fig. 7.4). In the *prophase*, the centrioles move apart to opposite sides of the cell and organize

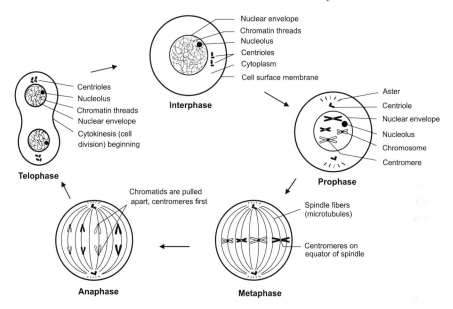

Fig. 7.4. The mitosis cycle (clockwise) of cell division in eukaryotic cells (after Taylor et al. 1997)

a division spindle. Later, in the *metaphase*, the nuclear envelope dissolves and the chromosomes align themselves in the equatorial plane, perpendicular to the spindle. In the *anaphase*, the copies of the chromosomes split at the centromere and each part is pulled toward its side of the spindle. Finally, in the *telophase* a new nuclear envelope forms around the chromosomes, the spindle fibers disintegrate, and the centrioles replicate. Subsequently, in *cytokinesis* the cell splits in two.

7.5 Sexuality and Meiosis

Sexuality is one of the most important inventions in the course of the evolution of eukaryotic cells. It permits the combination of the hereditary material of two parents. For example, consider two cells, one able to swim faster, and the other to feed more rapidly. A daughter cell that inherits the faculties of both parents has a much greater selective advantage than either parent. Sexuality was made possible by a modification of mitosis. Eukaryotic cells which contain two sets of homologous chromosomes are called *diploid cells*. *Meiosis* is the cell division, which separates the homologous chromosome pairs of the diploid cell into *haploid cells* with only a single set of chromosomes. The haploid cells are gamete cells; that is, egg cells or sperm cells. At fertilization, two gamete cells fuse. Figure 7.5 shows that meiosis consists of two mitosis-like division cycles (M I and M II). But there are important differences. In the prophase of M I there is an important reshuffling of the chromosomes, called *crossing over*. Instead of splitting at the centromeres, when the spindles pull the chromosomes apart in the anaphase, the chromosome copies stay together (Fig. 7.5). In M II the interphase does not occur and the chromosomes are not replicated. The pulling apart in the anaphase creates single sets of chromosomes. Unlike the four diploid cells produced in mitosis, meiosis produces four haploid cells.

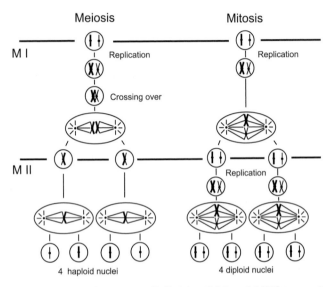

Fig. 7.5. The two stages of meiotic cell division (M I and M II) to produce gamete cells, compared with two ordinary mitotic cell cycles

Because the reshuffling of genes is so important for sexuality, the process of crossing over will be briefly explained. Figure 7.6a shows the chromosomes from two parents in replicated form, but each pair is still held together at the centromere. Note that the genes M, N, and Q of the parents differ, as indicated by upper- and lower-case letters. One chromosome of each pair crosses over to the other set (Fig. 7.6b), where they are cut and rejoined. For this, the homologous pairs have to be perfectly aligned in order to be clipped at the right positions (Fig. 7.6c). The result is that from two parental sequences, four different haploid gene sequences are produced (Fig. 7.6d).

To summarize, due to sexuality a particularly rapid change of the genome, and therefore more speedy adaption to the environment, can be achieved. This is made possible first by various combinations of chromosomes to form haploid cells in meiosis, second by crossing over, in which individual genes are reshuffled, and third – particularly for higher life forms – by the chance fertilization of an egg by one of many sperm.

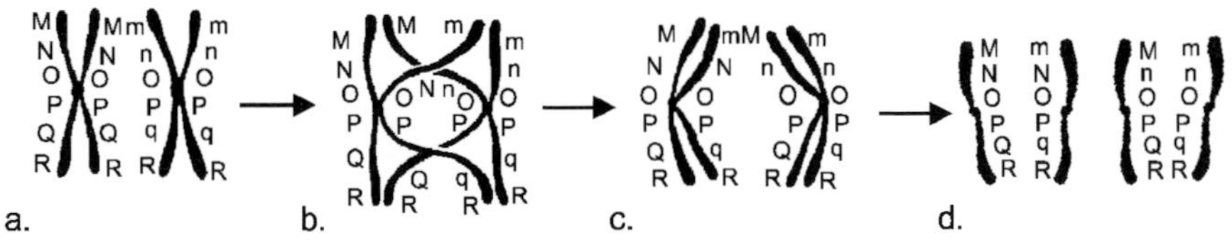

Fig. 7.6. The process of crossing over to reshuffle genes (after Novikoff and Holtzman 1976)

7.6 Genetic Evolution

The growth in the sophistication and fighting power for the struggle of survival that characterizes the development of eukaryotic cells into superbly equipped hunters was essentially an accumulation of information. That is, the increased complexity in terms of both tools (organelles) and processes (vacuole formation, mitosis, and meiosis) went hand in hand with a large increase in the number of instructions (genes) stored on the DNA. We would thus expect to see a correlation between the level of sophistication and the size of the genome. In a rough manner, this is indeed true, as seen from the growing size of the genome in DNA base pairs with the height of organization in Table 6.4. However, this table also shows that, for example, among the flowering plants a large variation ($1.2 \times 10^8 - 1.2 \times 10^{11}$ Bp) in the size of the genome occurs, which is unrelated to the level of sophistication. A much better correlation is found when the introns are subtracted from the genome size, and only the exons are counted (Table 6.4). Moreover, in some genomes, individual genes can appear in large numbers of identical copies. Counting only nonidentical genes, a much clearer correlation is found between the number of genes, the organization height, and the development time, as seen in Fig. 7.7.

Assuming that chemical evolution started with complexes of about 10 memory molecules, one can hypothesize that the Progenote, the first living cell, had perhaps around 50 genes, and the later, more sophisticated, RNA-world organisms of the order of 100 genes. We have already seen (Sect. 6.7) that the Last Universal Common Ancestor (LUCA) with its DNA machinery (dot in Fig. 7.7) could have had about 300 genes. The important invention of photosynthesis, which made life independent of the external food supply and is traced in the Isua Formation (Sect. 6.6), might have added perhaps another 300 genes.

Figure 7.7 shows that this amounts to a rather rapid development on the environmentally aggressive early Earth. However, after photosynthesis was around, the intense evolutionary pressure was considerably reduced, since a steady and abundant food supply was now available. This was true not only for the photosynthetic bacteria, but also for those that lived on the generated waste products or from hunting photosynthetic bacteria. The abundant food supply very likely explains the smaller rate of genome growth in later times.

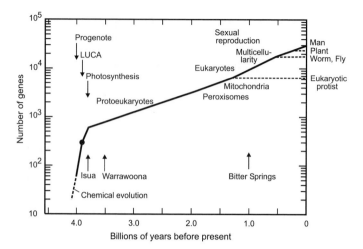

Fig. 7.7. The growth in the number of nonidentical genes with time in eukaryotic evolution (modified after Kaplan 1972). LUCA (*dot*) marks the Last Universal Common Ancestor of the DNA-world, Progenote the first living organism

The subsequent evolution can be reconstructed by using Table 6.4, which gives the number of genes in today's organisms. It has been found that the eukaryotic protist *S. cerevisiae* (yeast) has 6300 genes. This organism can be taken as a model for a typical eukaryote which lived about 1 billion years ago, before the invention of multicellularity and cell specialization. With Table 6.4 and the known dates of the appearance of plants, animals, and man, the rapid increase in the number of genes with time due to multicellularity and the development of organs can be displayed in Fig. 7.7. Not shown in the figure is the evolution of the prokaryotes which, after Table 6.4, carry between 468 genes for mycoplasmas and 4289 genes for *E. coli*. Since some prokaryotes have organelles, and others have lost genes because of their highly specialized lifestyles, the number of genes in prokaryotes varies. Yet their gene numbers agree nicely with those expected for the protoeukaryotic ancestors of the eukaryotes.

Figure 7.7 shows that over about 2.5 billion years the number of genes increased roughly by a factor of 10 (from about 600 to 6000). Yet, after the invention of sexuality, multicellularity, and cell specialization, this growth rate became more rapid. In an additional billion years, the number of genes multiplied by another factor of five, from 6000 to about 30 000. This shows the great importance of these inventions for life. It also explains the explosion of life forms at the Ediacaran period, roughly 570 million years ago, and the breathtaking speed of the evolution toward higher animals and to man. However, this also teaches us to appreciate what nature accomplished in the preceding 3 billion years: the production of the fabulously complicated eukaryotic cell, with its many well integrated organelles.

There is another interesting point that can be deduced from Table 6.4 and Fig. 7.7. Note the large variety in the number of genes of today's organisms. Since information is a decisive means of increasing the survival rate, we can conclude that the more information is available, the better a given organism should be equipped for its struggle for survival. But why, then, did organisms with many fewer genes not become extinguished long ago in this competition?

The fact that organisms with a small genome have survived to our times proves that they must have been very successful. Although information clearly leads to success, it is not only the sheer volume of information that is decisive. Success can also be achieved by specialized know-how. Some organisms, such as the bacteria, work on the principle "lean and mean". By avoiding unnecessary ballast (with a small genome and a tiny body size), they achieve a fast reproduction rate. This is combined with a high mutation rate, which, as has recently been discovered, can be deliberately increased by allowing the proofreading process during DNA replication to produce more errors. With high reproduction and mutation rates, the likelihood of a successful adaption to a new environment is greatly increased, since with a larger number of trials, an organism that optimally fits into a new environment can be found more quickly. We also discover this, unfortunately, from the fact that these organisms so quickly become resistant to medication.

7.7 Multicellularity, the Formation of Organs, and Programmed Cell Death

Multicellularity, the joining of many individual cells to form a common organism, together with cell specialization leading to true organs, is another landmark invention of the biological world. The decisive advantage that the organelles brought about for eukaryotic cells is equivalent to the large selective advantage that organs have brought about for multicellular life forms. We know today that the embryonic development of multicellular organisms recapitulates the historical evolution that occurred about 1 billion years ago. In a first so-called *blastula* stage (see Fig. 7.8a), the cells arranged themselves in a sphere, which created a protected internal space.

Such simple blastula-like organisms still exist today in *Volvox*. Like all multicellular life forms, *Volvox* has two cell types, *somatic (or body cells)* and *germinal (or reproductive cells)*. *Volvox* has about 2000 two-flagellate somatic cells which form the spheroid, and 16 large germ cells in its interior. These germ cells divide and build juveniles. Reaching maturity, the blastula ruptures and the juveniles are released. After this, the ruptured parental blastula *dies*. This points to another important fact of life, which came about with multicellularity, *programmed cell death*, called *apoptosis*. Unicellular organisms, which reproduce by simple cell division, are essentially immortal, except for mishaps or being eaten. That the death of the parental *Volvox* is programmed, and not a necessity, has been demonstrated by the discovery

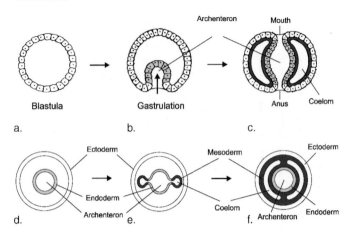

Fig. 7.8. The development of a multicellular organism. **a.** to **c.** Lengthwise cuts. **d.** to **f.** Perpendicular cross-sections (modified after Campbell 1996)

that a mutation in a single gene abolishes apoptosis, rendering immortality to that organism (Gilbert 1997). Why all multicellular organisms have vastly different programmed lifetimes is presently not fully understood.

The next step in the development to higher life is *gastrulation*, by which the blastula folds in on itself to build a simple gut, the *archenteron*, together with the ancestral anus (Fig. 7.8b). Later, the archenteron connects with an opening on the other side, the ancestral mouth (Fig. 7.8c), forming the alimentary tube through the body. The infolding of the surface layer, called the *ectoderm* (white), produces the *endoderm* (light gray), which lines the archenteron. In the cell line, which eventually leads to higher animals and man, this endoderm sprouts in a bilateral way (Fig. 7.8e) and produces the third germ layer, the *mesoderm* (black). The mesoderm expands and eventually fills the inside of the organism, forming a true internal body cavity, the *coelom* (Fig. 7.8f). In our ancestors *deuterostomes* the coelom developed from the endoderm. These important internal developments of multicellular organisms took about 500 million years and are shown in a phylogenetic tree in Fig. 7.9. It can be seen there, that during this time-span many branching points occurred, where different choices eventually led to a rich variety of different life forms.

The coelom, which is generally filled with fluid, has important functions. It acts as a hydrostatic skeleton, enables separate activities of the body surface and the alimentary tract, provides material, fluid, and gas circulation, and offers a safe space for enlargement of organs. The important advance of multicellularity is the specialization of cells into organs. It is from the three germ layers, ectoderm, mesoderm, and endoderm, that the organs develop. From the ectoderm develops, for example, the skin and its derivatives (glands, hair, feathers, and scales), the nervous system, and the sense organs. The mesoderm produces, for example, the skeleton, the muscles, the vascular

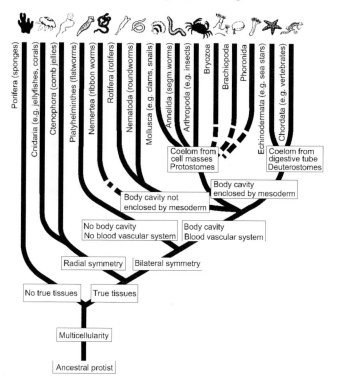

Fig. 7.9. The radiation of animals following the invention of multicellularity, roughly 1 billion years ago (after Campbell 1996)

system, the excretion organs, and the gonads. From the endoderm the digestive tract develops, for example, the liver, the pancreas, the thyroid gland, and the lungs.

There is also geological and paleontological evidence for this evolution. It has been discovered that suddenly, 630–542 million years ago, an extraordinary variety of new life forms appeared (Ediacaran period). The cause for this variety has been attributed to the fact that multicellular organisms had now developed master genes that controlled different parts of the body separately, and that therefore allowed the development of a very high degree of variation, specialization, and environmental adaption. Such genes are also called *homeobox genes* or *HOX genes* (see HOX 2005) and were discovered while studying the embryonic development of the fruit fly Drosophila melanogaster. The genes encode proteins that tell the cells in the various segments of the developing embryo what kind of structures they should make: antennae for the head and legs for the three thoracic segments. This all contributed to the large radiation of animals seen in Fig. 7.9.

The subsequent evolution toward the higher animals and man proceeded with the invention of the *chorda*, a cord-like central fiber that became the pre-

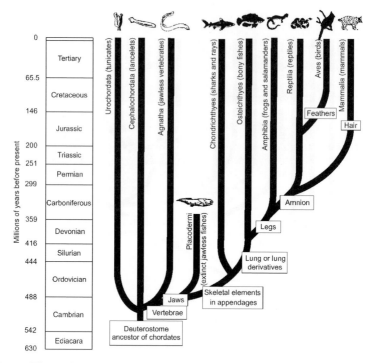

Fig. 7.10. The evolution of the chordates, after Campbell (1996), dates after Rhode (2005)

cursor of the internal skeletal structure of the vertebrate animals. The organisms with a chorda, the *chordates*, are the branch on the right side of Fig. 7.9. Their evolution is shown in more detail in Fig. 7.10. About 520 million years ago (in the Cambrian, 542–488 million years ago, and the Ordovician, 488–444 million years ago) the backbone, and later the jaw, developed. At the start of the Silurian, 444–416 million years ago, the chondrichthyes (cartilaginous) fish, and thereafter the osteichthyes fish (with a bony skeleton) evolved. Also at that time the first primitive land plants developed and arthropods (millipedes, insects, and spiders) conquered the land. The first amphibians appeared in the Devonian, 416–359 million years ago, together with the first flying insects.

7.8 Problems of Life on Land

A crucial event on the road to our type of intelligence was the conquest of the land that led to tetrapod (four-footed) animals, where eventually one pair of legs became arms and subsequently were liberated to permit manipulation and tool use. The essential basis for this development was the evolution of plants that proceeded on the land and provided the necessary alimentation and environment (rainforest) where the development of primates took place. The conquest of the land took place in the Ordovician (488–444 million years ago) where after 3.5 billion years of life in the oceans the descendants of the first bacteria had finally acquired the ability to conquer the land.

That this enormous expanse of virgin territory, covering 30% of the Earth's surface and providing very favorable light conditions for photosynthesis, had not been conquered before can be attributed to the truly formidable obstacles that land posed to the expansion of life. It was the multicellularity of eukaryotes, by allowing for cell specialization and permitting the development of completely new organs, that finally provided the tools for meeting these challenges. What were these obstacles that an empty, barren Mars-like environment, consisting of rocks, gravel, sand and clay plains, crossed by meandering rivers, studded with lakes and overblown by winds, inundated by sporadic and only seasonal rain posed to life?

There are four major difficulties. From seaside vacations we can recall that gravity is particularly felt when getting out of the water. In water, buoyancy working on a body with air-filled pockets (for example the air-bladders of fish or air-sacs of brown algae) greatly decreases weight. A second severe problem on land, other than in the ocean or even on the sea floor, is that it is difficult to move on dry soil. Here plants chose to avoid movement entirely and remain stationary. But how could then a plant at a fixed position on land obtain its necessary supply of water and essential nutrients? Some animals such as the arthropods (spiders, centipedes, insects, crustaceans), that probably had evolved from walking worms on the sea floor, did not have to modify their appendages much to move on land, while the need of locomotion for vertebrate animals (chordates) required more extensive skeletal developments.

Another problem on land is the mortal danger of drying out due to an ever-present evaporation, because even in the absence of winds unprotected plants and animals suffer from rapid desiccation. Finally, if all these problems are solved, how would a plant or animal procreate on land, that is, how would the gametes of eukariotic sexual organisms meet in the absence of open bodies of water? That this latter problem is particularly formidable can be seen from the fact that the evolution of plants from bryophytes to seedless vascular plants to gymnosperms and finally to angiosperms is a story of constant development towards greater independence of water and of greater control over sexual reproduction. It is not surprising that precisely the same can be said about the evolution of land animals from amphibians to reptiles and finally to mammals.

7.8.1 Conquest of the Land by Plants

The earliest life forms on land would most likely have been microbes, algae and lichens that populated the soil of lake shores and river banks long before the first land plants. Indeed, from DNA and RNA sequencing one finds that the ancestors of land plants are not, as one might expect, seaweed-like marine algae that occur in the tidal zone of the oceans, but rather microscopic fresh water forms, closely related to green algae of the class *Charophyceae*, living in lakes and rivers (Raven et al. 2003, Kenrick and Davies 2004). The reason for this fresh-water origin could be due to the fact that drying sea water leaves behind layers of salt that, by osmosis, suck out water from the sensitive plant tissue.

In the Middle Ordovician (475 million years ago) microfossils of the first spores appear, microscopic airborne cells that are characteristic of true land plants. Spores are assemblages of one to four usually haploid cells that are easily dispersed by the wind over long distances due to their microscopic size. Because of their robust construction and prodigious numbers, spores are ideal candidates for fossilization. These early fossil spores very likely derive from ancestors of the bryophytes (mosses, liverworts, hornworts), tiny plants from which only minute fragments are found (Kenrick 2003, Raven et al. 2003, Wellman et al. 2003).

The first step in the conquest of land by plants apparently consisted in sticking to minute size (to battle gravity) and to form sturdy spores (against desiccation). But since the first bryophytes – similar to green algae – require that their sperms swim through water to reach the eggs, these first land plants had to live in close association with water. That tendency is still seen

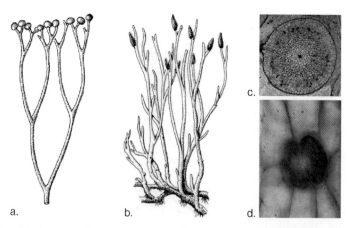

Fig. 7.11. Early land plants ca. 420 million years ago had leafless stick-like stems with simple branchings. **a.** Cooksonia, was about 2 cm tall. Stem-tops have terminal sporangia (spore sacks). **b.** Aglaophyton about 18 cm tall, growing from prostrate rhizomes with hair-like precursors of roots. **c.** Cross-section through a stem of a similar plant Rhynia showing central water conducting vascular cells surrounded by a cell system that transports the products of photosynthesis (bright). **d.** Stomata, openings at the stem surface, where gas exchange takes place, regulated by two kidney-shaped guard cells (after Raven et al. 2003, Kenrick and Davies 2004, Kerp 2005)

in modern bryophytes that most often grow in moist locations in forests or along the edges of streams and wetlands.

A next step was to battle gravity by the formation of sturdy stalk-like stems and development of a capillary system that is driven by evaporation at the top. One of the earliest so-called *seedless vascular plants* (club mosses, horsetails, ferns) with upright, gradually tapering shoots that bore terminal sporangia is *Cooksonia* from the Silurian (about 420 million years ago). Of this only about 2 cm tall plant (Fig. 7.11a), fossils are found worldwide. Other well-documented spore-producing seedless vascular plants are fossilized in the famous Early Devonian *Rhynie Cherts* of Scotland. Discovered near the village of Rhynie, 40 km north-east of Aberdeen, these roughly 410-million-year-old hard rocks were produced by hot springs that periodically flooded their environment with silica-rich water. The precipitating silica preserved the internal structure of plants and animals as well as their three-dimensional shape in situ, and in exquisite detail down to the microscopic level.

Like Cooksonia, early land plants such as *Aglaophyton* (Fig. 7.11b), *Rhynia* or *Psilophyton* were single stick-like organisms with bifurcating leafless stems that arose from a prostrate *rhizome* (horizontal growing root-like stem). By the Middle Devonian about 390 million years ago the tallest plants were of less than knee height but most were much smaller. Some of the plants now had stems that bore minute hairs and some, like the oldest club moss (lycopod) *Asteroxylon*, had even small scale-like leafs. The larger plants had a vascular system (tubing for the transport of water, see Fig. 7.11c) located in the core of the stems and *stomata* (openings in the stem surface, see Fig. 7.11d) to

Fig. 7.12. The first land plants at the shores of fresh water lakes and river banks in the Lower Devonian 416–398 million years ago (after Schaarschmidt 1968)

take up carbon dioxide for photosynthesis and to regulate the evaporation of water, thereby drawing it up from the rhizome. Some or all of the stem-tops had terminal *sporangia*, spore sacks that used the wind for the dispersal of spores. Figure 7.12 shows the first land plants growing near fresh water in the Lower Devonian.

7.8.2 New Organs of Land Plants

No other time period showed such huge changes in the evolution of plants than the Devonian (416–359 million years ago), a development rightly called the *Devonian explosion*. At its beginning one had bryophytes and simple seedless vascular plants. From horizontal rhizomes with hair-like forerunners of roots (Fig. 7.11b) in unfertile sandy soils near bodies of water these plants grew to heights of less than 10 cm. At the end and in the following Carboniferous (359–299 million years ago) there were extensive forests of canopy-forming club mosses and seed plants with tree heights of up to 40 m. These trees equipped with massive trunks and diameters reaching 2 m were sending their deep roots into by now highly processed and fertile soils and formed forests that extended over large areas away from the immediate contact with water. Associated with these external changes there was an equally significant internal revolution of the rooting system, of stems, leafs and the reproductive apparatus.

Transport Mechanisms

Anchoring and maintaining the upright position of a plant is one function of roots, but their role in the acquisition of water and nutrients is even more important. What are the nutrients of plants? Clearly the central role of plants as *autotrophs* (self-sufficient by photosynthesis in their nutritional need for organic compounds) and as the basis of the food supply of the *heterotrophs* (eaters of organic material produced by other organisms) requires the availability of light, carbon dioxide (CO_2) and water. While carbon (C) in the form of CO_2 is supplied from the air, there are many other nutrients that together with water have to be taken up by sufficiently specialized roots.

Foremost and required in large amounts are six so-called macronutrients: nitrogen (N), phosphorus (P), potassium (K), calcium (Ca), magnesium (Mg), and sulfur (S). While N, S, and P are essential elements for proteins as well as nucleic acids, the elements Mg, Ca and K are important for the function of chlorophyll, the basic molecule involved in photosynthesis, as well as in starches, sugars and in cell walls. Less important but still essential are a host of additional trace elements. Many of these nutrients have to be recovered by splitting and weathering of rocks, by dissolving decaying organic matter, or obtained with the help of microbes and symbiotic fungi at the roots. Finally, by their ploughing and mixing the ground, their chemical action and their decay, roots transformed the land from a barren accumulation of sand and clay into fertile soil suitable for life.

Relatively early in their history, as already discussed above, the plants had the problem of efficient transport of water and nutrients along their body. This was solved by the evolution of a system of fluid-conducting tissues called *xylem* and *phloem* (Fig. 7.11c) that permitted specialization of the plant in roots, stems and leafs. Of these, the xylem consists of cells that strengthen their walls with a very sturdy organic compound called *lignin* and subsequently die, thereby leaving behind an empty interconnected tube structure that serves a double function, to conduct water and inorganic nutrients upward, and to provide rigidity for the growth of a large organism. The phloem is a system of living cells that specializes in the upward and downward transport of internally manufactured compounds.

Such tissues permitted the plants not only to develop leaves to more efficiently harvest light, but also to grow to great height and thereby increase their surface area and photosynthetic efficiency in the competition with neighbors. This process has sometimes been called *colonization of the air* by which tree tops with maximum height levels of about 113 m (coastal redwood *Sequoia sempervirens* in the US northwest) are reached today.

Stems

For reasons of static stability, such growth demanded an ever-increasing stem diameter and more biomass. To save mass different solutions were tried. Ancient club mosses developed tough outer cylinders encasing soft inner tissue. Palaeozoic tree-horsetails had hollow tube-like stems, while tree ferns bound together several strands of smaller diameter, which were buttressed by roots growing downward to strengthen the base of the plant.

While a mass-saving rapid growth of a stem was advantageous in the competition for light, the disadvantage was that such stems more easily buckled under wind forces and could not develop extensive crowns. For this reason, seed plants such as conifers preferred solid trunks. The extinct club moss tree *Lepidodendron*, for example, reached heights in excess of 30 m. It had a slender branchless stem where the branching for the crown started 5 m from the top and the crown diameter was only 6 m. Because of the small final crown size and due to the lengthy growth time in which the plants consisted only of shoot-like stems, the light conditions and tree density in the primeval Devonian and Carboniferous forests were very different from our forests today. While now one has about 1–2 trees per ar (an area of 10×10 m^2), the different light conditions in the Carboniferous forests permitted 10–20 trees in the same area.

Reproduction

One of the biggest changes in plants and most beneficial to animals and humans were those associated with reproduction: the development of seeds, flowers and fruit. While today one has about 28 000 species of bryophytes (mostly mosses) and seedless vascular plants (mostly ferns) the number of *seed plant* species is ten times larger: about 800 species of *gymnosperms* (conifers, cycads, ginkgo and gnetophytes) and 235 000 species of *angiosperms* (flowering plants). Because the seeds of angiosperms are concealed inside a fruit, the other seed plants that do not develop a fruit are collectively called gymnosperms, which in Greek means plants with naked seeds.

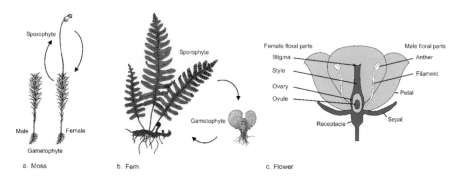

Fig. 7.13. Alternating diploid (sporophyte) and haploid (gametophyte) plant generations **a.** in the moss Polytrichum, **b.** in ferns. In the moss the sporophyte grows on top of the female gametophyte. **c.** Flower (schematic) with a cup of petals covered by sepals resting on the receptacle. Filaments and anthers represent the male, stigma, style and ovary the female organs. The ovule or embryonic sac contains the female egg, and the anthers the male pollen grains

Very significant for animal life on Earth has been the evolutionary drive in plants to make its reproduction process more efficient and to decrease the role of water in that process. The development started from a state found today in unicellular green algae. They have a life cycle in which a diploid thick-walled *zygospore* after some dormancy shows meiosis and haploid cells emerge that then reproduce asexually. Subsequently, two haploid cells of different mating strains come together to form a diploid zygote that once more develops a zygospore.

a. Bryophytes and Seedless Vascular Plants

The earliest land plants, the bryophytes and seedless vascular plants, have a life cycle of an alternating sequence of diploid and haploid generations that often are individual plants that differ substantially in appearance (see Fig. 7.13a,b). In this life cycle, which was already present in their ancestors the green algae, the bryophytes accentuated the gametophyte, while in ferns the sporophyte became the dominating generation (Fig. 7.13a,b). The diploid *sporophyte* develops sporangia (spore-producing organs) where meiosis happens and from where haploid spores covered by thick walls are dispersed by the wind. Most of the seedless vascular plants are *homosporous* where the haploid spores generate haploid bisexual *gametophytes* that produce both types of gametes (sperms and eggs). To prevent self-fertilization the two types of gametes are not produced at the same time. By swimming to the eggs in the female organs on a gametophyte of a different plant, the sperms fertilize the eggs and create a zygote that grows to a new sporophyte. Crucial in this process is the presence of water to allow the flagellated sperms to find the eggs. *Heterosporous* seedless vascular plants and bryophytes show a variant of this process, they generate male and female spores that subsequently produce male and female gametophytes.

b. Seed Plants (Gymnosperms)

This heterosporous life cycle was further modified in seed plants where the diploid and haploid life phases become bound into a single plant. In gymnosperms the male and female sporangia, called *microsporangia* and *megasporangia*, respectively, are usually carried on the same plant. After meiosis, the microsporangia form microspores, out of which so-called *pollen grains* develop that are dispersed by the wind, while the female megasporangium after meiosis develops megaspores. The megasporangium together with its protective layer, the *integument*, forms the *ovule*. Usually in an ovule only one functional megaspore is developed and is not released in the environment but remains captured inside and forms a female gametophyte that contains an egg. At the top of the ovule the integument leaves a gap. Here the pollen attaches and grows a pollen tube into the ovule to deliver the sperms to the egg without much need for water.

After fertilization, the integument forms a sturdy seed cover around the ovule together with attached wings, and the ovule becomes a seed that is released into the wind for dispersal and to grow a new plant (sporophyte). Because of the reduced dependence on water and the increased protection of the gametophyte together with its assured nutritional support, this new mode of reproduction proved to be very successful – a selective advantage that was also assured because of the much bigger size of the seeds that compared to the spores also carried a food supply for germination. This evolutionary development led to an overwhelming role of gymnosperms in the Mesozoic era 251–65 million years ago (Fig. 7.14), and is still seen today by the great extent of the conifer forests in the mid-latitude and subarctic climatic zones.

c. Seed Plants (Angiosperms)

Although angiosperms (flowering plants) because of their eye-catching flowers and fruits look very different, their reproduction, in principle, is rather similar to that of gymnosperms. The male and female sporangia are now concentrated in a hermaphroditic flower, the basic parts of which are shown in Fig. 7.13c. On the *receptacle* of the flower one often finds a cup of colorful *petals* protected by the *sepals*. The *stamens* with *filaments* and *anthers* producing the pollen grains constitute the male parts, while the *carpels* (with *stigma*, *style* and *ovary*) are the female parts of the flower. After a pollen grain falls onto the stigma it grows a pollen tube down through the style and ovary into the ovule where the sperms are delivered to the egg, again without much need for water. While the ovule and integument now form a seed the ovary grows a massive fruit that completely conceals the seed.

The decisive difference of angiosperms compared to the gymnosperms appears to be their reproductive strategy of how flower, pollen, seed and fruit are employed. While gymnosperms rely on the wind to disperse their pollen grains and seeds, the angiosperms developed mutually beneficial relationships with animals to deliver pollen and seeds. The difficulty to develop such relationships was apparently the reason why it took 230 million years from the start of the gymnosperms until angiosperms first appeared (see Fig. 7.14). Flower petals are powerful signaling devices to insects and birds to advertise nectar as a reward for the transfer of pollen, and ripe fruits are the compensation for birds and mammals for dispersal of the seeds. The fabulous numbers of different angiosperm species ranging from grasses to trees, shrubs, herbs and cacti in all climatic zones on Earth demonstrate the overwhelming success of this strategy.

Figure 7.14 displays the evolution over time of the number of species of the different plant phyla (plant subgroups) since they first appeared on land in the Ordovician. In this figure the bryophytes, the first land plants, are included gray because of the fragmentary fossil record their number of species over the ages is not known. For the display, today's number of species is taken. It is seen that the figure also includes extinct plants such as the seed ferns (pteridosperms) and cycad-like benettitales. Figure 7.14 is modified to follow Raven et al. (2003) who place the invasion of the land by plants in

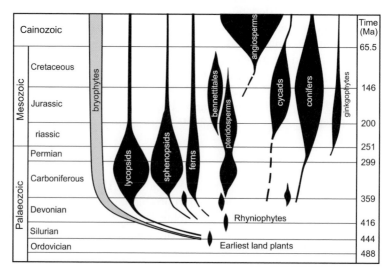

Fig. 7.14. Evolution of the number of species of plants since their conquest of the land in the Ordovician, modified after Skelton et al. (2003). Lycophytes are club mosses and sphenopsids horse tails. For lack of fossils, today's number of species are used for the bryophytes (*gray*). The time periods are in million years

the Ordovician around 488–444 million years ago and the development of gymnosperms in the late Devonian at 365 million years ago. Note that the oldest fossil angiosperms are about 135 million years old, appearing first in the Early Cretaceous. The geological time periods given in the figure (as used throughout this book) follow Rhode (2005).

7.8.3 Conquest of the Land by Animals

Associated with the plants in the Early Devonian Rhynie Cherts, mentioned above, several groups of freshwater animals like arthropods were found such as spiders, mites, and centipedes. These animals were *carnivores* (eating other animals) and *dentritivores* (eaters of dead and partly decomposed plant parts) but not *herbivores* (eaters of living plant tissue). While most of the organs and tissues of the early land plants had to be newly developed, early land animals resembled closely their aquatic ancestors. In their multiple transition to the land they retained most of their complex organs, although in modified form. However, animals following the plants onto the land had to face similar difficulties (gravity, motion over rough soil, respiration, desiccation, reproduction) as the plants.

Arthropods, because of their sturdy chitinous exoskeleton and limbs, did not have to modify their body much for support against gravity and to move around, but had to solve the problems of breathing and of desiccation. It was the vertebrate animals (chordates), in pursuit of plant and insect food, that particularly had to modify their anatomy for life on land. For support against gravity, three important tools had to be developed: legs, girdles, and a supportive backbone. In fish, the spine is essentially used for holding the body tightly together, acting like a taught spring for their wavy swimming motion (see Fig. 7.15).

On land this backbone was now required to take on the function of a crossbeam, from which the entire body is suspended. While the backbone of the crossopterygian in Fig. 7.16 was well adapted for the wavy motion of a fish, the spine's support function in a more advanced amphibian such as Mastodonsaurus was accomplished by modifying the individual vertebrae such that they mutually underpinned each other (Fig. 7.16). In addition, as

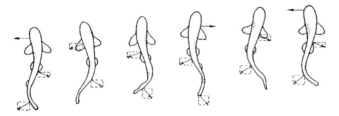

Fig. 7.15. The wavy motion of a fish (after Campbell 1996)

seen in Fig. 7.17, the backbone had to be anchored in strong girdles supported by powerful legs. It can be seen that the fins of the fish where not strong enough to support their body weight. They were sufficient for the vigorous wiggling by which a crossopterygian could escape over short stretches of land out of a desiccating lagoon. As shown in Fig. 7.17 the backbone and girdles of amphibians represent the same support mechanism for their body as in a garden swing, where crossbeam and stands support the weight of children in their seats.

Another important modification required for life on land was breathing, which in water is done through gills, while on land lungs are used. At the same time, important changes had to be made to the skin to battle against the threat of desiccation. This problem also concerns the eggs, and the food supply for the newly hatched juveniles. The amphibians (frogs, and salamanders) solved these problems by leaving both their eggs and the juvenile development in the water. The oldest amphibian fossils, showing the fish-tetrapod (animals with four limbs) transition, date back to Late Devonian times 370–360 million years ago (Clark 2004).

A great evolutionary step represents the *amniotic egg* in which the embryo is enclosed in a fruit-sack (amnion) and surrounded by a hard calcified shell. Together with parental care for the young, this allowed the reptiles (lizards, snakes, turtles, and crocodiles) to conquer land far away from the sea. While crocodiles stayed close to the water, lizards and snakes ventured

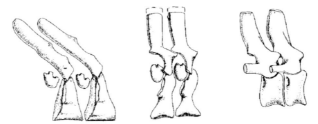

Fig. 7.16. The backbones of a fish (crossopterygian), a primitive amphibian (*Ichtyostega*), and an advanced amphibian (*Mastodonsaurus*) (after Romer 1974)

Fig. 7.17. The differing anatomies of a crossopterygian (*left*) and a primitive amphibian (*center*) (after Campbell 1996). Backbone and girdles represent the same support mechanism as in a garden swing (*right*)

even into the driest desert regions. The first reptiles with their tough skins appeared in the Carboniferous, 359–299 million years ago, when plants became massive land-dwellers as seen in today's extensive coal deposits. The reptiles were the dominant animals on land and in the sea in the Permian (299–251 million years ago), and stayed dominant all the way through the Triassic (251–200 million years ago), the Jurassic (200–146 million years ago), and the Cretaceous (146–65 million years ago).

Starting in the Triassic, dinosaurs supremely ruled the Jurassic and the Cretaceous. The reptiles were also the first vertebrate animals to conquer the air. The first birds appeared in the Jurassic and recently it has become certain that they have descended from coelurosaurian dinosaurs, a family of the suborder theropoda belonging to the order saurischian dinosaurs. Birds can therefore be seen as surviving saurischian dinosaurs (Padian and Chiappe 1998, Norell and Xu 2005). Although mammals with their great discovery of the *placenta* (where the embryo could develop within the maternal body) originated in the Triassic, they did not really flourish until the Tertiary because of intense competition by the dinosaurs.

The important innovation of the birds and mammals was body insulation, which covered the naked reptile and mammalian skins with feathers and hair, respectively. This allowed these animals to become warm-blooded; that is, to keep their body temperatures at elevated levels, at which the biochemical and sensory reactions are maintained at an optimum level. Since for large dinosaurs the body surface was comparatively small for the given body volume, the energy loss from the surface was minor. Thus, despite their naked skin, they could lead a very active life. This situation, however, was different for small animals, where radiation loss and thermal conduction from the body is most acute. For this reason, it was the small animals (birds and their ancestors as well as mouse-sized mammals) that first developed body insulation. This adaption proved to be a major advantage for survival, and its lack was a major reason for extinction, when a most unusual external catastrophe, the great K/T boundary event, occurred.

7.9 The Great K/T Boundary Event

The great K/T boundary event terminated the Cretaceous and marked the beginning of the Tertiary, which extends from 65 to 2 million years ago. On the basis of recent data, this event is characterized by the extinction of between 40% and 75% of all animal species. A similar event occurred at the Permian–Triassic boundary (called *P/T event*), 251 million years ago, where between 60% and 95% of all animal species became extinct. While the K/T event was most likely caused by an impact of a large extraterrestrial body, possibly a comet, an impact origin has also been proposed for the P/T event (Becker et al. 2004).

But there are also alternative theories. Some paleontologists believe that the extinctions occurred due to natural causes. However, as the geological evidence is mounting that the K/T boundary event must have happened in an even shorter time-span (less than a few thousand years) than previously thought, this gradualist theory is increasingly difficult to maintain. Another alternative theory has been put forward by geologists who believe that the extreme volcanism that was responsible for the formation of the Deccan Traps (an extensive mountain ridge in southwestern India, see Chap. 3) caused the K/T boundary event. However, this period of volcanism seems to have happened 67–68 million years ago, considerably earlier than the K/T boundary event. But extreme volcanism associated with the formation of the Siberian Traps (Chap. 3) that occurred in close proximity to the Permian–Triassic boundary (Mundil et al. 2004) can not be excluded as a cause of the P/T extinction event (Ward et al. 2005).

The impact theory attributes the K/T boundary event to the infall of a large meteorite, which struck in the region of today's Yukatan peninsula in Mexico 65 million years ago (Fig. 7.18). The reality of this infall is not in doubt because the impact location, named the Chicxulub crater, has been found. Seismic and microgravity measurements reveal a diameter of about 180 km, which makes it not only the biggest known impact crater on Earth, but also the biggest impact crater on any planet or moon of the solar system since the end of the heavy bombardments, 3.8 billion years ago. The crater shows several rings. Half of it is buried under Yukatan, and the other half under the sea. It originally had a depth of between 15 and 20 km, but the crater walls subsequently collapsed. Today, the crater lies 300–1000 m below the land and sea level.

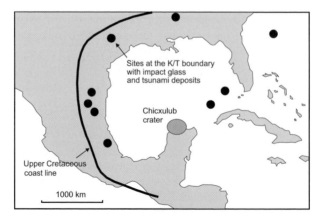

Fig. 7.18. The extent of the tsunami damage from the comet responsible for the Chicxulub crater (after Claeys 1996)

The estimated impact energy of the bolide was between 3×10^8 and 3×10^9 megatons of TNT (1 megaton of TNT is 4.2×10^{15} J or 50 times the Hiroshima bomb). Its diameter ranged between 9 and 17 km and its velocity between 15 and 30 km/s. The impact created a gigantic tsunami (sea wave) with a height of roughly 1 km, which raced over the ocean, depositing layers of sand and rubble of 2–3 m thickness in a 3000 km wide circle extending from Alabama to Guatemala (see Fig. 7.18). Shocked quartz uniquely testifies to the impact event. Argon isotopes in molten rock lead to a date of 64.98 ± 0.05 million years, while impact glass gives an age of 65.01 ± 0.08 million years.

Geologically, the K/T boundary is found in oceanic sediments over the entire Earth. For instance, the white deposits of chalk from the shells of sea organisms at a site near Biarritz (Fig. 7.19) are suddenly interrupted by a ca. 20 cm wide red and black clay layer in which the concentration of carbonate abruptly drops to 0–10% from the former 40% (Fig. 7.20a, solid). That this same layer shows a sudden jump in the iridium abundance to a 100 times greater value (Fig. 7.20a, dotted), with a mass of 3.6×10^{11} g worldwide, suggests a comet. This cometary origin is also indicated by the presence of unusual amino acids in the same layer. The mighty layers of chalk at the end

Fig. 7.19. The chalk and clay layer at the K/T boundary near Biarritz in the Bay of Biscay (after Rocchia 1996)

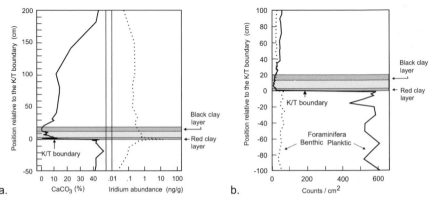

Fig. 7.20. a. The carbonate and iridium concentrations at the K/T boundary (after Rocchia 1996). **b.** Destruction and recovery of foraminifera (after Smit 1996)

of the Cretaceous consist of calcium skeletons of dead foraminifera, which are planktic (living near the surface) and benthic (bottom dwelling) unicellular organisms. Deposited at the bottoms of oceans, these layers normally grow at a rate of 1 cm in 500 years.

The following time-sequence of the events has been worked out. At the impact of the comet, a huge amount of mass, more than 10 times the cometary mass, was thrown high up into the atmosphere and distributed over the entire globe, from where it dropped back to Earth. This infall generated enormous heat, and extensive forest fires broke out. The large amount of dust subsequently blocked out the sunlight and the Earth's surface became completely dark for many months or even years. Due to the lack of solar radiation, it also became very cold. The infalling debris is thought to have generated the lowermost ca. 15 cm wide layer, which includes the red clay layer at the K/T boundary. Overlying that, the 5 cm black clay layer is attributed to the very impoverished population of surviving foraminifera of the millennia immediately after the K/T boundary event. The consequences of the impact affected the ecosystems in different ways.

On land: Small animals that could shield their bodies against the cold with fur or feathers survived as did animals that where able to hibernate, or burrow in the soil. The plants withered in the total darkness. This caused a mass extinction of large animals, which could not find food, and of those with naked skin, because of the lack of solar heating (dinosaurs). The remaining land animals survived by digging for worms, larvea, roots, and other plant remains. By erosion, decaying biological material seeped into the rivers, where it served as food for animal life (fish, turtles, and crocodiles).

In the ocean: The complete blocking of the sunlight led to a mass extinction of the planktic foraminifera (see Fig. 7.20b, solid). The benthic foraminifera (Fig. 7.20b, dotted), were not affected in the same way, as they live on the decaying residue of dead animals which accumulates in the depths of the oceans. The death of the plankton led to a mass extinction of shrimps and fish, which in turn caused a mass extinction of the great maritime reptiles. At the ocean surface, the entire food chain was disrupted. The extinction of planktic and benthic foraminifera is shown by the actual counts per cm^2 in a horizontally cut stone layer. The planktic counts decreased from typically 550 animals per cm^2 to less than 30 at the K/T boundary. The benthic foraminifera also decreased, but only from 60 to 20. They survived much better in the depths of the oceans and dominated immediately after the K/T boundary event. Yet 20 cm (about 10 000 years) above the black clay layer, the planktic foraminifera dominate again, although the total number of these animals remained only a small fraction of the abundance seen at pre-K/T times.

7.10 The Tertiary and the Evolution of Mammals

At the beginning of the Tertiary, no large land animals were present and the surviving amphibians, reptiles, mammals, and birds found the land surface covered with lots of plants and an abundance of insects. Figure 7.21 shows the evolution of placental mammals. It can be seen that various mammalian

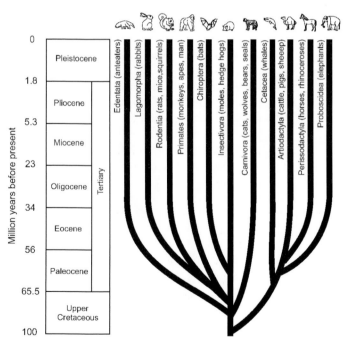

Fig. 7.21. The evolution of placental mammals (modified after Condie and Sloan 1998), dates after Rhode (2005)

orders had already radiated from the oldest order, *Insectivora* in the upper Cretaceous, while the remainder departed in the Paleocene. The important innovation of the mammals was to let their embryos grow in an uterus inside the body, in contact with the mother's blood supply over a placenta. Together with milk-feeding of the infants and intensive child care (which is also done by reptiles and birds), this invention led to a high survival rate of the offspring.

7.11 Primate Evolution

Primates branched off from the mammalian insect eaters in the upper Cretaceous period (see Fig. 7.22), and today there are six natural groups of nonhuman primates, located in the following regions:
1. Lemurs: Madagascar
2. Lorises: Africa, South and Southeast Asia
3. Tarsiers: certain islands in Southeast Asia
4. New world monkeys: South and Central America
5. Old world monkeys: Africa, South and Southeast Asia
6. Lesser and great apes: Africa, South and Southeast Asia

These primate groups are classified into two suborders. The first three groups belong to the *prosimians*. They are more distantly related to the last

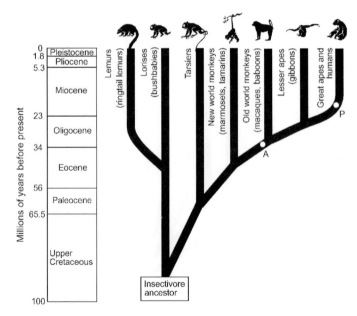

Fig. 7.22. The phylogenetic tree of the primates (after Martin 1990 and Steiper et al. 2004), dates are after Rhode (2005). A marks the *Aegyptopithecus*, P the *Proconsul* stages

three ones, called *simians* or *anthropoidea*. Lemurs and lorises split from the main primate line in the upper Cretaceous period (see Fig. 7.22), and the tarsiers branched off in the Paleocene. New world monkeys departed from the path leading to the great apes and man in the Eocene, old world monkeys in the early Oligocene, and the lesser apes (gibbons) in the middle Oligocene periods.

The geographical distribution of today's primates (see Fig. 7.23) roughly coincides with the band of the Tropics, from 23.5° northern to 23.5° southern latitude (dotted), reflecting their preference for a warm and moist climate, especially the rain forest. This is essentially the environment in which they evolved from their insectivore ancestors.

Throughout most of the Tertiary period, temperatures were much higher than today and the range of the tropical forests extended to much greater latitudes. This can be deduced from the isotope ratios of oxygen found in skeletal deposits of deep-sea dwelling organisms (benthic foraminifera), recovered from tropical ocean drilling cores (see Fig. 7.24). Because ocean water containing the lighter ^{16}O evaporates more easily than with the heavier ^{18}O isotope, the water with the lighter isotope gets deposited in glaciers and ice sheets in the polar regions, while the remaining ocean water (which the organisms take up to build their skeletons) gets enriched with the heavier oxygen isotope. A persistent increase in the $^{18}O/^{16}O$ ratio from the early Eocene up to the Pliocene (see Fig. 7.24) therefore indicates increasing glaciation and steadily declining temperatures during primate evolution. The present extent of the tropical forest is therefore only a fraction of what it was in Eocene and Oligocene times, when the extensive ice sheets of the Arctic and Antarctic regions had yet to develop. This explains why most fossils of early primates have been found in central Europe, and at similar latitudes in North America.

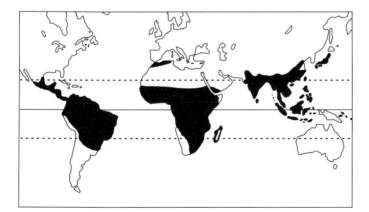

Fig. 7.23. The geographical distribution of the major groups of living nonhuman primates (after Martin 1990)

While insectivores typically led a life on the ground, the ancestors of primates began to follow insects up the trees, where also a great abundance of plant materials – leaves, fruit, and nuts – could be found. From the early Cretaceous period onward, as discussed above, newly evolving *angiosperm* (flowering) plants increasingly utilized insects, birds, and mammals for the fertilization and the dispersal of their seeds. In warmer regions, these flowering plants soon replaced the *gymnosperm* (conifer-like) plants, which let the wind carry their seeds. A new mutualistic relationship became established between these animals and the angiosperm plants, which in return for fertilization and seed dispersal were provided with nourishment.

Having climbed one tree, the problem for an early small primate was how to get to another. This was essential because plants, in order to maximize

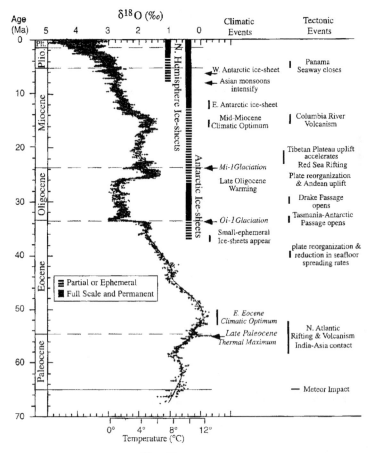

Fig. 7.24. The increase of the $\delta^{18}O/^{16}O$ isotope ratio in carbonate deposits from deep-sea dwelling foraminifera, correlated with the development of extensive glaciers and ice sheets (after Zachos et al. 2001). The *scale at the bottom* indicates the temperature increase compared to today

the service from animals, typically provide their nectar and fruit sparingly. By letting the fruit ripen at different times, they also ensure that the limited number of collecting animals will not be overwhelmed. With the forest floor infested with superbly adapted and camouflaged predators, getting all the way down from one tree to transit on the ground and climb up the next one was not just tedious but also dangerous. By jumping from branch to branch, primates were able to completely avoid the hostile forest floor, but to do this their mental equipment first needed to be improved. It was this arboreal environment in a tropical forest that stimulated the growth of the primate brain.

Two primate species represent important examples in this evolution. Both are now extinct simians, which lived near branching points of the evolutionary tree. Figure 7.25 shows the flora and fauna in the tropical forest in early Oligocene times in Egypt, some 35 million years ago. Here *Aegyptopithecus*, the common ancestor near the point (marked A in Fig. 7.22) at which old world monkeys departed from the lesser and great apes, is shown in the lower left-hand corner of the figure. Weighing about 500 g, it led the life of a quadrupedal tree-dweller.

In this environment, *stereoscopic vision* and a capability for fine-tuned hand and foot movement were essential, as false estimates of distance or branch thickness, or an imprecise grip could be life-threatening when jumping from tree to tree. To perfect stereoscopic vision, the eyes moved into a more frontal position, allowing a greater overlap of the visual fields. But

Fig. 7.25. The tropical forest and animal life in the Oligocene of Egypt, 35 million years ago with the *Aegyptopithecus* at the *lower left corner* (after Simons 1992)

this forward vision also increased vulnerability to predators. This, however, was compensated by the fact that the primate's arboreal environment was far less dangerous than the forest floor.

In addition, these animals required a better memory, as well as an improved pattern and color recognition capability, since the trees carrying ripe fruit in the rain forest at any given time are often widely separated. Moreover, unlike their ground-dwelling ancestors roaming about on a two-dimensional surface, monkeys had to remember pathways and locations in a much larger three-dimensional space. The greater mental capabilities required for this arboreal lifestyle and a more intense social interaction meant that primates developed a much larger brain, of roughly twice the size compared to that of the other mammals of the same body weight.

The other important stage (marked P in Fig. 7.22), past the branching point at which the lesser apes (gibbons) left the line leading to the great apes,(orang-utan, gorilla, and chimpanzee) and humans in the middle Oligocene period, is represented by *Proconsul* (see Fig. 7.26). This early Miocene ape, which lived about 20 million years ago, serves as a useful model for the common ancestor of the great apes and the hominids, especially since it had already acquired a weight of about 5 kg. This tenfold increase in weight necessitated a very different lifestyle compared to that of the much lighter *Aegyptopithecus*, and like today's orang-utan, it had to climb and move on sturdier tree branches and avoid jumping in the tree canopy, where the thinner branches might no longer bear its weight. This more sturdy mode of arboreal movement had led to a loss of the tail, which in the smaller monkeys serves as a steering rudder for jumping and as an additional gripping instrument. It also became necessary for *Proconsul* to visit the ground more frequently.

Ground-dwelling thus became increasingly important along the evolutionary path to man, as the apes grew and acquired the weight that characterizes today's chimpanzees and gorillas. About 14 million years ago, orang-utans branched away from the evolutionary line leading to man. The recently discovered 12.5–13 million-year-old middle Miocene ape *Pierolapithecus catalaunicus* (Moyà-Solà et al. 2004) is a likely model of the last common ancestor of the great apes and humans, which means that the departure of orang-utan

Fig. 7.26. The ape *Proconsul* lived in the Miocene, 20 million years ago (after Kelley 1992)

could be more recent (Fig. 7.27, dashed). Roughly 9 million years ago gorillas separated from the line of the great apes and some 6–7 million years ago
man's closest living relatives in the animal world, chimpanzees, became separated as well (see Fig. 7.27). Another reason why ground dwelling became
more important was because the tropical forest receded due to the climatic
evolution toward the ice ages (Fig. 7.24).

The creatures leading to us from this last branching point are called *hominids*. For an up-to-date account of our knowledge on hominids see Foley
(2005). Their line of development began with the *australopithecines*, comprising *Ardipithecus ramidus*, which flourished some 4.4 million years ago, *Australopithecus anamensis*, which existed at around 4.2 million years ago, *Australopithecus afarensis*, at 3.9 million years ago, and finally *Australopithecus
africanus*, which lived approximately 3 million years ago. Note that Fig. 7.27
includes the recently discovered earliest 5.7 to 7 million year old hominid fossils *Ardipithecus ramidus kadabba*, as well as *Sahelanthropus tchadensis* which
are very close to the branching point to the chimpanzees (Haile-Selassie 2001),
(Brunet et al. 2002). The australopithecines, weighing roughly 10% more than

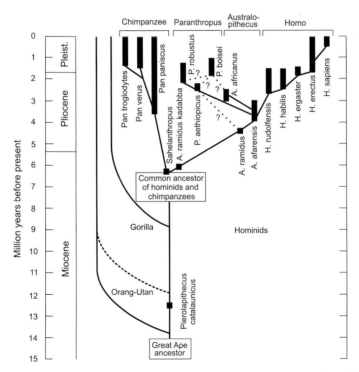

Fig. 7.27. The phylogenetic tree of the great apes and humans (modified after
Kelley 1992, Wood 1994). The recently discovered Miocene ape Pierolapithecus
could change the branching time of orang-utan (*dashed*). *Heavy lines* indicate the
time spans over which fossils were found

chimpanzees, were all upright walkers, but despite their striking similarity in appearance to later humans and modern man (see Fig. 7.28), they did not possess a significantly larger brain (and probably mental capability) than chimpanzees or gorillas (Fig. 7.33). Accustomed to ground-dwelling, they nevertheless remained very accomplished climbers and retired to the trees to avoid danger and to sleep at night.

Their upright walk (Figs. 7.28, 7.29), a development for ground-dwelling, was not just an energy-efficient way of covering large distances. More importantly, it freed the arms and hands for the transport of food and infants, and the use of tools. With an upright gait, enemies hidden in high grass and bushy environments are more easily spotted, and plant food on high branches better obtained. All these factors support the view that our ancestors left the dense rain forest and entered more open places, particularly the wide savannahs, while gorillas and chimpanzees, which normally use all four extremities for walking (knuckle walk), remained behind in the receding rain forest, where they survive to this day. As clear proof of the upright walk, Fig. 7.29 shows tracks preserved in moist ash from a volcanic outbreak which afterwards dried and became cement-like. These were made by *A. afarensis* about 3.7 million years ago, in the Laetoli area of northern Tanzania.

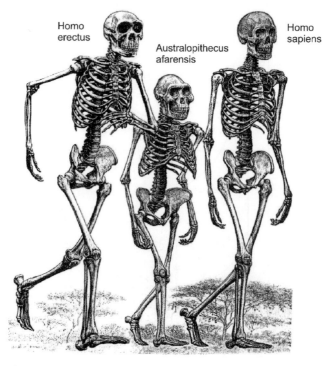

Homo
erectus

Australopithecus
afarensis

Homo
sapiens

Fig. 7.28. A comparison between *A. afarensis*, *H. erectus*, and *H. sapiens* (after Gore 1997)

Perfection of the upright walk, which is sometimes also clumsily performed by gorillas and chimpanzees, necessitated a number of important anatomical changes (Fig. 7.28). The gripping toe, for example, was modified into an acceleration tool for striding with longer legs, which were straightened and positioned more directly below the pelvis. This supported the body weight better and avoided spending energy to hold it upright. The fingers and toes changed from long curved phalanges, specialized for climbing, to the short straight forms of modern man. The pelvis was modified for the different attachment of the muscles for upright walking, the spine acquired an S-form to act as a shock absorber for the walking motion, and the head became balanced straight on top.

The first humans developed out of australopithecines and flourished some 2.5 million years ago as *Homo rudolfensis* and *Homo habilis* (see Fig. 7.27). Then, about 2 million years ago, a particularly efficient upright walker, *Homo erectus*, appeared (Figs. 7.28, 7.29). Recent fossil discoveries on the island of

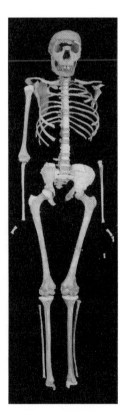

Fig. 7.29. *Left:* Tracks of *A. afarensis* made at Laetoli in Tanzania, discovered by Leakey (1979). *Right:* Almost complete skeleton of the "Turkana boy", a 1.6-million-year-old *Homo erectus* (after Tattersall 1997)

Java in Indonesia suggest that this ancestor, whose chest had become barrel-like to improve breathing, was still alive possibly as late as 27 000 years ago (Swisher et al. 1996). About 1–1.5 million years ago, *Homo erectus* began to use fire as a tool (Brain and Sillen 1988).

Finally, about 500 000 years ago, the *archaic Homo sapiens*, and roughly 200 000 years ago *modern Homo sapiens* appeared (McDougall et al. 2005) who, in the past 30 000 years, developed from a Stone Age collector and hunter to a farmer, builder, and producer of machinery. The *Neanderthal man*, or more properly called *Homo sapiens neanderthalensis*, was a side line of human evolution. He lived throughout Europe and the Middle East from roughly 230 000 to about 30 000 years ago and became extinct due to the intensive competition with modern H. sapiens (Mellars 2004). It has been shown using mitochondrial DNA sequencing that there was no interbreeding with modern H. sapiens (Serre et al. 2004).

As an example of an archaic Homo sapiens, also called *Homo heidelbergensis* (Balter 2001), Fig. 7.30 shows the skull of the *Tautavel man*, found in 1971 in the Arago cave at the village of Tautavel, near Perpignan, France. It has a brain volume of 1150 cm^3 and an estimated age of 450 000 years.

As the demands of fine-tuned hand manipulation and the social interaction between our ancestors became more complex, culminating in the development of speech, the volume of the brain increased rapidly, from an average of 400 cm^3 in *A. afarensis* to 900 cm^3 in *H. erectus* and finally to 1350 cm^3 in *H. sapiens*. This dramatic threefold growth of the brain is all the more staggering when compared with the body weight, which went up by only about 30% during the same period (see Fig. 7.33). Figure 7.28 compares

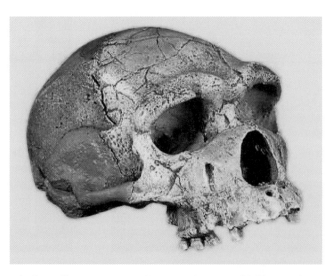

Fig. 7.30. *Archaic Homo sapiens*, the 450 000 year old Tautavel man (after de Lumley 2001)

Australopithecus afarensis, *Homo erectus*, and *Homo sapiens*. Given a time separation of roughly 2 million years between each of these hominids, both the similarities and the differences are striking.

7.12 DNA Hybridization

Before discussing human development in more detail, it is necessary to explain how the evolutionary relationship between the various apes and humans and the branchings of the phylogenetic tree were worked out. The heavy lines in Fig. 7.27 indicate time-spans which are known from dateable fossils, but such fossil evidence for the older branching points near 9 and 14 million years ago has yet to be found. In the absence of fossils, the molecular clock has been used to measure the ancestral relationship between monkeys, apes, and man. As discussed in Chap. 6, this method is based on counting the number of mutations of DNA of different species. While, because of a lack of sequenced genomes, precise counting of the mutations is not yet possible, there are already cruder methods that can be used to approximately estimate the number of mutations.

Fig. 7.31. Combined single strands of DNA of two species, 1 and 2

The procedure of *DNA hybridization* combines single strands of DNA of two different species into the double-stranded form of a hybrid DNA (Fig. 7.31). Since species 1 and 2 are closely related, the single DNA strands fit together at most places. However, as the DNA of species 1 differs from that of species 2 at points of mutation, there are regions that do not fit. In these sections the single strands cannot combine and therefore form loopings (see Fig. 7.31). The DNA hybridization method works by carefully heating samples containing double-stranded DNA and measuring at what temperatures the two strands separate (melt). The more distantly species 1 and 2 are related, the greater is the number of such loopings and the easier it is to melt the hybrid DNA compared to regular DNA. Because hybrid DNA melts at lower temperature than normal DNA, the differences in the melting temperatures are largest for the most distant species.

The DNA hybridization results for humans, great apes, lesser apes, and old world monkeys are shown in Table 7.1. The temperature difference between old world monkeys and all the other species is about 7.3 °C, and between the lesser apes (gibbons) and the great apes and humans 4.8 °C. Between the orang-utan and gorilla, chimpanzee, and man it falls to about 3.6 °C, and among the latter three it varies from 1.6 °C to 2.3 °C. These differences were used to construct the phylogenetic trees shown in Figs. 7.22

Table 7.1. Melting temperature differences (in °C) in DNA hybridization experiments for various monkeys, apes, and man (after Sibley 1992)

DNA–DNA distances	Go	Hu	Ch	Or	Gi	Owm
Gorilla (Go)	–	1.6	2.2	3.6	4.8	7.4
Human (Hu)		–	2.3	3.6	4.8	7.3
Chimpanzee (Ch)			–	3.5	4.8	7.1
Orang-Utan (Or)				–	4.8	7.4
Gibbon (Gi)					–	7.1
Old world monkey (Owm)						–

and 7.27. Note, however, that although undoubtedly correct in its rough outline, the details of the evolutionary tree are constantly being modified to incorporate new scientific evidence. A more detailed recent comparison of partial DNA sequences of man, gorilla, and chimpanzee has revealed that, different to Table 7.1, chimpanzees are our closest relatives in the animal world. Also, it has recently been proposed that human evolution might go from *A. afarensis* to *A. africanus*, and from there to *Homo*.

7.13 Brain Evolution and Tool Use

Returning to the outline of hominid evolution, what are the reasons for the impressive growth of the brain? Figure 7.32 shows a comparison of the skulls of *Australopithecus afarensis*, *Homo erectus*, and *Homo sapiens*. It can be seen that *A. afarensis* had a short skull, a protruding upper jaw, a low forehead, and a small brow ridge, while *H. erectus* had a longish skull, a heavy brow ridge, a less protruding upper jaw, and a higher forehead. *Homo sapiens*, by contrast, has an even larger skull, a high forehead, but no brow ridge and no protruding upper jaw. This morphology reveals a rapid growth of the brain. Indeed, direct measurements of the endocasts (casts of the contents of the brain cases) displayed in Fig. 7.33 reveal a astounding threefold increase of the average brain volume, which occurred while the body weight grew very little. This evolutionary development of hominids lies in stark contrast

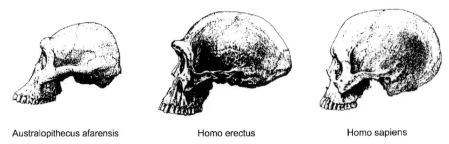

Australopithecus afarensis   Homo erectus   Homo sapiens

Fig. 7.32. A comparison of the skulls of *A. afarensis*, *H. erectus*, and *H. sapiens* (after Gore 1997)

to the brain evolution of the great apes. In chimpanzees, the brain volume remained at the level of the early australopithecines, while in gorillas it stayed at a similar level, despite the far greater growth in body weight.

Why did the human brain increase in size so much? A full explanation is presently hampered by the fact that we know so little about the brain's workings, particularly of the difference in brain function between man and chimpanzee or gorilla. Recent landmark advances in nuclear magnetic resonance (NMR) imaging and positron emission tomography (PET), which provide high spatial and temporal resolution for noninvasive investigations of the living brain, will reveal much in the future, particularly for the study of higher mental functions. Experiments and investigations of brain-damaged patients have already identified an impressive number of control and information-processing centers, and allow some partial answers to be given.

Figure 7.34 shows in cross-section areas of the human brain, overlaid with pictorial representations of the body parts that they control. As the brain devotes larger areas to those parts that receive more attention, these have been magnified out of proportion to produce a distorted human figure, known as a *homunculus*. Figure 7.34a shows the areas where the senses are especially fine-tuned, and Fig. 7.34b the motor areas where the body movements are particularly delicately controlled. The most important areas in both figures are roughly identical: the face, lips, tongue, and hands.

This strong emphasis on the mouth, face, and hands reveals the outstanding specializations of man: communication and tool use. Both played a fun-

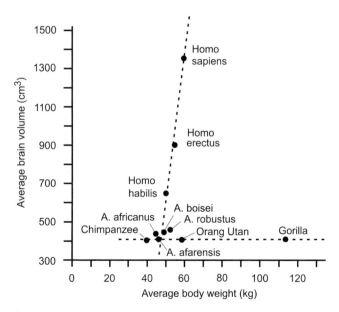

Fig. 7.33. A comparison of average body weight and average brain volume between ape and man (after Lewin 1989)

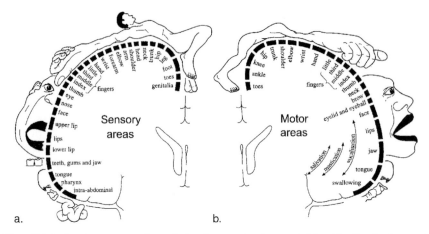

Fig. 7.34a,b. A cross-section of the human brain overlaid with the body parts that the respective brain areas control (after Taylor et al. 1997)

damental role in human development. In the early phases in particular, when language had yet to develop to its present level, the correct interpretation of nonverbal facial communication, such as happiness, sadness, aggression, compliance, pain, and sexual interest, provided a critical survival role in the tight social groups in which *Homo* lived. Another important growth area of the brain in hominid evolution is the neocortex, the front part of the brain where the higher mental functions such as thinking and planning take place, and where self-consciousness presumably resides.

7.14 Stone Tool Culture

Stone tool culture and its development provide more insights into the dramatic growth of the brain in hominids. Toolmaking has long been regarded as a dividing line between humans and nonhumans – the distinction between man as *toolmaker* and ape as *tooluser* having been made since the 19th century. Extended field studies on chimpanzees, gorillas, and other apes, however, have recently challenged the basis of this division. It is now clear that chimpanzees do not just use tools, in common with gorillas and orang-utans, but they also have the ability to make them as well. For example, some chimpanzees have been observed modifying branches by stripping off leaves to make a fine probe that they can use to extract termites from mounds. In addition, there is strong evidence of their ability to solve problems by thinking. In captivity, for instance, chimpanzees without previous experience have been observed stacking crates on top of each other in order to reach inaccessible food placed near the ceiling. It is also increasingly clear that chimpanzees possess a powerful ability to learn through observation – some have even been taught to communicate with their human captors in a language based on simple symbols (lexigrams). The defining differences between the mental faculties and functions of ape and man will undoubtedly prove to be a subject of intense investigation in the future.

Yet despite this uncertainty, language and the carrying of carefully constructed tools remain the undisputed characteristics of humans. There is an obvious lack of evidence for the former, but detailed studies of the mouth region of fossil skulls have recently shown that Neanderthal man did indeed possess speech. Similar studies will probably reveal much more in the future. For the moment, however, and leaving aside brain research, the only way to understand the mental development of ancient hominids is to study their tools. And since most early tools soon perished, being made from bones, horn, teeth, wood, and skin, it is mainly via the remains of the stone axe culture that man the toolmaker can be identified in the early fossil record.

The oldest known stone artifacts are 2.5 million years old and were discovered in Ethiopia (Ambrose 2001). They consist of very crude choppers and flakes, which were easily made by striking cobblestones together. Such implements, distinguishable from ordinary stones by their form, their frequency of occurrence, and their unusual material (for example, flint), are classified according to their level of perfection. More refined early tools are described as *Oldowan industry*, named after the famous East African prehistoric site Olduvai Gorge. Figure 7.35a shows a simple 2-million-year-old chopper and flakes, along with a 1.5-million-year-old crude Olduvai biface from the Olduwan industry (Fig. 7.35b). These early tools, which served as hammers, choppers, scrapers, and knifes, were common for a million years, during which time they underwent little additional refinement. About 1.5 million years ago, however, there appeared the superior *Acheulean industry*, which continued to be common until about 100 000 years ago. Figure 7.35c shows a carefully manufactured hand axe from the Acheulean period of 300 000 years ago.

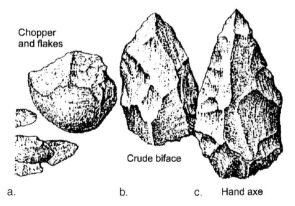

Fig. 7.35a. Chopper and flakes. **b.** Olduvai biface axe. **c.** Advanced Acheulean stone axe (after Gore 1997)

Such implements made new types of food accessible. It was now possible to deflesh animals, partition hides, chop nuts, and crush bones for marrow, which even the formidable teeth of hyenas could not crack. In addition, stone axes and points could be used as weapons. From approximately 100 000 years ago, the stone tool industries of the early, middle, and late Paleolithic periods became increasingly refined, making possible the production of superbly fashioned flint blades. By then, the variety of materials used in construction had clearly increased to include horn and wood – recently discovered hunting spears with carefully configured tips have been dated to 400 000 years ago (Thieme 1997).

7.15 Diet and Social Life

How did the hominids live? Australopithecines and early *Homo* formed small groups, and the considerable weight and height difference between the sexes in *A. afarensis* suggests the existence of a harem-type society similar to that of today's gorilla. Hominids foraged for food (fruit, shoots, roots, tubers, nuts, and eggs), and hunted small animals (birds, reptiles, rodents, fish, and insects). They also scavenged on the kills of the larger predators, such as leopards, who typically hide their prey from lions, hyenas, and vultures in trees. Here, the anatomically superior climbing ability of these hominids remained essential, with trees being frequently visited as ready shelters. *Homo habilis* was the first hominid to make simple stone tools, which were used for food preparation and as weapons.

Homo erectus, with its much larger brain, was a far more accomplished toolmaker. Moreover, its barrel-like chest and accomplished walking ability (Fig. 7.28) enabled it to forage over a much wider territory, taking on bigger game as it became an increasingly successful group hunter. Its larger brain, however, also caused problems. At birth, the bigger head required a wider birth channel in the pelvis. But as the pelvis could not be arbitrarily enlarged without interfering with its function as the basic platform for upright walking, *H. erectus* had to be born earlier, in a less developed and more helpless state, leaving it far more dependent on its parents.

It is believed that this development necessitated far more intensive child care and a closer bond between the parents, which nature then enforced by increasing the mutual sexual attraction of the parents through anatomical changes (large breasts and nonseasonal sexual receptiveness), and mental stimuli (an intense partner bond and monogamy). Greater infant dependency may also have stimulated a division of labor; short-range food collecting, rearing, and teaching of the children being performed by women, and long-distance collecting, scavenging, hunting, and fighting by men. More complex social interaction during group living, hunting, and child care also led to the development of sign language and the first forms of speech. With greater handling skills, and an increasingly shared knowledge about hunting, food

Fig. 7.36. A hypothetical alien with eyes on stalks

collecting, food treatment (to destroy toxic substances), toolmaking, building, medicine, and culture, the fight for survival became far more successful, but also necessitated a much longer apprenticeship for juveniles.

These developments continued up to *archaic Homo sapiens*, who in terms of anatomy was already pretty similar to us today. Very likely, the appearance of *modern Homo sapiens* about 200 000 years ago was associated with the full anatomical capability of complex language (Holden 1998). The perfection of language permitted an increasingly sophisticated sharing of knowledge and, crucially, its transfer over the generations by verbal education. The extent of such nonwritten expertise is impressive. The remains of a 5000 year old Copper Age man named Oetzi, found in the snowfields of the Alps, for example, revealed that he had an astonishingly sophisticated knowledge of animals, plants, hunting methods, clothing, weapon technology, tools, knowledge of geographical locality, architecture, medicine, food and social behavior.

7.16 The Logic of the Human Body Plan

Do the outlined reasons which led to modern man help to give us ideas about extraterrestrials? It is very difficult to predict how an extraterrestrial intelligent being might look. Nature is extremely inventive and imaginative, as the many life forms on Earth testify. There are, however, a number of physical and biological restrictions that limit pure fantasy, which would also apply to extraterrestrials at a similar stage of biological development as us (prior to the onset of conscious manipulation of DNA and the arrival of a completely artificial nonorganic existence, discussed in Chap. 9). Looking more closely at our own anatomy can reveal much about those parts that are there for an overwhelming internal reason, and those that exist due to mere chance.

Consider the picture of a hypothetical alien in Fig. 7.36. Like a snail, it has its eyes on stalks. Such eyes might be considered advantageous, allowing the alien to effortlessly survey its surroundings without much need to turn the head. The existence of such aliens, however, can largely be discounted. One of our most important abilities, inherited from our tree-dwelling ancestors, is stereoscopic vision, which remains absolutely essential for judging distances in our day-to-day existence. The physical basis of stereoscopic distance determination is to measure angles to an object from two different viewpoints (our two eyes). However, these angles can only be converted to a distance scale if the basis-length (the distance between the eyes) is kept meticulously constant. With eyes on freely moving stalks this could never be assured, which shows that stalk-eyed aliens are unrealistic.

Let us consider what parts of the human anatomy are there by necessity and dictated by broader physical and biological laws:

1. Eyes and ears: on the surface of our planet, light and sound are easily available as media that can be used by living organisms to recognize the environment. On a typical land surface with a complex geometry, stereoscopic vision and directional hearing constitute a decisive advantage for survival. Eyes and ears are thus set in a solid structure, the head, to ensure a constant basis-length for stereoscopic distance determination. Keeping the head on a movable neck has the added advantage of providing a turnable range-finder.
2. Upright walk and hands: an upright posture has many advantages, including freeing the hands for transport, manufacture, and tool use.
3. Mouth and anus: this is the old alimentary body tube, dating back to the gastrula stage of simple multicellular organisms.
4. Brain: learning, thinking, rational acting, and quick adaption all constitute intelligence, which gives man a decisive selective advantage.
5. Language: intelligent beings communicate, which on planets (in atmospheres and oceans as well as solids) is most easily accomplished by sound. This is because sound can be easily generated and controlled by muscle movements.
6. Economy: unnecessary expenditure of energy, or to grow nonessential body parts, is disadvantageous in the competition for survival. Thus more than two eyes or more than four extremities are superfluous, as they require greater brain control and produce extra body mass without providing decisive additional advantages. In the relentless struggle for survival, natural selection always opts for the more efficient and economical course.

This rudimentary list of essential biological and physical constraints suggests that extraterrestrial beings might not differ greatly from us in their general bodily appearance, since on other planets the same basic biological and physical laws will operate if a similar environment is present on a habitable planet. How the environment on Earth determined the architecture of our body is intensely investigated in developmental biology. For instance, would a planet completely covered by oceans allow the appearance of the above listed traits?

Clearly the gut developed in the oceans. The same can be said about the eyes and ears, where for instance the eyes have developed independently in many animal phyla (nematodes, mollusks, arthropods, chordates, see Arendt and Wittbrodt 2001). The ocean environment also led to the development of specialized appendages (hands) for gripping and manipulation by invertebrates such as mollusks (octopus) and arthropods (crabs) but the development of such organs in vertebrates (e.g. velociraptor, squirrel, otter, monkey, ape) occurred on land.

What about the brain development? As discussed above in Sects. 7.8 and 7.11, the mutually beneficial relationship with mammals and birds of the angiosperms, the flowering and fruit plants, was probably instrumental in increasing the primate and bird brains. In monkeys and apes, brain development, moreover, was greatly accelerated by the evolution of hands to multipurpose organs where walking was no longer the main function. But only after the complete freeing of the hands in hominids did our type of brain develop, made possible by the upright walk based on two legs when our ancestors left the rain forest. It is therefore very likely that the development of the human type of intelligence is intimately tied to the existence of continents, their plant cover and to the evolution of angiosperm plants.

If intelligence is so important for survival, why did it not appear at earlier stages in the history of the Earth? Clearly, for a high level of intelligence a sufficiently complex brain with specialized centers must be available, and such a brain could only develop over many steps from earlier stages. But why are there so many successful organisms with low intelligence? Here we see a similarity to the situation in Sect. 7.5, where the success of bacteria with relatively small genomes was compared to that of multicellular organisms having large genomes: clever behavior, highly specialized know-how, superb adaption, and specialization are very successful. *Cockroaches* are such examples, with body plans and behaviors that have changed little since the Devonian times. Yet it is clear that in the evolution from fish to amphibians, from reptiles to primates and man, one sees a progressive increase in intelligence: predatory dinosaurs are believed to have been more intelligent than amphibians or fish, the primates more so than reptiles and other mammals, the great apes more so than monkeys, and humans more intelligent than them all.

Finally, the innate logic of the human body plan can help us to hypothesize what might have happened to the development of intelligence if the K/T boundary event had not occurred. The high demand on mental capability required for survival in the angiosperm rain forest may well have triggered the appearance of primate-like intelligent tree-dwelling dinosaurs. The anatomical appearance of human-like creatures eventually based on them might not have differed much from our body plan in general layout. Future computer simulations might possibly give more answers here, and also provide clues to the appearance of extraterrestrial beings under alternative planetary parameters, including different gravity and other land to sea area ratios.

7.17 Evolution, Chance, and Information

Unlike physics or chemistry, evolutionary biology has been plagued since its beginning, and particularly since the 19th century after Darwin, by ideological battles (Taylor et al. 1997). The fighting is about how the phenomenon of evolution comes about, and to what degree the appearance of higher life forms and of humans must be attributed to chance. In 1809, Jean Baptiste Lamarck proposed a theory of evolution based on the use of inherited acquired characteristics. According to Lamarck, changes in the environment require more extensive use of certain organs, which leads to their increased size or efficiency. These newly acquired traits, in his view, are then inherited to the offspring. This hypothesis, called *Lamarckism*, was used, for example, to explain the evolution of the giraffe. The short-necked and short-legged ancestors feeding on tree branches of ever greater height would, after many generations, produce the long-necked, long-legged modern giraffe.

This hypothesis was challenged in 1859 by Charles Darwin, who published his theory in the famous book *On the Origin of Species by Means of Natural Selection*. As discussed in Sect. 7.1, this theory in today's language describes two processes: undirected chance mutations generating new organisms, and natural selection leading to the survival of the fittest. This theory is now universally accepted in the scientific community. Ernst Mayr (2000), for instance, describes Darwin's theory as follows: "Only the first step in natural selection, the production of variation, is a matter of chance. The character of the second step, the actual selection, is to be directional".

The heated discussion in evolutionary biology, mentioned in Sect. 7.1, involves two questions in particular. First, does the direction of evolution have a *teleological purpose*; that is, does nature have an innate aim to create higher life forms? Second, does Darwin's theory, and in particular its directedness, apply to a long range development, and does it therefore predict the eventual appearance of intelligence?

Concerning the first question, Mayr (1988, 2000) points out that evolutionary biology, unlike physics and chemistry, is a historical science and thus is particularly plagued by teleological explanations. He outlined that "starting from Aristotle's demand that everything must have a final cause, teleological explanations have a long tradition up to recent times, supported by great thinkers like Newton and Kant". In addition, we have to bear in mind that, particularly in the first half of the 20th century, outstanding researchers in the field were either priests such as Abbé Henri Breuil (an authority on Paleolithic art) or members of religious orders, such as the palaeanthropologist Pierre Teilhard de Chardin. Both certainly believed in God's grand plan for mankind. It is therefore not surprising that many scathing remarks are found in the literature of evolutionary biology, denouncing teleological arguments, while in the other sciences teleological arguments have long ago been discarded.

Here astronomy, another historical science, might serve as a guide. There is not even a remote chance that astronomers would assume that the solar system came about because the gas and dust cloud, the progenitor of the solar system, had an innate teleological aim to form planets. And yet accretion disks, as computer simulations show, invariably form planetary systems, complete with central stars, as discussed in Chap. 2. Astronomers attribute this persistent outcome of the evolution of accretion disks to the laws of nature and the properties of the environment which, modeled with a computer code, lead to a directed development without the need to invoke a teleological purpose. Because, like planet formation, biological evolution is also subject only to chance, to the laws of nature, and to the environment, we might suppose that one day it should be possible to simulate evolution on the computer. First steps toward a "virtual cell" – that is, to a complete computer simulation of cells – are already under way (Tomita 2001, 2005, Gavin et al. 2002). The directedness of biological evolution might then turn out to be no more mysterious than the directedness of the evolution of accretion disks.

The second question, of how intelligent life formed, also has a long history of intense debate. A.R. Wallace (1904), the cofounder with Darwin of the theory of natural selection, has already written about the existence of extraterrestrial intelligence: "An identical specific evolution cannot take place a second time ... Even less probable, then, is a whole series of identical developments from the dawn of life to the development of the human organism ... On another planet where conditions are even more diverse, the evolution of a species identical to humanity would be infinitely improbable".

This view is forcefully reiterated by evolutionary biologists up to the present day (Mayr 1988, 2000). Mayr writes that "even though quantum mechanics had placed determinism in doubt", he is impressed by how the physical scientists "still think along deterministic lines", while evolutionists are "impressed by the incredible improbability of intelligent life ever to have evolved, even on earth". "For three billion years after the first prokaryotes were formed, higher life did not evolve. Then when eukaryotes originated in the Cambrian period, four kingdoms of life developed in quick succession, the protists, fungi, plants and animals". "However, in none of these kingdoms, except that of the animals, was there even the beginning of any evolutionary trends toward intelligence". "Even then, there were hundreds of branching points that led to humanity". These chance events implied not only that humanity was improbable, but also intelligence itself. "There were probably more than a billion species of animals on earth, belonging to many millions of separate phyletic lines, all living on this planet earth which is hospitable to intelligence, and yet only a single one succeeded in producing intelligence."

Steven Jay Gould (1989) remarks "Evolution is a staggeringly improbable series of events, sensible enough in retrospect and subject to rigorous explanation, but utterly unpredictable and quite unrepeatable ... ". "Wind back the tape of life to the early days of the Burgess Shale; let it play from an identical starting point, and the chance becomes vanishingly small that anything like human intelligence would grace the replay." Burgess Shale is a rock formation in western Canada containing very early fossils.

As mentioned in Sect. 7.1, these views do not dispute the directedness of Darwin's theory at short range, but dispute its directedness in the long-range development. Long-range evolution is seen by these authors as the outcome of a type of random walk process, where the direction of evolution is frequently changed in an unpredictable manner.

Precisely this view of Gould has recently been challenged by another evolutionary biologist and authority on the Burgess Shale, Simon Conway Morris (1999). He points out that it overlooks an evolutionary phenomenon known as *convergence*. "This is the phenomenon that animals (as well as plants and other organisms) often come to resemble each other despite having evolved from very different ancestors. Nearly all biologists agree that convergence is a ubiquitous feature of life. Convergence demonstrates that the possible types of organisms are not only limited, but may in fact be severely constrained. The underlying reason for convergence seems to be that all organisms are under the constant scrutiny of natural selection and are subject to the constraints of the physical and chemical factors that severely limit the action of all inhabitants of the biosphere. Put simply, convergence shows that in the real world not all things are possible." He points out that, for example, whales are "from the perspective of the Cambrian explosion no more likely than hundreds of other endpoints, (yet) the evolution of some sort of fast, ocean-going animal that sieves sea water for food is probably very likely and perhaps almost inevitable." Conway Morris points to the great similarity between the famous marine reptiles ichthyosaur and today's dolphin, or between marsupial and placental sabre tigers. The recognition of convergence "effectively undermines the main plank of Gould's argument on the role of contingent processes in shaping the tree of life and thereby determining the outcome at any one time."

We thus conclude that Darwin's theory not only explains that nature tends to fill all possible modes of existence, attainable at a given evolutionary stage, subject to the constraints of the environment, but that this theory, indicated by the phenomenon of convergence, also leads to a long-range direction of evolution. As noted above, this applies particularly to the case of intelligent behavior, which can be seen to rise persistently with time from fish to man.

An example of this persistent rise of intelligence with time has recently been presented by Emery and Clayton (2004) who found that not only primates but also other large-brained social animals, such as corvids (crows, magpies, jays, ravens) understand their physical and social worlds and solve problems with intelligent behavior, tool manufacture and social cognition on a level comparable with higher primates. As the corvids have a vastly different brain structure and since the evolutionary separation of birds and mammals goes back at least as far as the late Triassic, 210 million years ago, this represents another fine example of convergent evolution.

Finally the perpetual increase in the number of genes in the eukaryotic cell line leading to man can be taken as another plausible example for a long-range directedness based on Darwin's theory, the growth of information, the know-how to survive, with time.

7.18 Cultural Evolution

As pointed out, the biological growth of information with time continued at an accelerated pace in the evolution of primates, due to a marked growth of their brain size relative to their body weight. In the hominids, the brain volume increased more than threefold over 4 million years. Even more striking, however, was the spectacular growth of the mental functions associated with knowledge, learning, and language.

Since the first appearance of man, there has been a rapid expansion of human knowledge, as can be seen from the increasing sophistication of tool-making, improvements in hunting, the use of fire, and migration into climatic zones that were previously uninhabited. This evolution is termed *cultural evolution*. In the past 400 000 years, biological evolution has largely ceased in *Homo sapiens*. It is cultural evolution that has taken over and developed exponentially, especially since the advent of writing.

Knowledge is the key commodity that divides us from our similar-bodied ancestors of 400 000 years ago. Knowledge of agriculture, after it was invented 13 000 years ago, became a fully established way of life a few thousand years later (Pringle 1998). It created the division of labor that, for example, enabled ancient societies to construct spectacular buildings such as the pyramids. But it was the invention of writing by the Sumerians about 5000 years ago that made possible a particularly rapid growth of information. Until then, knowledge had been stored in human memory and transmitted by long apprenticeship, but now it could be recorded. The invention of writing enabled human knowledge to be connected and accumulated across the centuries.

What will the future hold? By completing the sequencing of the human genome within the next few years (final draft), and decoding the functions of its 30 000 odd genes – which will take some time – man will eventually be able to discover how a self-conscious thinking brain is constructed. The number of possible building instructions for this is finite. Once the building blocks of an intelligent organism have been fully understood, it will become possible for humans to resume their biological development, in what will effectively constitute an unprecedented combination of cultural and biological evolution. This is discussed in more detail in Chap. 9.

8 The Search for Extraterrestrial Life

Looking at the nature, origin, and evolution of life on Earth is one way of assessing whether extraterrestrial life exists on Earth-like planets elsewhere (see Chaps. 6 and 7). A more direct approach is to search for favorable conditions and traces of life on other celestial bodies, both in the solar system and beyond. Clearly, there is little chance of encountering nonhuman intelligent beings in the solar system. But there could well be primitive life on Mars, particularly as in the early history of the solar system the conditions on Mars were quite similar to those on Earth. In addition, surprisingly favorable conditions for life once existed on the moons of Jupiter. Yet even if extraterrestrial life is not encountered in forthcoming space missions, it would be of utmost importance to recover fossils of past organisms as such traces would greatly contribute to our basic understanding of the formation of life. In addition to the planned missions to Mars and Europa, there are extensive efforts to search for life outside the solar system. Rapid advances in the detection of extrasolar planets, outlined in Chap. 4, are expected to lead to the discovery of Earth-like planets in the near future. But how can we detect life on these distant bodies?

8.1 Life in the Solar System

If life exists outside Earth in the solar system, where would we look for it? Are there places that have the required conditions for life (discussed in Chap. 5), such as a suitable temperature, an aquatic environment, and sufficient energy? Because neither Mercury nor our Moon have atmospheres or oceans, there is little chance of finding life there. For the same reason, asteroids, comets, and small moons can also be discounted. Venus is an inferno with surface temperatures of 480 °C. Shrouded under dense aerosol clouds of sulfuric acid, it has an atmosphere consisting of 96.5% CO_2, 3.5% N_2, and traces of SO_2, H_2S, and H_2O. However, because of its solid rock surface, Venus is another unlikely place for life. The same can be said of the giant planets Jupiter, Saturn, Uranus, and Neptune, which are completely covered by cold ($-100°C$ to $-190°C$) and deep oceans, consisting of liquid H_2 and He. As every inorganic or organic molecule is heavier than these elements, it will sediment down to the bottom of these oceans.

P. Ulmschneider, The Search for Extraterrestrial Life. In: *Intelligent Life in the Universe*, P. Ulmschneider, Adv. Astrobiol. Biogeophys., pp. 201–218 (2006)
DOI 10.1007/11614371_8   Springer-Verlag Berlin Heidelberg 2006

Remaining possible locations for extraterrestrial life in the solar system are therefore Mars, the large moons of Jupiter (Io, Europa, Ganymede, and Callisto) and maybe Saturn's moon Titan. Although the moons of the giant planets are far outside the habitable zone, it is the nearby planets that could supply the necessary energy for life by tidal heating. Since these moons have accumulated from planetesimals beyond the ice-formation boundary, it is not surprising that there was plenty of ice at their formation. Gravitational and magnetic field measurements by NASA's *Galileo* spacecraft have enabled the interior structure of these moons to be unraveled. We now know that the three inner moons of Jupiter have iron cores overlaid by a mantle consisting of silicate rocks. Above the mantle, Ganymede, with a radius of 2634 km, has a 1000 km deep surface layer of ice, while Europa (radius 1565 km) has a 350 km ice layer. No ice has been found on Io (radius 1821 km), probably because it was lost from that moon during its evolution, while Callisto consists of a relatively uniform mixture of ice and rock.

The existence of large amounts of ice on the surfaces of Ganymede and Callisto can also be seen from their white craters which, due to underlying fresh ice, are conspicuous by the sharp contrast with the surrounding dark dust-covered regions. However, it is doubtful whether moons such as Callisto and Titan can be seats of life, as the available energy appears to be too small because both moons are too far away for tidal heating.

Closer to Jupiter, however, much more energy is available, as demonstrated by the very active volcanism of Io. While Earth's Moon (radius 1738 km) shows no trace of volcanic activity, the similar-sized Io has very active volcanism, which is attributed to the heating by tidal friction exerted from the nearby Jupiter. The surface of Io consists largely of sulfur and sulfur dioxide, with lakes and lava streams composed of liquid S and SO_2. Hot spots on Io show temperatures of $17\,°C$, while the surroundings are as cold as $-178°C$. Here one would not expect life based on organic chemistry. Because the tidal interactions decrease rapidly with distance, the second moon Europa receives much less heating, and Ganymede even less.

8.2 Europa's Ocean

Surprisingly, Europa has a surface that is practically devoid of craters (see Fig. 8.1). The figure shows that large plates or rafts of ice seem to have been sliding over deeper layers on Europa, in much the same way as Earth's continents move over our planet's oceanic crust. The parallel linear ridges (Fig. 8.1, left panel) between the plates have many similarities to the mid-ocean ridges on the Earth's sea floor, where new crust forms due to upwelling material between the separating plates. As the pieces of plate fit together like a jigsaw puzzle (Fig. 8.1, right panel), plate-tectonic-like activity might be occurring on Europa. The material between the cracked and separated plates looks like slush, which is now frozen solid at the very low surface

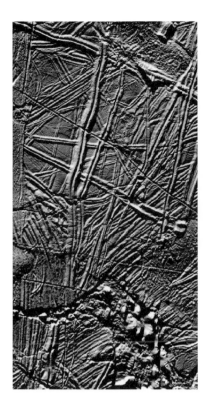

Fig. 8.1. The surface of Jupiter's moon Europa, observed from the Galileo space-craft (courtesy of NASA)

temperatures ($-143°$C). Strangely shallow impact craters and basins indicate that subsurface ice was warm enough to fill in the deep holes. From the appearance of a reworked surface and the fact that there are so few craters on Europa, it appears that the surface is very young, and that in places the ice surface is possibly only about 3–4 km thick (Turtle and Pierazzo 2001).

All this evidence suggests that there could be an ocean of liquid water under the surface ice of Europa and that the above-mentioned 350 km ice layer could be partly melted. Moreover, observations of surprisingly strong magnetic fields also point to subsurface oceans of liquid water on Ganymede. Such bodies of water with a frozen surface are well known from the ice lakes of the Earth's Antarctic regions (see Fig. 8.2). As these lakes teem with life, this might also be the case for Europa and Ganymede. Although life usually needs energy in the form of light to carry out photosynthesis, it was discussed in Chap. 6 that at hydrothermal vents on the ocean floor and at geothermal hot springs life is sustained in complete darkness, using heat and chemical energy from volcanic fluids, and that some of these organisms – such as thermophilic (heat-loving) bacteria – are among the most ancient life forms on Earth.

Fig. 8.2. Lake Bonney, a dry valley ice lake in the Taylor Valley of the McMurdo Sound region, Antarctica (after Priscu 2001)

It is not surprising that these observations have generated intense interest in Jupiter's Galilean satellites, which has resulted in a series of space missions (see LP missions 2005). The Galileo spacecraft, that arrived in 1995 and till its planned fiery end in the atmosphere of Jupiter in Sept. 2003 had completed 34 orbits around Jupiter in which it had numerous close encounters with the four large moons. A dedicated mission to search for the existence of oceans of liquid water on Europa, Ganymede and possibly Callisto, to determine their depths and investigate their nature and global extent is planned. The spacecraft *Jupiter Icy Moons Orbiter* (JIMO) to be launched in 2015 will orbit these three moons.

8.3 Life on Mars

Since it orbits in the habitable zone, Mars represents the greatest hope for finding traces of extraterrestrial life in the solar system. In its early history, this planet had a dense atmosphere and liquid water on its surface. At this time, primitive life could well have formed, traces of which might still exist today or might be found in fossilized form. To look for such traces is one of the main aims of the Mars missions planned in the near future.

8.3.1 Early Searches

In the year 1877, G. Schiaparelli, an astronomer from Milan, Italy, discovered canals (canali) on Mars, which were immediately attributed to intelligent Martians. This let to an outbreak of public excitement about Mars.

The American amateur astronomer P. Lowell confirmed Schiaparelli's discovery and, observing from a specially built telescope at Flagstaff, Arizona, drafted maps of these canals, an example of which is shown in Fig. 8.3. Ingenious methods for establishing contact with the Martians were proposed. For instance, it has been suggested that large mirrors should be built, which could direct sunlight from Earth to Mars, in order to catch the attention of the Martians and send messages. But with the ability to measure the low Martian surface temperatures from its infrared emission, doubts have slowly grown about the existence of waterways on Mars and of Martians.

Fig. 8.3. A map of Mars (after Lowell 1909)

Mars has seasons like the Earth, because the rotational axis of Mars is inclined against the axis of its orbit around the Sun (Table 5.4). They are about twice as long as on Earth. One Mars year has 669 Martian days, while a Martian day (24h 39m) is not much longer than an Earth day. In the Martian winter the temperature varies from $-113°C$ at night to $-98°C$ in the day, while in the Martian summer it varies from $-100°C$ at night to $0°C$ in the day. The polar regions have extended ice fields which most likely consist of water ice.

In 1969, the spacecraft Mariner 4 returned the first pictures from Mars, while in 1971 Mariner 9 provided the first high-resolution images, together with photographs from the Mars moons Phobos and Deimos. These pictures and others obtained from the Viking Orbiters in 1976 unequivocally proved

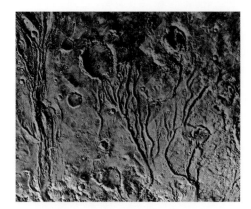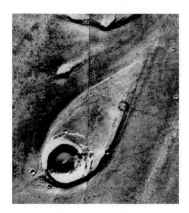

Fig. 8.4. Pictures from Mars, taken in 1976 by a Viking Orbiter (courtesy of NASA/JPL)

that there are no canals on Mars, and that these canals must be attributed to an optical illusion by which the human eye tends to connect faint features with lines. Yet Fig. 8.4 shows flow patterns and ancient river systems, which indicate that running water was once common on Mars. Canyons 100 km wide and 7 km deep were discovered, which contain river beds. The 900 km wide extinct volcano Olympus Mons rises 27 km from the surrounding plane. It is the largest volcano in the solar system. In addition to the remnants of ancient volcanism, Mars was found to be covered by a large number of impact craters. The atmosphere of Mars consists of 95% CO_2, 3% N_2, and 2% Ar, and at the surface has a pressure of 7 mbar, compared to 1013 mbar on Earth. Despite this thin atmosphere, Mars also shows weather, with clouds of haze and dust.

8.3.2 The Viking Experiments

In July and November 1976, leaving behind the Orbiters, the Viking 1 and 2 Landers descended onto Mars. A main purpose of these missions was to search for life. Three experiments (Horowitz 1977) were conducted to investigate the Martian soil (see Fig. 8.5):

1. The gas exchange experiment (Fig. 8.5a): Using a shovel, a sample of Martian soil was brought into a reaction chamber and a solution with a mixture of organic nutrients dissolved in water ("chicken soup") was added. Microorganisms were supposed to grow and produce gases such as O_2, CO_2, and H_2. The resulting enhanced gas emission would prove metabolism and thus the existence of life.
2. The labeled release experiment (Fig. 8.5b): A radioactive nutrition solution containing ^{14}C was added to the soil sample. After metabolizing the nutrients, living organisms would release radioactively labeled gases such as CO_2 and CH_4, which would be identified with radiation detectors.

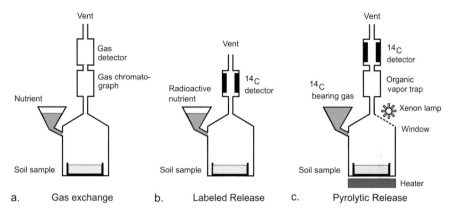

Fig. 8.5a–c. The three different experiments to detect life on Mars aboard the Viking mission

3. The pyrolytic release experiment (Fig. 8.5c): A soil sample was placed in a reaction chamber containing an atmosphere virtually identical to that of Mars, except that the gases CO and CO_2 were replaced by radioactive counterparts containing ^{14}C. After a time, the so-called incubation time, the organisms due to photosynthesis were supposed to have taken up some of these gases. The chamber was then flushed and the contents heated to a temperature of 750°C, which incinerated all organic material. The detection of radioactive ^{14}C in organic compounds would then indicate biological processes and thus the presence of life.

The following results were found:

1. Large quantities of O_2 were released, which today are attributed to oxygen-producing chemical reactions caused by the addition of water.
2. After a first wetting, a sudden rise in the radioactivity level of the gases was found. But after a second wetting, which should be equally nourishing for the organisms in the soil, the radioactivity level did not only fail to rise, but actually decreased. The explanation was that the nutrition solution reacted with oxygen-rich compounds in the soil, but after the first wetting all these compounds were used up.
3. Radioactive ^{14}C was taken up by the soil sample and after incineration could be detected as planned. A second experiment then heated the soil sample for a long time to 175°C, whereby all forms of life must have been destroyed, and then the experiment was repeated by adding radioactive gases. This procedure gave the same results as the previous experiment, indicating that biology had nothing to do with the translocation of ^{14}C from the gases CO and CO_2 to other compounds. On the basis of these negative results, there is universal agreement today that none of these three experiments indicated the presence of life on Mars.

8.3.3 Mars Meteorites

In August 1996, great excitement was generated when NASA scientists (McKay et al. 1996) presented a Mars meteorite which was supposed to show traces of fossil Martian life (see Figs. 8.6, 8.7). The meteorite, named ALH84001 after its finding place the Allan Hills, Antarctica, is one of 23 such Mars meteorites presently known (see Table 8.1). It has a weight of 1.9 kg and the size of a large potato (see Fig. 8.6). That this meteorite came from Mars was determined from its abundance ratios of elements (see below) and the following history could be unraveled. From isotope dating, this meteorite crystallized 4.0–4.5 billion years ago on Mars and was ejected from that planet into space by an impact 15 million years ago. This event is dated by isotopes of He and Ne produced by cosmic rays that bombarded the meteorite in space. About 13 000 years ago, it fell on Antarctica (dated from the decay of ^{14}C produced by cosmic rays) and roughly 20 years ago it reappeared at the surface of the ice layer, where it was found in 1984.

It was suggested that the tiny worm-like feature found on ALH84001 (see Fig. 8.7) showed a fossil bacterial colony consisting of members with a diameter of about 100 nm. Although claims that these objects are too small to be bacteria could be refuted by pointing out that in mammalian blood there are cells with a diameter of 70 nm, and in the soil bacteria with a diameter of 80 nm (McKay et al. 1997), most scientists regard the feature seen in Fig. 8.7 as an artifact, perhaps produced by the gold–palladium coating carried out to prepare the sample for viewing under a scanning electron microscope. However, it has been pointed out (Buseck et al. 2001) that magnetic material found in the meteorite could only have been produced by bacteria. Yet despite of this the present scientific discussion remains controversial, and so far Mars meteorites do not provide undisputed evidence of former life on that planet.

Fig. 8.6. The Mars meteorite ALH84001, found in Antarctica (Baalke 2005)

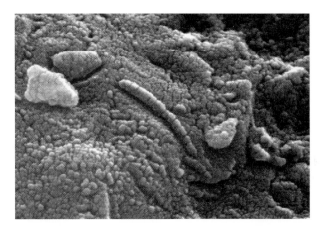

Fig. 8.7. Possible microscopic fossil traces of life on the Mars meteorite ALH84001 (after McKay et al. 1996)

Table 8.1. Mars meteorites (after Baalke 2005). Meteorites of the "find" category are noticed only by their unusual appearance, while meteorites of the "fall" group have been observed during the impact

Name	Mass (kg)	Mode	Location	Year
Cassigny	4.00	fall	France	1815
Shergotty	5.00	fall	India	1865
Nakhla	10.0	fall	Egypt	1911
Lafayette	0.80	find	Indiana USA	1931
Gov. Valadares	0.16	find	Brazil	1958
Zagami	18.0	fall	Nigeria	1962
ALHA77005	0.48	find	Antarctica	1977
Yamato793605	0.016	find	Antarctica	1979
EETA79001	7.90	find	Antarctica	1980
ALH84001	1.90	find	Antarctica	1984
LEW88516	0.013	find	Antarctica	1991
QUE94201	0.012	find	Antarctica	1994
7×Dar al Gani	10.4	find	Libya	1997−1999
2×Los Angeles	0.70	find	California USA	1996−1999
11×Sayh al Uhaymir	11.4	find	Oman	1999−2004
2×Dhofar	1.1	find	Oman	2000
16×Northwest Africa	4.2	find	Morocco, Algeria	2000−2001
6×Antarctica	15.8	find	Antarctica	1998−2004

How can one be sure that ALH84001 and the meteorites listed in Table 8.1 come from Mars? This question is answered by measuring the abundances of different elements (for example, Ca, Mg, Fe, and Si) in the meteorites. If, for example, the Ca/Si ratio is plotted against Fe/Si as in Fig. 8.8a, or Mg/Si against Al/Si as in Fig. 8.8b, it is found that the rocks of Mars (filled symbols) and Earth (open symbols) occupy different regions of the diagram. Why this is the case is presently not well known, but it is attributed to variations of the element abundances with distance from the Sun in the planetary accretion disk at the formation of the solar system. Observationally, this variation is well established and recent investigations of Martian soils by Viking, and of

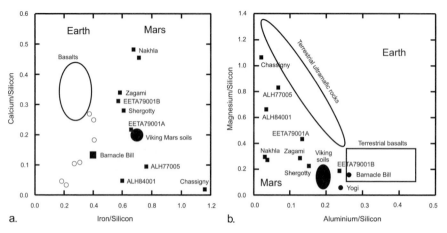

Fig. 8.8a,b. Element ratios allow the identification of Mars meteorites, after Lorenz (1997). Material from Mars is indicated with *filled symbols*, that from Earth with *open symbols*

rocks (named Barnacle Bill or Yogi) in the vicinity of the landing site of the spacecraft *Pathfinder*, fit into this picture (see Fig. 8.8).

8.4 The Early Atmosphere of Mars

Meandering river valleys, canyons, and delta-like flood planes (see Fig. 8.4) show that in ancient times running water must have been present on Mars and the temperatures must have been much higher. From the Pathfinder mission, which landed in 1997, close-up images of the rocks at the Martian surface have provided independent evidence of a benign early climate. These images revealed conglomerates (rocks made of sand and water-worn pebbles), as well as pebbles that could have been rounded in flowing water. That Mars probably had a relatively warm climate, starting with a dense (1–5 bar) CO_2 atmosphere prevailing over hundreds of millions of years is also indicated by simulations of the planetary atmosphere. Figure 8.9 shows a comparison of the surface temperatures of Mars and Earth, covering the time when the first oceans formed up to the present and into the future, taking the rise of the solar luminosity into account.

That Mars was once covered by a large ocean (Head et al. 1999) has been suggested by measurements of the laser altimeter (MOLA) aboard the *Mars Global Surveyor* which orbits at an altitude of about 400 km above the Martian surface. Mars Global Surveyor has found that much of the planet's northern hemisphere is a low-lying plain, roughly centered on the north pole (Fig. 8.10). A comparison of these Martian lowlands with the floor of the Atlantic Ocean near Southwest Africa (see Fig. 8.11) suggests, that for a time, Mars might have had something like terrestrial plate tectonics, which leads to

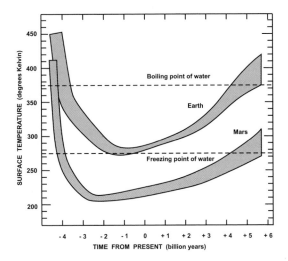

Fig. 8.9. Surface temperatures in the early atmospheres of Mars and Earth (after Sagan 1977)

continents composed of lighter rocks that float on top of the heavier material of the ocean basins.

Plate tectonics would also produce the heavy volcanic activity (exemplified by Olympus Mons) that brings CO_2 back into the atmosphere, thus counteracting its removal by weathering and carbonate deposition on the ocean floor. A that distance from the Sun, CO_2 is particularly important for the greenhouse effect to raise the temperature sufficiently for the existence of oceans. However, since Mars has only $1/10$ the mass of the Earth, the cooling of its core and mantle proceeded much more quickly, so that plate tectonics and volcanism probably stopped more than 2 billion years ago. The decreasing amount of atmospheric CO_2 would then have led to an irreversible glaciation. Moreover, as the Martian atmosphere is not protected by O_3, photolysis of water by solar UV radiation would have generated hydrogen and oxygen (Chap. 5). Hydrogen easily escaped the low-mass planet and O_2 oxidized the abundant iron, giving the planet its red color and its highly oxidizing soil.

If life formed on Mars, it probably did not survive the irreversible glaciation (Chap. 5), because for life an aquatic environment is essential. However, the fact that life might have died out after several hundreds of millions of years would be a great advantage for the investigation of the chemical evolution of early life, because on Earth these early traces are drowned in the large number of chemical signatures that are the result of 4 billion years of biologically aggressive terrestrial evolution. A Mars mission would therefore be of great importance for the investigation of the earliest steps of the formation of life, even if no surviving life is found on the planet.

Fig. 8.10. Mars observed with the laser altimeter MOLA on board the Mars Global Surveyor at 90° longitude, color coded for height. The figure shows the deep (ocean?) basin of the northern hemisphere (*blue*), the north polar ice cap (*green*), the volcano Olympus Mons (*leftmost white rise*), and the canyon Valles Marineris (*horizontal blue-green channel below the equator*) (after Smith et al. 2001)

8.5 Future Mars Missions

At the end of 2005 five missions are in progress on and around Mars (see Mars missions 2005), the *Mars Global Surveyor* orbiter just had its 8th anniversary (see Fig. 8.10) was accompanied in October 2001 by *Mars Odyssey*, an orbiter designed to study the composition of the planet's surface, to detect water and buried ice, and to study the radiation environment. It succeeded to detect subsurface ice (Boynton 2002) (see also Chap. 5). In 2003 the *Mars Express* was launched to look for subsurface water (ice) and study the Martian atmosphere, structure, and geology. Figure 8.12 shows an unnamed impact crater located on Vastitas Borealis, a broad plain at Mars's far northern latitudes. The crater is 35 km wide and has a maximum depth of about 2 km beneath the crater rim. The circular patch of bright material located at the center of the crater is residual water ice. That such ice fields are only the tip of the iceberg and that there is abundant subsurface ice on Mars has recently been shown using radar observations from Mars Express that reveal that 2 km of

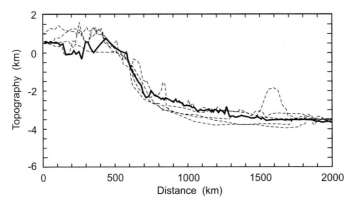

Fig. 8.11. Matching lowlands on Mars with the South Atlantic ocean floor off Southwest Africa (*solid line*) (after Kerr 1998)

Fig. 8.12. Unnamed impact crater with central water ice field located on Vastitas Borealis, a broad plain at Mars's far northern latitudes (courtesy of ESA)

layered deposits rich in pure water ice underly the North Polar Cap (Picardi et al. 2005).

Also in 2003 two powerful Mars exploration rovers *Spirit* and *Opportunity* were sent to Mars. These are robotic explorers that carry all their instruments with them and were able to travel up to 100 m per day across the Martian surface. They still rove over the Martian landscape to analyze rocks, soils and the atmosphere. A more advanced surveyor mission has been launched in 2005 and is expected to arrive on Mars in March 2006, called the *Mars Reconnaissance Orbiter*. It is scheduled to perform very high resolution imaging of the Martian landscape. To be launched in 2009 a smart lander and long-range rover mission called *Phoenix Mars Lander* will serve as a mobile science laboratory. Beyond 2009 these missions are planned to be extended in 2011 by a *Mars Scout* that might involve airplanes or balloons. Further plans are for a *Mars Sample Return* to bring back soil and rock samples and to test the return flight to Earth from the Martian surface, a *Mars Astrobiology Field Laboratory* and a mission that aims to test deep drilling and other technologies. All these missions primarily have the aim of understanding the Martian geography and geology, and of finding water in order to eventually produce fuel using robotic manufacturing plants, which is essential for extensive return missions and later human expeditions to Mars. For a summary of these and other space missions, see Table 9.1.

8.6 Life Outside the Solar System

An even more promising way of searching for extraterrestrial life is to look for Earth-like habitable planets outside the solar system. As discussed in Chap. 4, the search for extrasolar planets has so far resulted in more than 90 discovered planets. Unfortunately, up to Dec. 2005 they were all in the Jupiter size range and not of the terrestrial type, where the chance for life is highest (Chap. 5). Great advances in instrumentation, however, will eventually allow direct observations of such Earth-like planets (Chap. 4).

How can we detect life on these planets? Clearly, the presence of water (H_2O), indicated by absorption bands in the infrared spectrum near 8 μm, will be an important indicator, because life requires an aquatic environment (Chap. 5). Yet by the time the Viking missions were being planned, it had already been pointed out by Lovelock (1965) that life should be recognizable by detecting compounds in the atmosphere that are not in thermal equilibrium; that is, incompatible with each other on a long-term basis. For instance, because methane (CH_4) is quickly oxidized, there should be no methane in the Earth's atmosphere, and yet one actually finds an abundance, 140 orders of magnitude higher than predicted by thermal equilibrium (Sagan et al. 1993). This methane comes from methanogenic bacteria living in anaerobic environments, such as rice paddies or cow guts.

The presence of nitrous oxide (N_2O) is another indicator of biological processes arising from bacterial activity in soils and the ocean. But the most important indicators are free oxygen (O_2), a waste product of photosynthesis, and ozone (O_3). While the A-band of O_2 at 0.76 μm in the visible spectrum would be difficult to observe, because of the very large contrast between the planet and the central star, the infrared ozone absorption band at 9.6 μm

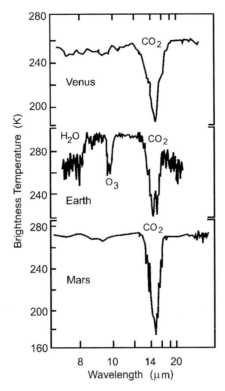

Fig. 8.13. The infrared spectra of Venus, Earth, and Mars (after Angel and Woolf 1996)

would be ideal as it is the next strongest absorption feature in the Earth's atmosphere after the 15 μm band of carbon dioxide (CO_2). Figure 8.13 shows the infrared spectra of Venus, Earth, and Mars. While all three planets show the dominant CO_2 absorption band, only the Earth spectrum has bands of water and ozone.

In the absence of life, the production of O_2 is due to photolysis of H_2O and CO_2. Yet under such conditions the abundance of O_2 is very low, because of the many oxygen sinks such as the volcanic gases H_2, CO, and SO_2, which are easily oxidized and overwhelm O_2 by 100 to 1. O_3 is produced photochemically from O_2. Thus the presence of O_2 and O_3 is a good indication of life (Kasting 1997), and through it life has been detectable on Earth for over 2 billion years (see Fig. 7.3). But, from a great distance, how could life have been detected on Earth before that time? Kasting and Brown (1998) pointed out that methanogenic bacteria are very ancient, producing methane via the process $CO_2 + 4H_2 \rightarrow CH_4 + 2H_2O$. Their atmosphere models of the early Earth indicate that it would have shown strong detectable 7.6 μm CH_4 absorption features. Therefore, strong absorption bands of H_2O, O_3, and CH_4 in the infrared spectrum would be good indicators of life on Earth-like planets. Naturally, the detection of radio emissions would be an even better and more direct indicator. This type of search for extraterrestrial intelligence, SETI, is discussed in Chap. 10.

8.7 UFOs

Since we have mentioned SETI, another likely place in the solar system where we might look for extraterrestrial life comes to mind: the Earth. Some readers might say that the search for extraterrestrials is unnecessary, because they are already here, as demonstrated by the presence of UFOs ("Unidentified Flying Objects"). This term was coined in 1952 by the US Air Force to name what were previously called "flying saucers". Modern interest in UFOs apparently started in 1947, when a private pilot, Kenneth Arnold, reported that from his plane near Mt. Rainier in Washington State he had observed nine disk-shaped objects flying in formation at speeds of over 1000 miles per hour. These were named "flying saucers" by the press and within a year of Arnold's report, 850 sightings were claimed by various people. In 1948 the US Air Force, worried about violations of the US air space in the Cold War, began to investigate these UFO reports, first under Project Sign and then Project Grudge, which in turn became Project Blue Book and continued until 1969. At the end of 1969, the Air Force decided to terminate the UFO investigation project. This decision was based on the final report prepared by the eminent physicist Edward Condon (1969), who summarized his and his colleagues findings: "Our general conclusion is that nothing has come from the study of UFOs in the past 21 years that has added to scientific knowledge".

In addition to the many deliberate UFO hoaxes that have consistently plagued not only the general public and the press but even the scientific world (a phenomenon also found elsewhere – as for instance, the Piltdown man as a human ancestor), UFO investigators have pointed out that there are a huge number of natural and psychic phenomena that can fool even an experienced observer. Table 8.2 gives a list of objects that have erroneously been taken for UFOs. These consist of a great number of different material artifacts, classified experimental and spy planes (such as U2 and Blackbird) or balloons, atmospheric events, astronomical objects, physiological and psychological phenomena, defects in the recording media, and radar malfunctions. In addition, the question of UFO sightings soon developed into a field where intense ideological views were expressed, even among normally rational scientists. One camp of scientists has strongly attacked Condon as being one-sided and for apparently suppressing suspicious sightings, while the other camp has claimed that UFOs cannot possibly be of extraterrestrial origin, because it would be impossible for extraterrestrials to come to us over interstellar distances.

Table 8.2. Natural phenomena mistaken for UFOs (after Menzel 1972)

A. Material objects	spider webs	B. Immaterial objects	reflection from bright
1. Upper atmosphere	insects	1. Upper atmosphere	sources
meteors	swarms	auroral phenomena	electric lights
satellite re-entry	moths	noctilucent clouds	street lights
rocket firings	luminous	2. Lower atmosphere	flashlights
ionosphere experiments	(electrical discharge)	reflections of searchlights	matches
sky-hook balloons	seeds	lightning	(smoker lighting pipe)
2. Lower atmosphere	milkweed, etc.	streak	autokinesis
planes	feathers	chain	stars unsteady
reflection of sun	parachutes	sheet	stars changing places
running lights	fireworks	plasma phenomena	falling leaf effect
landing lights	4. On or near ground	ball lightning	autostasis
weather balloons	dust devils	St. Elmo's fire	(irregular movement)
luminous	power lines	parhelia	eye defects
nonluminous	transformers	sundogs	astigmatism
clusters	elevated street lights	parselene	myopia (squinting)
clouds	insulators	moondogs	failure to wear glasses
contrails	reflections from windows	reflections from fog and mist	reflection from glasses
blimps	water tanks	haloes	entopic phenomena
advertising	lightning rods	pilot's halo	retinal defects
illuminated	TV antennas	ghost of the Brocken	vitreous humour
bubbles	weathervanes	mirages	E. Psychological phenomena
sewage disposal	automobile headlights	superior	hallucination
soap bubbles	lakes and ponds	inferior	F. Combinations and special effects
military test craft	beacon lights	C. Astronomical objects	G. Photographic records
military experiments	lighthouses	planets	development defects
magnesium flares	tumbleweeds	stars	internal camera reflections
birds migrating	icebergs	artificial satellites	H. Radar
flocks	domed roofs	sun	anomalous refraction
individual	radar antennas	moon	scattering
luminous	radio astronomy antennas	meteors	ghost images
3. Very low atmosphere	insect swarms	comets	angels
paper and other debris	fires	D. Physiological phenomena	birds
kites	oil refineries	after-images	insects
leaves	cigarettes tossed away	sun	multiple reflections
		moon	I. Hoaxes

The debate about the extraterrestrial origin of UFOs is compounded by the huge commercial value that aliens represent for the publishing, TV, and film industries, independently of the question of whether or not they exist. Movies showing extraterrestrials are our everyday TV experience, as are reports claiming government or armed forces conspiracies. Many of our contemporaries have become rich by writing outrageous revelations about historical visits of aliens to the Earth, about the mysterious connection of the "face of Mars" with the Pyramids, or the tracks in the deserts of the Andes identified as landing strips of aliens – extraterrestrials who were powerful enough to travel over interstellar distances but nonetheless needed landing strips in the Andes! All of this has made the general public quite prepared to accept people who have claimed to have observed UFOs and even become victim to alien abductions.

From all this, it was clear to the UFO investigators that they could accept only the most rigorous proof and reliable evidence, since a convincing and unequivocal proof of a sighting or a contact with an extraterrestrial being would undoubtedly be one of the greatest single scientific discoveries in the history of mankind. The astronomer Allen Hynek (1972) classified possible encounters with extraterrestrials as "Close encounters of the first kind" (seeing UFOs), "the second kind" (hard evidence of UFO landings), and "the third kind" (actually meeting aliens). But despite the fact that such a documented contact would bring instant fame and fortune to the discoverer, nobody has ever come forward and presented such an irrefutable proof. Clearly, a nonhuman artifact from a vastly superior extraterrestrial civilization, which could be sold at an absolutely astronomical price at international auctions, has never been found.

Yet this does not mean that we should close our mind to the UFO phenomenon. As the great majority of scientists accept that there must be extraterrestrial intelligent beings in our galaxy, and certainly in the universe (Chap. 10), it has to be admitted that their technical capabilities could be enormous. There is thus a nonvanishing chance that such beings are able to visit Earth, and that they might already be here.

Part III

Intelligence

9 The Future of Mankind

The questions "Why don't we have contact with extraterrestrial intelligent life?" and "How will mankind evolve in the near future?" are intimately connected. Clearly, civilizations that are far behind our technological state would not be capable of communicating with us. But even societies more advanced than us would have difficulties in making contact, as radio waves or spacecraft take a long time to cross the huge distances in our galaxy. In addition, such advanced societies might no longer exist. They could have fallen victim to external or internal dangers, or they might not wish to communicate with us. The only way to gain some insights into the possible dangers afflicting extraterrestrial intelligent societies and their likely mode of behavior is to consider our own future development, because these civilizations are expected to have gone through our own technological state long ago.

Although it is very speculative to predict the likely evolution of human society, it is impossible to overlook three current developments that have particularly far-reaching consequences for our future: the advances in information technology, the conquest of space, and the mastering of the biological world. While the conquest of the solar system is expected to open up truly limitless economic prosperity and an unimaginable diversity of human society, information technology and the mastering of the biological world will not only enable us to change our own bodies and minds in a fundamental way, but may even result in the evolution of entirely new forms of intelligent beings. Yet this constant increase of knowledge and explosive growth of technological power also carries with it the rapidly rising danger of complete annihilation of the human life form. Due to this it is absolutely essential to develop survival strategies to counteract these dangers.

9.1 Predicting Mankind's Future

Predicting the future, as every stockbroker will readily admit, is notoriously difficult. Numerous wrong predictions, even by professionals in their fields, have demonstrated this. It suffices to recall the "scientific" prediction in 1895 by the famous physicist William Thomson (Lord Kelvin), president of the

P. Ulmschneider, The Future of Mankind. In: *Intelligent Life in the Universe*, P. Ulm-schneider, Adv. Astrobiol. Biogeophys., pp. 221–253 (2006)
DOI 10.1007/11614371_9   Springer-Verlag Berlin Heidelberg 2006

Royal Society, that "heavier-than-air flying machines" will never be possible. Others include the 1977 prediction by Ken Olson, president of Digital Equipment Corporation, that "there is no reason anyone would want a computer in their home", or the financially devastating assessment of the managers of IBM a few years later, that personal computers (PCs) would never be important in terms of sales compared to mainframe computers. We are not only notoriously bad at predicting the future, but often even lack imagination when new inventions have been introduced. In addition, there is a tendency to overemphasize negative developments. This has a long tradition, going back to antiquity and medieval times, when "prophets of doom" delivered their end-of-the-world type messages.

Is it possible, despite these warnings, to make predictions about the future development of mankind? I think that by extrapolating far-reaching and fundamental trends in human history, it should be possible to make reasonable predictions. Let us consider the advances in information technology. In less than a quarter of a century, computers have mutated from specialized machines for some scientists to universal and absolutely essential tools in manufacture, process management, administration, commerce, communication, education, and recreation. Because of their usefulness, these machines have modified everyone's lives both at work and at home, and we no longer want to live without them. They allow us an almost instant contact with millions of people and databases around the world, and they permit access to the remotest places on Earth. There is little doubt that this progress in information technology will continue by making computers even more sophisticated and intelligent. They could be viewed as evolving external organs of our body, which greatly help to improve our interaction with the environment and other human beings. The full consequences of this evolution are impossible to predict, but the development in the other two areas mentioned above could turn out to be even more fundamental for mankind.

9.2 Settlement of the Solar System

The second, easily foreseeable development in the near future is the conquest of the solar system. But why go into space? Why not stop all growth and live in harmony with our environment here on Earth? The expansion of the Greeks into the western Mediterranean, the attack of the Huns on the Roman Empire, the search for the routes to the spice islands, the colonization of America and Australia: Have these activities all been driven by hunger and overpopulation in the homeland? Or were they not frequently initiated by the excitement of adventure, the possibility of untold treasure, and by the chance to shed the old and make a new beginning? I think that these adventurous aspects of our nature are deeply implanted in our minds in order to assure the survival of our species.

9.2.1 The Space Station

Since *Sputnik*, the first Russian satellite orbiting Earth in 1957, the conquest of space has steadily progressed. Technology has advanced in a staggering way: the first commercial communications satellites launched in 1965 could handle only 240 simultaneous telephone connections, but they now carry one third of the world's telephone traffic and essentially handle all TV broadcasts between countries. It is now common to receive TV directly from powerful satellites in geostationary orbits, 36 000 km above the Equator, parked at a fixed spot in the sky. With a series of new communication systems such as *Iridium*, which use cross-linked satellites that orbit the Earth at heights of less than 1000 km, our planet now has instant telephone connections and high-rate data transmissions from every location on Earth. Low orbits help to avoid the noticeable time-lags in voice communications currently encountered over geostationary satellites, and caused by the finite speed of radio waves.

While for telecommunications the conquest of space is now commercially rewarding, other Earth-observing satellite ventures are also on their way to becoming economically successful. Such space activities include weather prediction, navigation (with the *Global Positioning System*, GPS, and the forthcoming European navigation system *Galileo*), resource management, studies in the Earth sciences, and military defense.

The greatest endeavor to date, however, is the *International Space Station* (ISS). In 1998 (see ISS 2005), the first two modules of the ISS were launched and joined together in orbit. Other modules soon followed and the first crew arrived in 2000. Over the past five years construction proceeded and the station was continuously inhabited by a staff of 2–3 astronauts. After a considerable delay following the disaster of the space shuttle Columbia in Feb. 2003, the station is now scheduled to be completed by 2010 with an expected final weight of 460 t.

At a present height of 350 km above ground, in a so-called *Low Earth Orbit*, the ISS circles around the Earth in 92 minutes. This height varies because of atmospheric drag. As it has decreased by 50 km since 2003 it needs to be periodically reboosted to greater altitude. When completed, the station will provide a platform in space (see Fig. 9.1) that can house seven permanent staff in a pressurized living and laboratory environment the size of a large jumbo jet. Because of its huge cost, 16 nations are participating in the construction of the station.

Admittedly, many of the present tasks of the station could have been carried out more cheaply by unmanned satellites. Yet there are important future applications with humans being personally able to carry out investigations in situ, which make the ISS the next fundamental step in the conquest of space. If we want to conquer space, we must know how man reacts to weightlessness, how to combat undesirable changes in our bodies, how plants grow in space, how limited biospheres must be controlled to provide a suitable environment for life, how unhealthy or offensive emissions from machines and

Fig. 9.1. International Space Station ISS, photographed from the Space Shuttle Discovery in Aug. 2005 (courtesy of NASA)

materials can be managed, and how pests and microbes can be kept in check. In addition, the ISS is the place where industrial applications in zero gravity and space manufacturing will be investigated, with the aim of making these activities commercially successful.

Clearly, however, one main purpose of the space station is its future function as a way station to space: to serve as an extraterrestrial base and supply depot from which satellites and space vehicles can be repaired and serviced, and where large interplanetary missions can be given last minute checkups before departure. The ISS will also function as an emergency center from where help can be launched quickly when space projects develop difficulties.

Finally, in the more distant future, the ISS and similar stations in low Earth orbit will be the seat of industrial processing of materials brought from asteroids or the Moon. Here the most important industrial product will be rocket fuel. Its manufacture in the future will turn the ISS into a vitally important refueling station for space travel.

The essential importance of the ISS for the conquest of space was also recognized by the US space policy announced in Jan. 2004 under the name *Vision for Space Exploration* by President George W. Bush (see Space Policy 2005). The aims of that policy are to complete the International Space Station by 2010, retire the Space Shuttle by 2010, develop a so-called *Crew Exploration Vehicle* CEV by 2008 patterned after the *Apollo* capsules that were used in the manned Moon project 1963–1972, conduct a first manned mission with it by 2014, develop two launch vehicles for cargo and for the CEV using Shuttle components, explore the Moon with unmanned missions by 2008 and manned missions by 2020, and eventually explore Mars and other destinations with unmanned and manned missions.

Table 9.1. Dates of launch and start of operation, as well as purposes of lunar and planetary missions carried out by various space organizations (NASA USA, ESA Europe, JAXA Japan, ISRO India, CAST China) (see LP missions 2005)

Launch	Mission	Start	Organization	Purpose
2001	Mars Odyssey	2002	NASA	Mars Orbiter, operating 2005
2003	SMART-1	2004	ESA	Lunar Orbiter, operating 2005
	Hayabusa (Muses-C)	2005	JAXA	Asteroid Itokawa Lander, Sample Return 2005
	Spirit (MER-A)	2004	NASA	Mars Exploration Rover, operating 2005
	Opportunity (MER-B)	2004	NASA	Mars Exploration Rover, operating 2005
	Mars Express	2003	ESA	Mars Orbiter, operating 2005 (Lander failed)
2004	Rosetta	2008	ESA	Asteroids Rendezvous, Comet Churyumov-Gerasimenko Lander
	MESSENGER	2008	NASA	Mercury Flyby and Orbiter
	Cassini, Huygens (1997)	2004	NASA/ESA	Saturn orbiter, Titan Lander
2005	Deep Impact	2005	NASA	Rendezvous, Impact on Comet Tempel 1
	Mars Reconnaissance Orbiter	2006	NASA	Mars Orbiter
	Venus Express	2006	ESA	Venus Orbiter
2006	New Horizons	2008	NASA	Pluto/Charon and Kuiper Belt Flyby
	Dawn	2007	NASA	Asteroid Ceres and Vesta Orbiter
	Lunar-A	2006	JAXA	Lunar Orbiter and Penetrator
	Stardust (1999)	2000	NASA	Coma Comet Wild-2, Sample Return 2006
2007	Selene	2007	JAXA	Lunar Orbiter and Lander
	Chandrayaan-1	2007	ISRO	Lunar Orbiter
	Phoenix	2008	NASA	Small Mars Scout Lander
	Chang'e 1	2007	CAST	Lunar Orbiter
	NEAP	2008	NASA	Asteroid Nereus Rendezvous
2008	Planet-C	2009	JAXA	Venus Orbiter
	Lunar Reconnaissance Orbiter	2008	NASA	Lunar Orbiter
2009	Mars Science Laboratory	2010	NASA	Mars Rover

9.2.2 Moon and Mars Projects

The International Space Station, with all its expected benefits, must not be regarded as an aim in itself, but rather as a means of accomplishing more ambitious space projects. Investigations using the satellites *Clementine* and *Lunar Prospector*, which orbited the Moon in 1994 and 1998, claim that surprising amounts of ice have been detected at the polar regions of the Moon, in quantities estimated to be similar to the amount of water stored in the North American Great Lakes.

Such ice is expected because at the polar regions of the Moon, the Sun never rises much above the horizon, so that in deep craters sunshine never enters. These regions are very cold and water vapor released from meteorites that impact on the Moon freezes out and accumulates there. Because there is some controversy about the validity of these findings (Bussey et al. 2005), further lunar missions are necessary. Particularly important for the search for ice will be the specially equipped *Lunar Reconnaissance Orbiter* mission to be launched in 2008 (Table 9.1).

All future expeditions to the Moon, as well as the manned missions to Mars, face the problem of how to obtain the large amounts of rocket fuel necessary for a return flight to Earth. To bring this fuel all the way from the

Earth's surface to Moon or Mars is prohibitively expensive. The only logical way is to manufacture it in space. For this aim ice is most important because, with the help of solar or nuclear energy, it can be split by electrolysis into hydrogen and oxygen for rocket fuel. But local manufacturing and building operations will also make sense for other products such as, for example, food and shelters. It is therefore of great importance to assess in detail what materials are available on Moon, Mars, and the asteroids.

Table 9.1 presents a list of 24 executed and planned lunar and planetary missions up to the year 2009, comprising seven flights to Mars, five to asteroids or comets, six to the Moon, two to Venus and one each to Mercury, Saturn, Pluto and the interplanetary space. For an up-to-date summary on each of these missions, see LP missions (2005). The purpose of most missions is a detailed surface exploration of the targets by orbiting or even landing on them. The missions to Mars have already been discussed in Chap. 8. Selecting the Moon, Mars, the asteroids, and comets as prime targets indicates that one main purpose of the missions is to search for ice, and other suitable materials that can be used in the space industry.

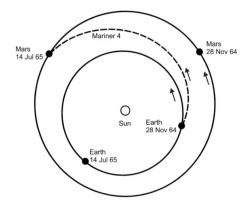

Fig. 9.2. The Mariner 4 flight to Mars in 1964/65

The next great steps in the exploration of the solar system will be a manned mission to Mars and a semipermanent Moon station. Figure 9.2 shows the orbits of Earth and Mars and the flight path of Mariner 4, which was the first flyby mission to this planet, launched in 1964. The trip to Mars, over a distance of 523 million km, lasted for 228 days. Because of this large distance, a manned return trip will take several years. For such a mission, where rescue in case of failure is not possible, a new level of reliability and self-sufficiency of the spacecraft, the landing vehicle, and the Martian ground station will have to be attained. Before the astronauts are sent on their trip, automated fuel manufacturing plants will already have had to operate on Mars to ensure that the required fuel will be available for the safe return of the crew. While more extensive settlement of Mars cannot yet be foreseen, the planet will some day have at least semipermanent outposts from which extensive studies in the Earth sciences (polar ice, climate evolution, plate tectonics, volcanos, and continent and ocean basin formation) as well as the life sciences (searching for early traces of life) will be conducted.

Fig. 9.3. A permanent Moon station with two mass drivers (after Hartmann et al. 1984)

Because of its exceptional location, a mere 380 000 km and only a few days of flight away from Earth, it is very likely that the Moon will soon be the seat of a sizeable permanent, but at first semipermanent, settlement. In order to save costs, such a settlement will need to be completely self-sufficient in terms of energy, air, food, and fuel. It will contain repair shops for maintenance of the habitat, its extensive equipment, and the surface vehicles. To provide sufficient energy, large solar or nuclear power plants will have to be installed, as well as an industrial capability to electrolyze water for air and fuel. The water will either be transported from the polar regions or retrieved from crystal water in rocks. Facilities will be needed for manufacturing various equipment and goods. The production of building materials from lunar soil for the construction of pressurized and shielded frames, to be used for living quarters, industry, and farming, will be of great importance. In due time, the inhabitants of such a station will be expected to be able to enlarge their base without help from Earth. With important industrial production and mining operations as well as extensive farms, such Moon stations will eventually become commercially successful by exporting ice, rocket fuel, building materials, components for space vehicles, food, and soil to other stations.

Another very valuable commodity that could be mined on our closest neighbor in space is the helium isotope ^{3}He that has been brought to the Moon by the solar wind blowing since billions of years. This isotope, embedded in lunar soil and discovered during the manned Moon missions is extremely scarce on Earth. Potentially, ^{3}He could be used in fusion reactors such as those being presently realized by the *International Thermonuclear Experimental Reactor* (ITER) (see ITER 2005) fusion project. While in ITER one uses tritium fusion in the reaction ^{2}H + ^{3}H → ^{4}He + n, where a deuterium ^{2}H and a tritium ^{3}H fuse to produce helium ^{4}He and a neutron n, the ^{3}He fusion with the reaction ^{2}H + ^{3}He → ^{4}He + p, is similar except instead of a neutron now a proton p is produced. Because high-energy protons can be much better handled than high-energy neutrons, which degrade the shielding walls of the reaction chamber and lead to radioactive waste materials, the ^{3}He fusion would be much cleaner. While one Space Shuttle load of 25 metric tons of ^{3}He fuel would satisfy the entire yearly energy needs of the USA, the problem is whether this material could be effectively mined (see He3-mining 2005) on the Moon and brought down to Earth. This question can only be answered by conducting further lunar missions.

Figure 9.3 shows a planned lunar station, which would provide supplies for the space stations in orbit around the Earth. Because the lunar gravity is only one sixth that of the Earth, reaching the escape speed of 2.4 km/s from the Moon is much less expensive than attaining Earth's escape speed of 11.2 km/s. As the Moon has no atmosphere to slow down rapidly moving bodies, it is proposed that electromagnetic accelerators, so-called *mass drivers*, will be employed to bring lunar material to escape speed. Figure 9.3 shows two mass drivers of such a Moon station, powered purely by electricity. In these drivers, buckets filled with ice, for example, would be accelerated to escape speed by a long series of suitably phased magnets. By abruptly stopping the buckets at the end of the driver, the contents will be ejected into space, where they can be collected and transported to a space station for further processing. Clearly, the concept of mass drivers would also be applicable for the other materials produced on the Moon.

9.2.3 Space Travel

Before discussing the exceptional importance of the near-Earth asteroids for the future of space colonization, we will briefly discuss rocket flight. In Chap. 5 the concept of escape speed has already been explained, which is the speed any object must reach in order to leave a planet or moon. Table 5.2 gives the escape speed v_E from the Earth and v_M from the Moon. In order to reach such velocities, a rocket is accelerated by expelling gas at high speed through the nozzle of its engine. This gas is generated by burning large amounts of fuel.

Indeed, in modern space vehicles such as Ariane or the Space Shuttle, more than 85% of the weight is fuel, which explains the huge costs of space travel. For the Space Shuttle, for instance, each of the two solid rocket boosters carries 500 t of solid fuel in a frame of 85 t, the external tank stores 730 t fuel in a 27 t frame and the orbiter has a total weight of 114 t of which 25 t is the payload. This adds up to a total lift-off weight of 2041 t of which only 1.2% is payload.

The high exhaust speed is achieved either by combustion or by nuclear heating. For instance, hydrogen and oxygen can be burned to produce steam, or water can be heated by nuclear reactions. Which of these types of rocket propulsion systems is used is a question of availability during space travel. In addition, hydrogen and oxygen usually have to be refrigerated in order to condense the gases into a handy fluid, which adds yet more costs.

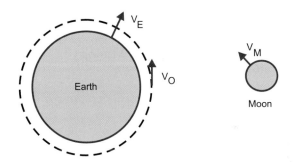

Fig. 9.4. The Earth, the Moon, and the low Earth orbit

Consider, for instance, a return trip to the Moon (see Fig. 9.4). From the Earth's surface, to reach the orbit of the International Space Station, the Low Earth Orbit at about 400 km height, a velocity of $v_O = 7.7$ km/s has to be attained. An additional speed $v_S = 3.5$ km/s has to be added to reach the escape speed from Earth, $v_E = 11.2$ km/s, if one wants to leave it for a space mission. As in falling from space onto the lunar surface the space craft would gain a speed equal to the escape speed, it has to brake this speed down to zero in order to land. Taking off from the Moon, it is necessary to once more attain the lunar escape speed v_M and, in order to brake at the ISS, it is necessary to reduce the speed again by the value v_S. Therefore, for the ISS–Moon–ISS trip, a total velocity change of $2v_S + 2v_M = 11.8$ km/s has to be achieved by expending rocket fuel.

Let us compare this trip with a return flight to a near-Earth asteroid at 1.1 AU. As the orbit of a near-Earth asteroid is very similar to that of Earth and as the escape speed from the asteroid is essentially zero, a velocity change of only about $2v_S = 7.0$ km/s is required. This shows that in terms of costs, it is much easier to travel to a near-Earth asteroid, than to the Moon. Note that considerable fuel could be saved in all these trips by using a heat shield and employing the Earth's atmosphere to transfer from space to Low Earth Orbit. This would be similar to the Space Shuttle, which uses the atmosphere to brake the descent to the Earth's surface without expending rocket propellant.

Table 9.2. The commercial value of a 1-km size M asteroid in billions of $, with metal prices of 2005

Metal	Fraction (%)	Mass (kg)	Price ($/kg)	Value ($billion)
Fe	94	7.5×10^{12}	0.06	450
Ni	5	4×10^{11}	16	4 800
Co	0.6	4.8×10^{10}	37	1 800
Pt	0.001	8×10^{7}	31430	2 500
Au, Ag, Cu, Mg, Ti etc.				2 500
			Total:	12 050

9.2.4 Near-Earth Asteroids and the Mining of the Solar System

Why are the asteroids so important for mankind? In the discussion of asteroids in Chap. 2 it was outlined that they greatly differ in their chemical composition. To show the commercial importance of the mineral resources that asteroids represent (Lewis 1997), we evaluate as an example the scrap value of a metal asteroid. Consider an M asteroid with a diameter of 1 km. With a mass of about 8×10^{12} kg, Table 9.2 shows that at current prices the metal value of such an asteroid would be $12 trillion. Moreover, as the metal already exists in the form of pure scrap, it would not need to go through a complicated reduction process from ore as happens on Earth. In addition, the low-gravity and high-vacuum space environment would permit the efficient operation of mass spectrometers to separate the various metal components.

There can be no doubt that the mining of M asteroids will one day be commercially important, as will be the exploitation of D and P asteroids for hydrocarbons. But our technical capabilities in space are so far very limited. At present, our primary interest will be directed more toward a C, P, or D asteroid where the most pressing fuel problems could be solved by bringing large quantities of ice to the ISS for the manufacture of rocket fuel. It might even be advantageous to produce the already finished electrolyzed fuel components on these asteroids.

With the fuel problem solved, more ambitious efforts in mining and in industrial manufacture could be undertaken. Expansion of industrial capabilities in space is a logical development, as for every imagined material, there would be a suitable asteroid where these materials could be simply and profitably obtained. In the more distant future, the next logical step would be the construction of more advanced space habitats. It is likely that by then bulk components for habitats, reentry- and space vehicles will all be manufactured in space, and only the more sophisticated equipment brought up from Earth. The transport of the processed materials down to the ISS and to Earth from space will become progressively cheaper, as it requires only a frame with a heat shield and a small amount of fuel for navigation, all of which could be manufactured in space. Because of this inexpensive reentry, the commercial success of mining in space will start to have an enormous impact on the terrestrial market.

9.2.5 Space Habitats

Living in a cramped, tightly controlled space, surrounded by a near-vacuum environment, is only one of the many inconveniences for humans dwelling in space stations. Much more serious is the effect of weightlessness. In that state, many familiar daily activities such as moving, sitting, sleeping, eating, washing, and other bodily functions become cumbersome. Moreover, there are severe physiological reactions by our body. Even on Earth, many types of degenerative change occur when prolonged bed rest takes the load of gravity off our bodies. It is therefore clear that extended stays in zero-gravity or low-gravity environments on the Moon and Mars will not be enjoyable. For the successful colonization of the solar system, it is therefore necessary to find ways of restoring gravity. This can be achieved by employing centrifugal forces. The use of rotating habitats was proposed almost half a century ago (see Billingham et al. 1979). Figures 9.5–9.7 show plans for large space colonies, constructed like wheels, rotating around a central axis. Humans would live inside the hollow pressurized tube, forced to the inner surface of the torus by centrifugal forces. Such a wheel, with a spoke radius of 10 km, for example, could provide a habitat of 63 km length. Rotating with a speed

Fig. 9.5. A torus-type space colony (courtesy of NASA Ames)

of 1100 km/h, which would be unnoticed just like travel in an airplane, the inhabitants would experience a weight similar to that felt on Earth.

Figure 9.6 shows a multi-level torus which has recreational, residential, and industrial areas, with trains servicing the length of the colony and elevators the different levels. By importing soil from the Moon, the landscape could be modeled in an Earth-like manner, with individual housing designs. Intensive and well controlled agricultural production (Fig. 9.7) could supply food for a population of many thousands of inhabitants.

Other designs of such space colonies, envisioned by O'Neill (1974, 1989), would have the shape of long cylinders with a diameter of 6 km and a length of 30 km, rotating around their axis once every 110 s. The axis of the habitat would be pointed toward the Sun. As seen in Figs. 9.8 and 9.9, the cylinder walls of such a colony would have three transparent sections where sunlight would illuminate three interior "valleys" via mirrors. By manipulating the mirror surfaces, the Sun could be made to rise and set in a natural 24-hour day-and-night cycle, and even the seasons could be mimicked. A problem with this type of space habitat is, however, that by orbiting around the Sun, the station's rotation axis will lose its orientation to the Sun. O'Neill therefore envisioned two identical cylinders rotating in opposite directions, connected by thin support structures (Fig. 9.10). In such twin systems, the rotation axes could be forced to always point to the Sun.

Fig. 9.6. A view of a residential section of a torus-type space colony (courtesy of NASA Ames)

Fig. 9.7. View of an agricultural section of a torus-type space colony (courtesy of NASA Ames)

It was found that inside an O'Neill station, clouds would form naturally at a height of 1 km above the inner surface. Soil provided from either the Moon or near-Earth asteroids would be the basis for landscaping and agriculture, generating environments similar to those on Earth. With a five times larger usable area, compared to a torus-type colony, the O'Neill station could have a population of from 200 000 to 20 million people. Moreover, such colonies would be surrounded by small service stations (Fig. 9.10), in which every environmental condition for optimal agricultural production could be realized. Because of the limited volume of such agricultural stations, control of the environment, pests, and plant diseases could be easily achieved.

In this most advanced stage, space colonies would become fully self-sufficient. With their own fleet of space vehicles and extensive asteroid mining operations, they could evolve into small independent worlds of their own, possessing a fully developed industry capable of manufacturing the whole range of commercial products (foodstuffs, household machines, tools, building materials, electronic devices, telecommunication equipment, motor vehicles and spacecraft). As a consequence of this industrial activity, they would become major trading partners of the Earth, by exporting metals, minerals, energy, equipment, and food in return for specialized high-tech products. By manufacturing reentry vehicles and heat shields in space, it will be possible to transport these goods cheaply to the terrestrial surface.

In addition, the O'Neill-type habitats would become major construction centers for all kinds of projects to build other habitats and outposts on planets, moons, and asteroids. They would possess workshops to build power plants and have dockyards for assembling space vehicles. One can safely envisage that, eventually, large numbers of such O'Neill-type space colonies with millions of inhabitants will float in the vast expanse of the solar system, producing every commercial product conceivable, while at the same time being fully self-sufficient and independent.

9.2.6 Cultural Impact of Space Colonization

As a consequence of the growing colonization of the solar system, important cultural developments will almost certainly occur on Earth. With the possibility of asteroid mining, supplies of raw materials would suddenly become essentially boundless. Moreover, environmental damage on Earth could be cut back dramatically and our planet returned to a state prior to the aggressive exploitation of all its limited resources.

However, the settlement of space would not only fundamentally change our economic situation; it very likely would also have a basic impact on our

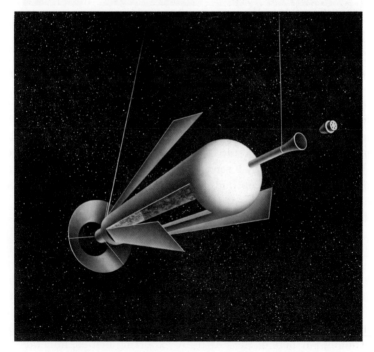

Fig. 9.8. A cylindrical space colony (after O'Neill 1974)

Fig. 9.9. The interior of a O'Neill-type space colony (courtesy of NASA Ames)

Fig. 9.10. Twin O'Neill-type space colonies (courtesy of NASA Ames)

own self-image and culture. As the human mind opens up to the limitless possibilities and opportunities of living in space, a great diversification of human society can be envisioned. Unlike Earth, where limited resources together with economical and geographical factors do not permit the maintenance of impermeable boundaries between nations, space colonies could develop into fully self-sufficient statehoods, which might individually regulate their physical contact and communication with Earth and other space colonies, and thus develop their own independent cultures and ways of life. As a huge number of asteroids, moons, Kuiper belt objects, and comets await exploitation, territorial behavior among these nations would be impractical, unnecessary, and unreasonable. Ultimately, this might lead to an unimaginable diversity of cultures and lifestyles of mankind. This cultural diversity, and the spread of mankind over the entire solar system, would be the best insurance against any fatal catastrophe in the future arising from external and internal threats.

9.3 Interstellar Travel

The exploration and conquest of the solar system is obviously an immediate aim, but mankind also dreams of exploring other solar systems. Four spacecraft (Pioneer 10 and 11 as well as Voyager 1 and 2) are already well on their way to leaving the solar system, with Pioneer 10 and Voyager 1 having reached distances more than twice as far away as Pluto by 2005. In 4 million years, Pioneer 11 will conduct a flyby mission of the star λ Aquilae. Table 9.3 lists the 29 closest stars to a distance of 11.9 Ly (1 Ly $= 9.5 \times 10^{17}$ cm). Except for two white dwarf stars (Sirius B and Procyon B), an A-star (Sirius A), and an F-star (Procyon A), it can be seen that essentially all stars are K- and M-stars. There are only two G-stars, α Centauri A, almost our closest neighbor, and τ Ceti. As in Chap. 5 it was argued that only G-stars should have Earth-like planets suitable for life, these two are therefore particularly interesting in the search for terrestrial planets. Yet with today's maximum velocities of about 100 km/s, a trip to α Centauri would take 12 000 years.

If we assume, for a working hypothesis, that in the future technical advances will allow an increase of this speed to at least 15 000 km/s, that is, to one 20th of the speed of light ($c_L/20$) this would decrease the travel time considerably: about 80 years to α Centauri A and 240 years to τ Ceti. To cross our entire galaxy with its diameter of 100 000 Ly, at a speed of $c_L/20$, would take 2 million years. Taking into account the large costs of such flights, interstellar manned flight is unlikely to be attempted in the next few centuries, except for journeys to the very closest stars. Yet this does not apply to unmanned missions, which could travel for hundreds of years and send back their reports by radio waves. Indeed, proposals have been made to send unmanned probes (so-called von Neumann probes) to extrasolar planets, where they would automatically manufacture new space vehicles and then launch them on to more remote planets. In this way, all known stellar systems of the galaxy containing Earth-like planets could be visited in less than 2 million years.

Manned interstellar travel would be very much easier if intelligent life were no longer to depend on organic beings. Conditions for long-lived androids, hundreds of times smaller than we are, would be very different. Energy requirements and costs for miniature spaceships would be greatly reduced and travel times of hundreds or even thousands of years might be quite acceptable,

Table 9.3. The 29 nearest stars in 19 stellar Systems. The distance is measured in light years, the last column gives the spectral type (after Henry 2002)

Name	Distance (Ly)	Spectral type
Proxima Centauri	4.22	M6V
α Centauri A	4.36	G2V
α Centauri B	4.36	K0V
Barnard's Star	5.96	M4V
Wolf 359	7.78	M6V
Lalande 21185	8.29	M2V
Sirius A	8.58	A1V
Sirius B	8.58	WD
Luyten 726-8 A (UV Ceti A)	8.72	M6V
Luyten 726-8 B (UV Ceti B)	8.72	M6V
Ross 154	9.68	M4V
Ross 248	10.32	M6V
ε Eridani	10.52	K2V
Lacaille 9352	10.74	M2V
Ross 128	10.92	M4V
EZ Aquarii A, B, C	11.27	M5V
Procyon A	11.40	F5IV
Procyon B	11.40	WD
61 Cygni A	11.40	K5V
61 Cygni B	11.40	K7V
Gliese 725 A	11.53	M3V
Gliese 725 B	11.53	M4V
Gliese 15 A	11.62	M2V
Gliese 15 B	11.62	M4V
ε Indi	11.82	K5V
DX Cancri	11.83	M7V
τ Ceti	11.89	G8V

particularly when large populations and essentially entire miniature colonies were to migrate. In addition, it would be easy to slow the clock speed for androids (as in computers today) so that physiologically they would experience a much faster trip. For speculations about the possible creation of androids, see Sect. 9.5.

9.4 Mastering the Biological World

A third foreseeable development for mankind is the mastering of the biological world. On the basis of the advances in molecular biochemistry, our current era has been called "Age of Biology". As discussed in Chap. 6, the complete information that defines a biological organism is stored in the DNA, the organism's master archive. This archive not only contains the building plan of the organism, but also the commands for its replication, instructions to read the genes, regulate their expression and transcribe them onto the mRNA blueprints, and the information about how to translate it from the mRNAs to the proteins, which serve as building blocks and tools to carry out the construction and metabolism of the organism. At present, completely determined sequences are only known for a few organisms (see Table 6.4) and the problem is that only for a fraction of the discovered genes has the exact function been worked out. However, it is only a matter of time before the DNA sequence of practically all important organisms will be known and the function of their genes understood.

This will eventually not only allow us to precisely understand what life is, but also to reveal how biological organisms carry out the manufacture of chemical compounds and body parts, how memory and information-processing works, the way in which immunological warfare is conducted against internal and external enemies, and even how conscious thought arises.

9.4.1 Creating Life in the Laboratory

The current efforts to create a minimal organism starting from mycoplasmas have been discussed in Chap. 6. As mentioned, the roughly 300 genes represent the complete information about a minimal life form based on the DNA machinery. The great challenge at present is to understand the precise function and mode of interaction of the translated products of these genes. This will provide us with the answer to the fundamental question of what constitutes life, and it will allow us to create life in the laboratory by assembling the necessary chemical building blocks.

While understanding the set of a few hundred mycobacterial genes appears to be a relatively limited task, a much bigger effort will be necessary to clarify what the few thousand genes of bacteria accomplish and to comprehend the functions of the 20 000–30 000 genes of higher life forms (see Table 6.4). Great progress in working out the function of systems of proteins has been made by Gavin et al. (2002), which gives hope that in the not too distant future the precise working of a cell will become understood, and even be modeled on a computer (Tomita 2001, 2005). This will allow us to simulate the reaction of cells to external stimuli and different environments, and eventually permit us to predict the consequences of mutations.

Comprehending the detailed building plan of organisms does not only clarify the working of body processes: it will also enable changes of the body plan. One particular change could concern, for instance, the process of *aging*, which is a property of all multicellular life (Chap. 7). Within the past decade, it has become clear that *apoptosis*, or programmed celldeath, is an important tool in the construction of organs during the growth of an embryo. Aging and programmed cell death are also global strategies of nature, developed a billion years ago to cope better in the battle for survival. As mentioned in Chap. 7, the human body, like that of other higher life forms, contains a small number of *germinal cells* that are essentially immortal, and an overwhelming number of *somatic cells*, equipped with a self-destruct mechanism that sets in after about 50 cell divisions. What is the exact purpose of this death program, and does it still make sense for us today? Is there a chance that programmed celldeath and aging could be halted? Here research concentrating on *telom-*

eres, noncoding appendages to chromosomes, provides the possibility that it may indeed one day be possible to become *immortal* (de Lange 1998). Such immortality, however, would not avoid accidental death: it would only mean that death would no longer be pre-programmed. However, at present we cannot even begin to comprehend a society in which the individuals are essentially immortal.

9.4.2 The Decoding of the Human Genome

We have recently witnessed the great effort of the *Human Genome Project* to chart the entire human genome (Table 6.4). A first draft had been published by 2001, and a final corrected version was released in 2003. Figure 9.11 illustrates a small portion of its continuous sequence, comprising a total of ca. 3.2 billion letters, using a four-letter (G, A, C, T) alphabet, and distributed over 24 different chromosomes (22+X+Y). Detailed knowledge of the DNA sequences will help to identify and eliminate inherited disorders, malfunctions, and susceptibilities to disease. As mentioned above, it is only a matter of time before we will fully understand the roughly 30 000 genes that comprise the human master building plan. There is the hope that subsequently it should be possible to activate parts of this plan on demand, and for example, grow a new finger or a replacement kidney. Unraveling the genome of all important organisms (bacteria, fungi, and parasites) and viruses that affect our health, and using this knowledge to develop specific medication, will eventually revolutionize medicine and allow us to lead a life of unprecedented quality.

```
gatctcttgagctcaggaggtcaagaccagcctgggcaacatggcgaaacccccgtctcta
ttaaaaaaaaaattaatacaacaattatcctggagtggtggtgcacacctgtagtcccag
ctacccaggacgctgagacgggaggatcgcttgatcccgggggatgtcgaggctgccatga
tcgcaccactgccctccagccagggtggcagactgagaccccatctcaaaaaataaataa
ataaaagcaaacaagaaaaaaaaaggcttgaaacatatctgatagataaagggctaatca
acacaatatataaagaactgcaaatcagtaaactaagagcaaataacccaatataaagac
attaaagggtagccacggacatctcagacgacgaaaaacaaaagacagtaaacgtataat
aaaacatgtaattgcaaggtgatccgggaatagtaagcgaaaagcaacaattaaatacta
ttttctcatccaccagaacgccaaaaattaaaaagcctaacaatgtccagggctggcgag
aatgtggcagaaggtgatgtcacataccctgcaagtgggaatctaaacagattcagggtt
ttggttttttttttaatcgcaattaggtggcctgttaaatttttttttcttgagacagagtt
```

Fig. 9.11. A sample of the human genome (courtesy of Sanger Centre, Cambridge UK). The total sequence consists of ca. 3.2 billion letters (g, a, c, t), distributed over the 24 different chromosomes (22 + X + Y)

9.4.3 Understanding Intelligence

The decoding of the human genome will not only provide us with the building instructions for body parts and organs. A certain finite stretch of the DNA will tell us *how a self-conscious thinking brain is constructed*. Just as the building plan of the simplest organisms will show us what life is, so the construction plan of the brain will allow us to understand the physical principles behind an intelligent self-conscious thinking mind.

Once the brain's construction plan and the function of the genes responsible for its components are understood, medical malfunctions and deficiencies could be repaired and its working made more efficient. As mankind has always pondered over an important building plan with the aim of perfecting it, one will find ways to improve, for example, our memory, the learning process, and the control of our emotions. Eventually, one might try even outright modifications. Such successful changes would result in a restart of the biological evolution of our body, with the consequence that eventually we could have descendents who would far surpass us in mental capabilities and power, and who would then take over the leadership in all fields of mankind. The interesting question is whether all of mankind will partake in this combined cultural and biological evolution, or whether, as has been the case for the past 4 billion years, only a small part of the total population will evolve. In the latter case we would have to envision a whole "zoo" of humans of different intellectual abilities occupying the solar system, similar to if *Homo habilis* and *Homo erectus* were still alive today.

9.5 Androids and Miniaturization

There is another foreseeable development that will come from the full understanding of the construction plan of our brain. Stripping that plan of its specialized instructions used for building blocks made from organic chemistry, and exchanging them with commands to assemble suitable modern computer components, a self-conscious, nonbiological brain may one day be built. As a result, many biological limitations – such as, for example, the isolation due to confinement in a body, the restriction to five senses, the slow speed of nerve signals, or mortality – would no longer apply. An intelligent brain in a suitable mechanical body, a so-called *android*, would easily connect with others, rapidly communicate by electronic means, and essentially be immortal, except for accidental death. In addition, androids would probably have very different sizes. Most likely, due to costs, they would be much smaller. Our brain, with a diameter of about 15 cm, contains roughly 10^{11} neurons with sizes of about 30 μm. By replacing these neurons, for instance, with 300 times smaller nonbiological nanostructures of 100 nm size, an android's brain could have a diameter of 0.5 mm.

At this point, it is foreseeable that mankind would have to face a tough problem, already envisioned by the famous science fiction writer and biophysicist Isaac Asimov (1982) in his great robot novel series: Can one accept androids as humans? To a lesser degree, this problem would already have arisen if our ancestors *Homo habilis* or *Homo erectus* were still alive today.

Since the body weight of *H. habilis* and *H. erectus* was not much different from ours, their average brain volumes of 650 and 900 cm^3, respectively, relative to our present volume of 1350 cm^3, certainly imply that their mental capabilities were much less developed. This despite the fact that as mentioned in Chap. 7, *H. erectus* already used fire and superbly manufactured spears as tools. The criterion "toolmaker", by which humans are defined, would certainly apply to androids. In terms of intellectuality, there would be no doubt that an android with a brain as complex and capable as ours would qualify to be included in the classification "human". Compared to the mindless unstoppable robots portrayed in some horror movies, androids would be very different. Yet will it matter that they are made out of inorganic materials?

The important issue here is that when constructing an android on the basis of the human body plan we would not *create* the phenomenon of self-consciousness. That self-consciousness exists is due to the laws of nature. These laws cannot be created; they can only be found. Once we know the building plan of a self-conscious brain, we will have found the law of nature governing self-consciousness. For example, when a certain combination of neural components is present and the neurons are connected in the right way, then self-consciousness arises. This law has existed since the formation of the universe. Self-conscious intelligence is therefore a possibility, which nature has eventually expressed in humans on the basis of organic components. In constructing androids, we would merely set the conditions such that this long-existing possibility is now expressed on the basis of inorganic components. We have to accept that since the beginning of time the laws of nature have presumably had a large number of possibilities in store, many of which became reality, and that others, which so far have not yet been realized, are going to appear in the course of our future evolution. From this perspective, it seems difficult to view androids as unnatural and nonhuman.

9.6 Connected Societies

One can speculate about an even more distant foreseeable development that mankind might face. This speculation concerns a synthesis of information technology and biochemical research. A lot of effort is presently put into experiments to generate interfaces between our body and the electronic world. Nerve fibers are being grown onto semiconductor chips, with the aim of establishing connections between nerves and electronic devices (Zeck and Fromherz 2001). Electronical bridging of severed nerves (for example, in the spine) would constitute miracles for the handicapped. In addition, one could establish direct contact with computers, machines, databases, androids, and even with other human beings.

At first, this contact would be only a quantitative enhancement of our existence, but it could well be that this stage signifies another fundamental step in the evolution of life. As multicellularity has led to higher life forms, so the connection of a multitude of intelligent brains could lead to a fundamentally new entity. Asimov (1991) has envisioned a society in which members share a common sphere of super-consciousness. One could call this a *connected society*.

We are not yet imaginative enough to picture the true possibilities of such an evolutionary step, and a whole series of questions arises from such a picture. As a sphere of super-consciousness requires quasi-instant contact, and as communication cannot surpass the speed of light, the individuals of a connected society could not be separated by distances larger than those on a planet. The distance to the Moon, which light traverses in about one second, would already be intolerable for an integrated thought process. A connected society would therefore be highly localized, evolving in a different direction than that of the large diversity and individuality envisioned for mankind in the space colonies.

9.7 Fear of the Future

For those who shudder when reading all this speculation about the foreseeable possibilities and perspectives, a word of consolation is in order. As humans, we cannot but be impressed when we see the orderly growth of organisms: the development of plants or animals, the growth of a child to adulthood, or even the evolution of inorganic objects, such as the birth, life, and death of stars. These observed developments proceed in such an ingenious, straightforward, and beautiful manner. This beauty is due to the laws of nature that govern our physical development, as well as that of the stars, of the biological world, and of the mental world. The future evolution of man is, likewise, the result of the laws of nature. Just as it would be wrong to prevent a child from growing up and to forbid it to realize its natural destiny, so would it be wrong to halt the evolutionary development of mankind. Our fear of the future is the fear of an adolescent concerning adulthood, and of the unknown.

9.8 The Dangers for Mankind

While the conquest of space and the mastering of the biological world are revolutionary and exciting foreseeable developments, there are other prospects that portray a much more sinister future for mankind. It has been suggested that an explanation for our lack of contact with extraterrestrial intelligent societies is that they no longer exist. This view believes that, due to disastrous external or internal events, every intelligent civilization will invariably come to a sudden end, perhaps soon after it reaches our technological stage and long before it can conduct extensive interstellar travel. But how likely is this scenario? Clearly, it is not possible to imagine all the many dangers that an advanced intelligent civilization will face. However, some external and internal dangers that threaten the survival of mankind can be foreseen, and one may ponder ways to avoid them.

Earth's history shows several episodes of mass extinction. Figure 9.12 illustrates the extinction intensity deduced from the changes in numbers of marine animal genera in the fossil record over the past 550 million years. It is not surprising that mass extinction events coincide with the ends of geological epochs, since these epochs have often been defined by precisely such abrupt changes in the paleontological record; that is, by the presence and absence of certain leading fossils. Episodes of massive volcanism, impacts of comets or asteroids, supernova or gamma ray events, global glaciations, and attack by deadly bacteria or viruses have all been proposed as possible causes for such events.

9.8.1 Bacterial or Viral Infection

Between 1347 and 1352, the plague wiped out one third of the estimated population of 75 million people in Europe. There are two types of plague. First, and more widely known, is *bubonic plague*, an infection that causes high fever and a painful swelling of the lymph glands called buboes. The Black Death of the 14th century was mainly of this type and is commonly spread through fleas that have fed on an infected rat. Second, and even more dangerous, is *pneumonic plague*, spread through aerosol droplets released by coughing and sneezing. Although not as common as the bubonic strain, it is more deadly, with an untreated mortality rate of nearly 100%, compared to 50% for bubonic plague. As witnessed by the contemporary Italian writer Giovanni Boccaccio, the disease struck and killed people with terrible speed, such that victims "ate lunch with their friends and dinner with their ancestors in paradise". In many towns the dead outnumbered the living, and bodies piled in the streets faster than nuns, monks, and relatives could bury them.

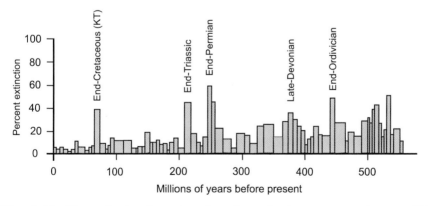

Fig. 9.12. The extinction intensity of marine animal genera over the past 550 million years (after Sepkoski 1995)

Earlier outbreaks have been recorded in Athens during the Peloponnesian war of 430–425 B.C., and in Constantinople in 541 A.D. After the 14th-century outbreak, which was the most violent, periodic reappearances of the disease from the 15th up to the 18th centuries caused widespread depopulation of towns and regions. However, after a particularly violent outbreak in Marseille and the Provence 1720–1721, the plague became extinct in Europe.

The history of the plague and other violent bacterial or viral infections demonstrates not only that diseases with a very high mortality rate have periodically attacked mankind, but there were always surviving individuals who were not affected or became immunized. Since such episodes of infection have occurred for millions of years, it is reasonable to assume that the immune system is versatile enough to withstand similar attacks in the future, and that there is no danger that the entire human population will be wiped out. However, this may be different when an engineered mixture of deadly bacteria and viruses is deliberately released by man himself. Yet this case of biological warfare can no longer be classified as a natural disaster: it represents an internal danger, discussed in more detail below.

9.8.2 Episodes of Extreme Volcanism

Extreme volcanism has been suggested as another reason for mass extinction. It has been proposed, for instance, that this might have been responsible for both the P/T and K/T boundary events (Chap. 7).

In the late Cretaceous, when India was still an isolated landmass drifting northward toward Asia, it moved over a hot spot at which huge amounts of molten rock rose from the Earth's mantle and flooded the land with basaltic lava, giving rise to the Deccan Traps (see Chaps. 3 and 7). It was proposed that the large amounts of CO_2 released during this period of extreme volcanism triggered a greenhouse effect warming of up to 30 °C, creating a heat stress that led to a large wave of extinction, both on land and at sea near the K/T boundary.

As noted in Chap. 7, the rapidity of the extinction event together with the more precise dating that suggests that the Deccan Trap volcanism occurred earlier, perhaps as early as 67–68 million years ago, now point to the impact of an extraterrestrial body as the explanation of the K/T boundary event.

An even larger episode of extreme volcanism occurred during the formation of the extensive Siberian Traps (Chaps. 3 and 7) that coincided with the P/T boundary event. The resulting or perhaps coincidental decrease of the atmospheric and oceanic oxygen (Payne et al. 2004, Grice et al. 2005) has been seen as a cause of the severest extinction in Phanerozoic times. However, as this volcanic episode and the perturbations in the oxygen and carbon cycles extended over millions of years it is difficult to explain the rapidity (Jin et al. 2000) of the P/T extinction event. Here again a large impacting extraterrestrial body (Becker et al. 2004) could be the prime reason for the extinction or might represent the decisive added calamity in an already stressed biosphere. This indicates that extreme volcanism could be an important cause of mass extinction but possibly only if additional catastrophic impacts occur.

9.8.3 Irreversible Glaciation and the Runaway Greenhouse Effect

Irreversible global glaciation is a threat to life, as already discussed in Chap. 5. Figure 5.13 shows that the greatest danger occurred about 2 billion years ago, when large amounts of the greenhouse gases were destroyed by biogenic oxygen and the Sun was much less luminous. This Huronian glaciation (see Chap. 5) might have been triggered by migration of continents and the formation of continental ice sheets in the polar regions, similarly to that indicated in Fig. 7.24 for the last ice age. Yet while these effects posed great dangers in the past, the threat of irreversible glaciation that could wipe out all life on Earth continuously lessens with time, because of the steadily increasing solar luminosity.

Another likely danger might be a runaway greenhouse effect due to global warming from natural causes and human activity. However, for millions of years in the earlier Tertiary, the temperatures were considerably higher than those of today, coinciding with teeming life and allowing for tropical forests as far north as Egypt. Significantly higher temperatures may actually be more of a norm for our planet and, as Fig. 7.24 shows, we are presently in an unusually cold period of planetary climate. Thus irreversible glaciation and

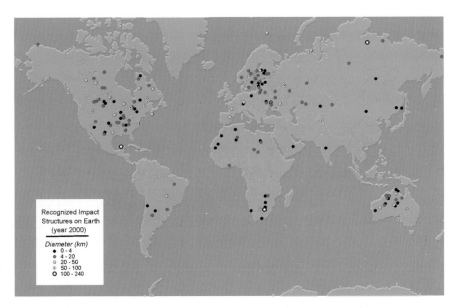

Fig. 9.13. Impact craters on Earth (see IMP 2005)

a runaway greenhouse effect do not seem to be overly convincing threats for the extinction of life on Earth.

9.8.4 Comet or Asteroid Impact

Among the greatest external dangers are perhaps the impacts of comets or asteroids. Evidence is now mounting that the most serious mass extinction events, such as the P/T boundary event at the Permian–Triassic boundary, about 251 million years ago (where 60% of the fossilizable marine animal genera and perhaps 95% of species were extinguished), and the K/T boundary event at the Cretaceous–Tertiary boundary (where 40% of the genera and maybe 75% of the species perished, see Fig. 9.12), were caused by the impact of comets or asteroids (Chap. 7). Figure 9.13 shows a map of large impact craters found on Earth resulting from such bombardments. Among the most famous are the Barringer Meteor Crater in Arizona (1.2 km diameter, impact 49 000 years ago) seen in Fig. 9.14, and the Chicxulub crater (180 km diameter, impact 65 million years ago, see Fig. 6.17), which is thought to be responsible for the K/T boundary event.

On average, bodies about 100 m in diameter strike several times per millennium. With a typical impact velocity of 20 km/s, they have an explosive energy equal to a 100 megaton TNT hydrogen bomb. Due to their chondritic consistency, most of the bodies with diameters less than 100 m split up high in the atmosphere. An example is the *Tunguska event* in the year 1908, in which an impactor in central Siberia split up at an altitude of 8 km, flattening all trees out to a distance of 20 km from the target site. Meteorites between

Fig. 9.14. The Barringer Meteor Crater, Arizona (after Hamilton 2001)

150 m and 1 km in diameter reach the ground and form craters. A 1 km body, for example, strikes the Earth about once every 300 000 years, forming craters of 10–15 km diameter. The amount of dust injected into the atmosphere could trigger severe global disruptions, such as worldwide crop failure. In contrast, a 5 km object arrives only about once every 10–30 million years, and a 10–20 km object, such as that responsible for the K/T boundary event, perhaps once every 100–200 million years.

Figure 9.15 shows the orbits of some of the known potentially Earth-impacting objects, together with those of Mercury, Venus, Earth, and Mars. The very crowded appearance of these orbits close to Earth might suggest that a large number of impacts is imminent. However, it must be kept in mind that the distances between the orbits of Venus and Earth, as well as between those of Mars and Earth, are 42 million and 78 million km, respectively, leaving huge amounts of space. In this vast volume, an asteroid, in order to impact, must encounter Earth at a precisely given moment to hit its comparatively tiny surface. Even for a large number of objects with near-Earth orbits, the likelihood of an impact is extremely small. Moreover, as illustrated by Fig. 9.15, by searching for the big near-Earth asteroids and computing their orbits, the uncertainty about an unexpected catastrophic impact can be greatly reduced.

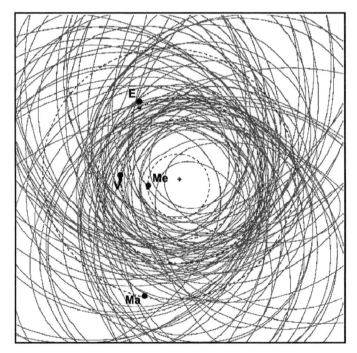

Fig. 9.15. The orbits of known potential Earth-impacting bodies, together with the orbits (*dashed*) of Mercury, Venus, Earth, and Mars (see Minor Planets 2005)

Although large impacts do occur, the great number of craters in Fig. 9.13 shows that they do not pose a serious threat to life, apart from the very largest impact events. It is even doubtful whether a K/T-type boundary event today would wipe out mankind. A few weeks or days of warning would be enough to get a fair fraction of the human population into shelters with enough supplies to survive the few months or years of the aftermath of an impact catastrophe.

To assess the danger of impacts by large bodies on Earth, NASA in 1998 initiated the *Spaceguard Survey* with the aim of finding 90% of the potentially hazardous near-Earth objects (NEOs) larger than 1 km diameter before the end of 2008 (see Spaceguard Survey 2005). *Near-Earth objects* (see Minor Planets 2005) comprise comets, near-Earth asteroids (NEAs, Chap. 2) and large meteorites. This was followed in 2003 by a NASA "Study to Determine the Feasibility of Extending the Search for Near Earth Objects to Smaller Limiting Diameters" (see NEA Search 2003) that recommended an extension of the survey to produce a catalog for hazardous objects down to 140 m diameter.

The Spaceguard Survey so far was very successful because by August 2005, NASA had already discovered 793 of an expected number of 1100 potentially dangerous NEOs larger than 1 km (see Spaceguard Survey 2005). Moreover, as expected, the total number per year (131, 91, 101, 69, and 57) found in the years 2000–2004 decreased despite improvements in the observational technology due to the approaching completeness of the sample, while the total number of discovered NEOs per year (362, 439, 485, 439, and 532) increased over the same time. To carry this survey to smaller objects and to long and short period comets will take more time (roughly 100 years).

Completion of the Spaceguard Survey will allow us to be forewarned of an impending large impact event by years and even decades, although the infrequent appearance of new hazardous comets cannot be predicted. For such chance events and the predicted impending impacts, despite our still limited technical possibilities in space at present, we would already be able to mount a decisive effort to avoid disaster by trying to manipulate the orbit of such a hazardous cosmic impactor, or by breaking it up using atomic weapons. In view of the great increase in the industrial capabilities in space expected over the next several decades, the threat of a giant impact is expected to decrease rapidly.

9.8.5 Supernova Explosions and Gamma Ray Bursts

Supernova explosions (discussed in Chap. 1) and the recently discovered *gamma ray bursts* share the distinction of being the most powerful explosions in the universe. Supernovae can rival the luminosity of an entire galaxy for months, while gamma ray bursts are by far the most luminous photon emitters in the cosmos, with peak luminosities of up to 10^{19} times the solar luminosity. Imagine what would happen if such events occurred close by. Luckily, these events are the end phases of stellar evolution of very rare stars with masses of more than 8 $M_\odot$. A high estimate for our galaxy is that one has about one supernova explosion every 15 years (Dragicevich et al. 1999), which amounts to 3×10^8 events over the Earth's lifetime. This may seem to be a large number, but distributed over the huge volume of our galaxy, the likelihood of one occurring nearby is small.

Actually, if one assumes a lethal distance of 10 Ly, one can expect only 0.17 life-threatening supernovae over the history of the Earth. Here the lethality is attributed to direct radiation from the supernova. In addition, supernovae eject clouds of energetic particles (cosmic rays), generating a so-called *super-nova remnant* that has a typical radius of 5 Ly. But only 0.02 supernovae are expected to be so close that the remnant engulfs the solar system and showers the Earth with deadly cosmic rays. Yet if one wanted to explain the mass extinctions at intervals of 40–140 million years as being due to super-nova explosions, one would need a rate of 33–115 close-by supernovae over the past 4.6 billion years. In addition, because 8 $M_\odot$ stars have such a short lifetime, the highest likelihood of a supernova in our vicinity occurred when the Sun was still close to its star cluster of birth (Chap. 2); that is, within the first 60 million years, when life did not yet exist on Earth.

But even if such an event happened, the irradiation by X and gamma rays from the supernova depends on the striking angle relative to the Earth's axis. If it went off from the direction of the poles, only half the Earth would be affected, while if it were to strike along the equatorial plane, all of the Earth would be bathed for about 60 days in the supernova radiation. But since for most of the Earth's history life has existed only in the water, a supernova going off in the equatorial plane would produce X and gamma rays which, near the poles, would penetrate the water at a very oblique angle and would be readily absorbed, not endangering life there.

Although the mechanism that produces gamma ray bursts is not well understood, it is thought (Burrows 2000) that these events, lasting from a fraction of a second to minutes, are produced by unusual supernovae at cosmological (millions of Ly) distances. Apparently, in these explosions one has very anisotropic conditions and extremely powerful jets are generated which are the source of the bursts. For the lethality of these jets the same argument about the striking angle applies. As gamma ray bursts are highly directional and constitute only a small fraction of the supernova events, they are even less likely than ordinary supernovae to threaten life on Earth.

To summarize, it appears that the known external dangers are not likely to be fatal to life and intelligent life. If our absence of contact with extraterrestrial beings can indeed be attributed to a fatal destruction, it therefore have to be the internal dangers that are the most formidable. Indeed, this is also what one is forced to conclude when looking at our own past and present history. Over the past few millennia there has been a clear progress in the amount of destruction that man has caused and that mankind is capable of doing. While man-made irreversible environmental damage might be still seen as a somewhat external threat, it is the likelihood of uncontrollable inventions – and particularly the possibility of deliberate man-made destruction that poses the most dangerous threat to the survival of mankind.

9.8.6 Irreversible Environmental Damage

Our generation has become acutely aware of the large amount of environmental destruction surrounding us. With the land, oceans, rivers, and atmosphere continuing to be polluted on a unprecedented scale, the long-term effects of this abuse are unknown. While environmental damage has already happened in the past, for example, in the classical Mediterranean regions of Lebanon, Mesopotamia, Turkey, Greece, Spain, and North Africa, which since antiquity have suffered from the large-scale destruction of the landscape due to overuse and deforestation of the land, the environmental destruction that we are witnessing today is at its absolute peak. Only a century ago most of sub-Saharan Africa, the Arctic regions, or the American and Asian rain forests were still largely untouched by man.

Yet while many of these destructions are still ongoing, there have been some signs of positive developments. With most developed countries finally becoming aware of the dangers to the environment, efforts are being undertaken to counteract bad practices. Improved environmental care is now executed in construction, manufacturing, and land management, while recycling of materials is starting to be common. Equally encouraging are the rapid increase of the forest cover in the temperate zone of the Northern Hemisphere, and the foreseeable limit to the growth of the human population due to increased fertility control. Despite the great environmental danger that we presently face, it is therefore likely that with tough measures and a big effort it can be mastered in time.

9.8.7 Uncontrollable Inventions

The story of the sorcerer's apprentice, who was able to start the broom but could not stop it, is a nightmare. There is no way of knowing whether or not an uncontrollable invention such as this will happen one day. Before the first detonation of their atomic bomb, Edward Teller and Robert Oppenheimer in 1942 carefully checked whether the huge temperatures generated would lead to a self-sustaining chain reaction in the Earth's atmosphere once nuclear fission had started. Luckily, it was found that the temperatures in the nuclear fireball would cool much too rapidly for such fusion processes to occur. Recently, people have become worried that the construction of new, ever more powerful particle accelerators such as the Large Hadron Collider (LHC) at Geneva, Switzerland, and the Relativistic Heavy Ion Collider (RHIC) at Brookhaven NY, USA, which are able to create extreme energies, might pose similar uncontrollable dangers (Jaffe et al. 2000). It has been suggested that these machines could create black holes that subsequently might swallow Earth. However, careful investigations of these claims have indicated that there is no danger. Moreover, it has been found that the energies involved are much less than those observed in very-high-energy cosmic ray particles and gamma ray photons that occasionally strike Earth, and which would have long ago triggered a collapse of our planet if such an instability existed. Yet all of this depends on the amount of energy generated, and one day we may well exceed the energies occurring in nature and abruptly meet unforeseen instabilities.

9.8.8 War, Terrorism, and Irrationality

The greatest danger to mankind and really the only one that is presently known to be absolutely life-threatening is *deliberate global destruction*. And it is a danger that becomes perpetually more formidable. In the Stone Age, tribal warfare occurred at a very local level, typically leading to only small numbers of casualties. But by the 8th millennium B.C. (Jericho), when agriculture permitted division of labor and political organization, war became nonlocal. In the third millennium B.C., the horse, and by the 17th century B.C., horse-drawn chariots, were introduced for warfare, while in Roman times, powerful catapults and siege machines were invented. By that time, the war dead numbered many thousands and in medieval times entire populations were slain.

With the Renaissance and the introduction of firearms, the number of war dead strongly increased. It is estimated that the Thirty Years' War (1618–1648) caused 7 million military and civilian deaths. Due to large-scale mechanization and further technological advances (machine guns, tanks, and airplanes) there were 20 million war dead in the First World War and 56 million in the Second World War. And because of associated social and political catastrophes (collectivization and the Holocaust) the total death tolls were easily doubled. During these wars and afterwards, even more formidable weapons of mass destruction were created: chemical and biological as well as atomic weapons. Judging from the enormous advances of weapons technology, mankind will have to face still more powerful and deadly inventions in the future.

But it is not only war that constitutes the great danger; it is the mounting concentration of power put into the hands of a few individuals. Weapons of mass destruction could be found in the possession of fundamentalists, revolutionaries, dictators, bosses of criminal organizations, and of insane individuals. Today, the pressing of a red button would destroy continents by thermonuclear strikes or by massive biological and chemical attacks. Tomorrow, a similar button may endanger the whole of human civilization.

These internal dangers, which arise from ourselves, are particularly worrying, since mankind is faced with an incessant and unstoppable development toward more knowledge, and consequently vastly greater power. In view of the achievements of past centuries, what will we know in a few hundred or a few thousand years? With mankind set in its present ways, it is clear that before long our world is heading for a nightmare of possible destruction. How can we control this irrationality and irresponsibility? Only if we learn how to solidly control these destructive tendencies will mankind be able to survive even the next few centuries.

9.9 Survival Strategies

How can the ever-increasing danger of destruction be avoided? As a first step, a concentrated effort must be made to study all the possible dangers. Programs are under way to develop methods for the prediction of earthquakes and volcanic eruptions by monitoring tiny surface motions, as well as magnetic and electric field variations.

The NASA Spaceguard Survey program that is currently underway (see Sect. 9.8.4) will by 2010 identify most of the dangerous impacting extraterrestrial bodies with sizes of 1 km and larger and most of the 100–1000 m sized objects a few decades later. At relatively low cost, the external threats to mankind can therefore be made much less formidable. By monitoring with advanced space telescopes and by distributing mankind over O'Neill-type space colonies, the external dangers would further decrease while the internal dangers would not.

To control the internal dangers is the greatest and most serious challenge that mankind is likely to face. This is much more difficult than the conquest of space, the renaturation of our planet, or even the construction of intelligent androids. Careful study, how to reliably control each of these perils will be necessary: how to contain power, to base important decisions on group consent, to effectively control weapons and means of mass destruction, and to discipline the irrationality and arbitrariness of any kind of fundamentalism. We have to learn how to combat criminal and terroristic organizations, develop methods to detect failures in persons entrusted with responsibility, and find ways to remove irresponsible and dangerous individuals from power. In addition, more effective procedures for teaching people how to act responsibly will have to be found.

Only time can tell whether this can be eventually achieved. To act selfishly and irresponsibly is deeply ingrained in us and probably helped our ancestors in the battle for survival. However, we no longer live in a Stone Age type of environment with its individual battle for survival. Extortionists pressing a good life out of others, well known from predator–prey communities, must become an outdated strategy of our biological past. Technical advances in the near future will certainly be able to provide for the basic needs of every individual. Already, automatic manufacturing plants are envisioned which will produce goods without human intervention and which will be able to manage their own repair.

But what about jealousy, spite, revenge, anger, mischief, and pleasure in the misery of others? What can be done against an instant outbreak of boredom, a fanatical missionary calling, or an intense death-wish? While the worst consequences of these mental disturbances can be avoided by widely distributing mankind over different space habitats, it may well be that these dark forces turn out to be unmanageable. In that case, one possibility for survival might be to change the hardware; that is, to modify our brains to achieve a more stable responsible society. Another possibility would be the connected society, where the dark side could be held in check by constant contact within the mental community. But again, a single disconnected individual could pose a mortal threat. While the road that will be taken in the future is unknown, there is one insurmountable fact: if our irresponsible side cannot be successfully controlled, then mankind is doomed, because sooner or later a mad powerful individual or movement will succeed in putting an end to it – a fate that might have befallen extraterrestrial intelligent societies that existed before us.

10 Extraterrestrial Intelligent Life

Having laid some extensive groundwork, we are finally ready to address the central questions of the existence and possible nature of extraterrestrial intelligent life. Although no trace of such life has so far been found, there are clearly compelling reasons for its existence. In Chap. 5 we estimated that Earth-like planets should occur quite frequently in our galaxy, and in Chaps. 6 and 7 it was shown how life, and intelligent life, have formed on Earth. Using the Drake formula, these results will now be combined to estimate the expected number of extraterrestrial intelligent societies. Since many probability factors entering this equation are controversial, the opinions of various authors will be discussed.

Because the first Earth-like planets were born about 10 billion years ago, and assuming that they have developed in a similar way to the Earth, extraterrestrial intelligent societies should already have been around for many billions of years. This raises the so-called *Fermi paradox*: "Where are they, and why don't we have contact with them?" Because advanced extraterrestrials would have gone through our stages of development, possible answers to these difficult questions should be attainable from our discussion of the foreseeable future of mankind in Chap. 9. Although the present chapter is very speculative, it will permit us to see the Fermi paradox, the search efforts for extraterrestrial intelligence (SETI), and even the future of our own civilization, in an entirely new light.

10.1 Does Extraterrestrial Intelligent Life Exist?

If the emergence of intelligent life were a unique chance event that has practically never occurred anywhere else, then estimates of the number of communicating intelligent societies, as attempted by the Drake formula (Chap. 5), would be pointless. Many of my colleagues and I believe, however, that reasonable arguments can be made that life, and intelligent life, should be quite common on other Earth-like planets. That the formation of life might be seen as the predictable result of chemical evolution has already been pointed out in Chap. 6, where Christian de Duve (1991) was quoted with his opinion that it is almost certain that the origin of life came about by a sequence of small steps, each with a plausible probability. He argued that this suggests that life

P. Ulmschneider, Extraterrestrial Intelligent Life. In: *Intelligent Life in the Universe*, P. Ulmschneider, Adv. Astrobiol. Biogeophys., pp. 255–277 (2006)
DOI 10.1007/11614371_10   Springer-Verlag Berlin Heidelberg 2006

will also occur on other planets if the chemical and environmental reasons for its development are not much different. Can a similar case be made for the appearance of human-type intelligence on other planets?

In Chap. 7 it was noted that the evolution of life is governed by Darwin's theory of mutation and natural selection, in which the mutations are chance events while the selection is directional, leading to the evolution of new organisms. It was also mentioned that there is a dispute over whether Darwin's theory only determines short-range evolution or whether it also governs the long-range development of the biological world. It was concluded that the phenomenon of convergence demonstrates that Darwin's theory must also hold for long-range evolution, because the survival of the fittest is due to the specific environment and the laws of nature that severely constrain the inhabitants of the biosphere.

The situation in biology, as suggested in Chap. 7, is not very different from that in astronomy, where one has the evolution of celestial bodies, which is also governed by chance events, by the properties of the environment, and by the laws of nature. If the evolution of interstellar clouds can be successfully modeled up to the formation of stars and from there to the final stages of white dwarfs or supernovae, it should also be possible to model biological evolution.

Certainly such simulations of the biophysics of cells, of organisms, of geophysical and climatological environments on terrestrial planets, and of ecological interactions between fauna and flora will be vastly more complex than that governing the evolution of the relatively simple astronomical objects, but there is no reason why this modeling should not be possible with future computational power. While such efforts so far represent as yet unrealized challenges, I attempted in Chap. 7 to show that there are definite reasons why long-range biological evolution eventually led to humans and intelligence.

Chapter 7 described how bacteria employ different strategies and how the strategy of eukaryotic cells rests on the accumulation of information. Large numbers of genes, which constitute blueprints for battle tools, enabled an ever more sophisticated fight for survival. By incorporating endosymbionts and using them as organelles, the eukaryotic cells became particularly efficient. This strategy of using massive amounts of information to fight the battle for survival continued with the creation of multicellular organisms, which allowed the formation of organs, and culminated in man with about 30 000 genes. As the laws of nature are the same everywhere in the universe, we can assume that Darwin's theory will result in similar strategies of nature on other Earth-like planets, and that the method to fight successfully by accumulating information will be a possibility that nature would not neglect elsewhere.

Does that lead to the development of human-type intelligence? Here we must note that intelligence is not a singular property that only happens in man, but that it also occurs in various degrees in animals. Intelligent behavior has always improved the chances of survival. That there is a noticeable growth of intelligent behavior from fish to amphibians, and from reptiles to mammals, has been pointed out in Chap. 7. Primates typically have twice the brain volume compared to other mammals of similar body weight, and the intellectual capabilities of the great apes far surpasses those of lower monkeys.

The development of high intelligence in corvid-type birds, on a level with that of monkeys, represents a particularly fine example of convergent evolution. Finally, the fabulous growth of brain volume (Fig. 7.33) from the australopithecines, *H. habilis*, *H. erectus* to *H. sapiens* implies great leaps in the level of intelligence. Here, a lot of research must be done in the future to understand these fundamental leaps and in what way we differ, for instance, from *H. erectus* who already used fire as a tool, 1–1.5 million years ago. Clearly, more intelligent behavior, employing tool use and group communication, gave our ancestors a higher chance of survival.

We can speculate that human-type intelligence might have developed from the dinosaurs, had the K/T boundary event not happened. From the logic of the human body plan, discussed in Chap. 7, we can even assume that these hypothetical intelligent beings would not have looked too different from us, and this argument should also hold for intelligent extraterrestrials in the state when they first appeared on their own planets.

Certainly, intelligence came about on the basis of Darwin's theory, by mutations and the relentless fight for survival. The cultural evolution associated with the accumulation of information in the form of "software" due to learning can be viewed as a natural continuation of the ancient survival strategy of eukaryotic cells, of amassing knowledge in the form of genes. In summary, therefore, it appears reasonable that intelligent life will also be an outcome of the biological evolution on other Earth-like planets.

10.2 What is the Hypothetical Nature of the Extraterrestrials?

What can be said about the nature of the extraterrestrials? In Chap. 5 it was argued that Earth-like planets that orbit in a continuously habitable zone are essentially found only around G-stars of the metal-rich population I. As the first population I stars appeared around 10 billion years ago, and assuming that the development time for human-type intelligence is about 4.6 billion years (as on Earth), the first intelligent societies could have appeared about 5 billion years ago. We thus have to face the possibility that there are intelligent societies billions of years older than ours.

Is there a way of visualizing these much more advanced life forms? As will be demonstrated below, there is absolutely no hope that our imagination is capable of providing an even approximately adequate picture of such beings, who are billions of years more advanced. Fortunately, however, as these societies must, at some distant time, have gone through our stage of technological development, we might gain some vague glimpse of their nature by considering our own future. In Chap. 9 we have already discussed some of the staggering changes that are expected in only the next few centuries due to the conquest of space and the mastering of the biological world, combined with the advances in information technology. We have speculated about humans with greatly superior mental capabilities, tiny artificial self-conscious androids, and connected societies. But I think that even our most daring futuristic views constitute a very poor preview of the true development of the next millennia, and will certainly be completely inadequate to predict the nature of our descendents, millions or billions of years in the future. There may well be qualitative changes. In addition, as mentioned above, nature may have more fundamental steps in store than the three basic steps of life already identified by Aristotle: vegetative, sensitive, and conscious life. To foresee the possible development of mankind over such time-spans is truly beyond imagination.

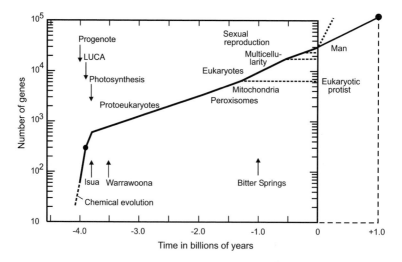

Fig. 10.1. Extrapolation of the number of genes of hypothetical organisms 1 billion years into the future

Yet there is an analogy that may help us to gain some appreciation of the magnitude and quality of the difference between these advanced stages and us. This method, called linear extrapolation, is often used with good results in science. In Fig. 7.7 the growth with time of the number of nonidentical genes of the most advanced life forms on Earth is displayed. Figure 10.1 shows a 1 billion year extrapolation of this curve into the future (solid line). Clearly, such an extrapolation is possible and likely, because molecular biochemistry has recently acquired the capability to artificially add and modify genes. Because these mutations, which in due time will result in the growth of the genome, are introduced deliberately, the accumulation of information could proceed even more rapidly (dotted line).

Although plotting an outlying dot in Fig. 10.1 does not tell us anything definite about the nature of such beings, what can be deduced from this figure is that the difference between such advanced extraterrestrials and us must be at least as great as the difference between us and life forms 1 billion years ago. That is, in magnitude and quality the difference must be comparable to that between a thinking, self-conscious multicellular being like us and a single-cell organism swimming in the Earth's oceans 1 billion years ago. In view of the perpetual growth of knowledge and power, observed in human history since the times of *Homo habilis*, we have to conclude that these extraterrestrial beings must be essentially God-like, with faculties that border on omniscience and omnipotence.

In Chap. 9 we have also discussed the rapidly mounting dangers associated with the incessant growth of knowledge and power, and indicated the need to develop strategies to avoid a fatal end for our species. Such developments must also be expected on other planets. As the history of an intelligent civilization is highly individualistic, characterized by outstanding persons, unique trends as well as chance, we may assume that the dangers faced by extraterrestrial societies during their histories will be very specific, as will the strategies to cope with them. There may be sudden catastrophic deaths of some civilizations, but probably there will also be a large number of successful solutions to the various crises. It is difficult to imagine that there should be an explicit law of nature that stipulates that all extraterrestrial intelligent civilizations inevitably come to a fatal end. It seems much more likely that some societies fail while others endure. Clearly, those societies that survive for millions or possibly billions of years must have developed a very *stable and responsible society*, because otherwise one irresponsible member or a single uncontrolled event would long ago have led to the annihilation of the entire civilization.

We have mentioned above that these extraterrestrials, in an evolutionary stage comparable to us today, may have had a body fairly similar to ours, based on organic chemistry. Yet in Chap. 9 it was speculated that in the future, after the construction plan of the human brain becomes known, there might be self-conscious human-like beings, androids, which have bodies that are no longer built on the basis of organic chemistry. Both in size and shape, androids could be very different from what we see in today's biological world. It is therefore obvious that when picturing extraterrestrial beings, millions of years older than us, we must expect them to look very dissimilar to us.

As advances in knowledge lead to an increase in power, it is not surprising that already, 40 years ago, the Russian physicist Nicolai Kardashev (1964) has envisioned extraterrestrial intelligent civilizations with technical capabilities vastly superior to ours. He has classified these societies into three categories: *Type I societies* have a technological level and an energy consumption similar to ours. This would mean that they control a power of 4×10^9 kW.

Type II societies are able to control an energy output equivalent to an entire star; that is, a power of 4×10^{23} kW.

Type III societies finally control the power output of an entire galaxy, or 4×10^{34} kW.

After looking for a detectable power output from such types of civilizations, it can be concluded that our galaxy does not possess a type III society, but type II civilizations are not ruled out. Since the energy dissipated or controlled is a poor, or even inappropriate, measure of such societies, it is doubtful whether the Kardashev classification is able to do justice to the nature of extraterrestrial civilizations.

10.3 The Drake Formula, the Number of Extraterrestrial Societies

After this highly speculative reasoning about the nature of extraterrestrials, we will now discuss the full Drake formula, which attempts to estimate the number N of communicating extraterrestrial civilizations in our galaxy. In Chap. 5 we have subdivided this formula into two parts, the estimated number of habitable Earth-like planets N_{HP} and the likelihood f_{IC} that, on such planets, communicating intelligent societies will develop. With these factors, the Drake formula can be written

$$N = N_{HP} f_{IC} = N_{HP} f_L f_I f_C L / L_S .$$

Here f_L is the probability that life will develop on an Earth-like planet, f_I the fraction of planets on which life evolves into intelligent life and f_C the likelihood that these intelligent societies develop into a technical society and are willing to communicate. L is the average lifetime of a technological society and L_S is the time during which Earth-like planets have existed in our galaxy.

Table 10.1 shows the remaining factors of the Drake formula used by various authors, together with the values for the number of habitable planets N_{HP} taken from Table 5.5. It can be seen that in the column that specifies f_L, except for the value adopted by Rood and Trefil (1981), there is a relatively good agreement among the authors that, given an Earth-like planet, life will surely appear on it. Rood and Trefil base their low value on two arguments: that it is difficult for isolated chemical systems to develop a reproduction process, and that other Earth-like planets do not have a large moon to create the high tides which they consider necessary for the generation of life in tidal basins. Note that the requirement of a large moon is demanded here for a different reason than its effect on the stabilization of the Earth's rotation axis, discussed in Chap. 5. However, these uncertainties have been largely overcome due to the progress of our understanding of the likely course of the chemical evolution toward life (Chap. 6). Therefore, Rood and Trefil's low value for f_L should be revised upward.

Table 10.1. Values used in the Drake formula by different authors. This formula estimates the number N of intelligent communicating societies in our galaxy. L_S and L represent lifetimes given in years, f_L, f_I, and f_C are probabilities for life, intelligent life, and communicating life, respectively, and N_{HP} is the number of habitable planets

Author	N_{HP}	f_L	f_I	f_C	L_S	L	N
Cameron (1963)	1×10^{10}	1	1	0.5	3×10^9	10^6	2×10^6
Sagan (1963)	1×11^{11}	1	0.1	0.1	1×10^{10}	10^7	1×10^6
Rood & Trefil (1981)	2×10^6	0.01	0.5	0.25	1×10^{10}	10^4	0.003
Goldsmith & Owen (1993)	2×10^{10}	0.5	0.75	1	8×10^9	10^6	1×10^6
The present author (2002)	4×10^6	1	1	1	1×10^{10}	10^7	4×10^3

With respect to the factors f_I there is little discrepancy, except for Sagan (1963), who supposes only a 10% likelihood. Most authors assume that once life forms, intelligent life will eventually develop. This view was discussed in detail in Chap. 7. Whether an intelligent society develops into a communicating one is described by the values f_C, which are also not in great dispute. Sagan, by pointing out that presently the high-technology part of human history is only a fraction of man's entire history, gave f_C a 10% probability. This choice of f_C, however, is inconsistent with his estimated value of a lifetime $L = 10$ million years for human society, which would significantly exceed the duration of the low-technology phase. His value of f_C should therefore be increased. Rood and Trefil (1981) multiply their 50% probability for f_C by another factor of 0.5 to allow for the possibility that some societies are capable, but do not want to communicate. The same abstinence factor has been applied by Cameron (1963) in his value $f_C = 0.5$. That extraterrestrial societies may not want to communicate is an important point, which was not taken into account in the original Drake formula. Based on experience with our own discovery and use of radio waves, it was assumed that if extraterrestrial societies were to develop this mode of transmission, they would use it to communicate with their own kind.

While there is rough agreement about the time L_S for which Earth-like planets exist, the final estimates N of communicating societies show large differences. From Table 10.1 it can be seen that there is the usual opinion that there should be of the order of a million communicating societies in our galaxy. In contrast to this, my present estimate and that by Rood and Trefil (quoting their "cautious estimate", disregarding the "optimistic" and "pessimistic" ones) give much smaller numbers. In fact, Rood and Trefil's estimate is so low that only one civilization would be expected in 330 galaxies. Their estimate would still be low if their value of f_L as discussed above were revised upward by a factor of 100. My own estimate leads to about 4000 high-technology civilizations, of which some might be noncommunicating, as will be discussed below.

The low values of N in Table 10.1 are mainly due to the estimates of the number of habitable planets N_{HP} and of the lifetime L. Here, advances in the theory of planet formation and in our knowledge about the necessary conditions for Earth-like planets, orbiting in narrow continuously habitable zones around relatively infrequent metal-rich G-stars, enter the picture. In addition, and very significantly, the new planet observations have made us aware of the need to avoid migrating Jupiters, a fact that has significantly lowered the probability of the existence of Earth-like planets. The very low value of N of Rood and Trefil is in addition due to their very pessimistic assumption of the lifetime L of a communicating society.

10.4 The Lifetime of an Extraterrestrial Civilization

In fact, the lifetime L of high-technology civilizations represents the greatest uncertainty in the Drake formula. For this value, we have no basis for a reliable estimate even for our own society, and every author's personal optimistic or pessimistic view acutely enters the equation. Mankind started with *Homo habilis* and *Homo rudolfensis* around 2.5 million years ago, yet it is possible that we might die in a nuclear and bacteriological holocaust in the next 100 years. Rood and Trefil's (1981) view is that mankind will not last for much more than another 10 000 years. As mentioned above, I doubt that there is an explicit law of nature that inevitably terminates intelligent societies. Here the finite lifetime of animal species such as, for example, the dinosaurs provides no counter argument. They were succeeded by their descendents the birds, or by others with comparable capabilities. Similarly, *Homo erectus* was followed by *Homo sapiens*, and we ourselves might some day be replaced by more advanced intelligent beings. But this will not diminish the capability to communicate, and thus it will not change L.

As we have no guide to selecting L, we may speculate that in our galaxy there are some very old intelligent societies together with many younger ones. An average lifetime L may thus be derived by taking a (logarithmic) mean somewhere between the lifetimes of the oldest (10^9 years) and the presumably youngest (10^4 years) civilizations, which suggests a value of about $L = 10^7$ years, or maybe as much as 10^8 years.

To select a mean lifetime L does not imply that this is a typical lifetime of an individual extraterrestrial society, but only that it is an average time when comparing societies that live for as little as another 10 000 years with those living as long as a billion years. Let us look at the consequences of such a choice. Assuming a value of $L = 10^7$ years, we have the following picture (see Fig. 10.2). I have estimated (Table 10.1) that there are 4 million Earth-like planets in our galaxy (the rising line in the figure), on all of which life and subsequently intelligence will probably develop. If one has a formation time of intelligence of 4.5 billion years, then the earliest intelligent societies have appeared about 5.5 billion years ago. Earth-like planets before that time

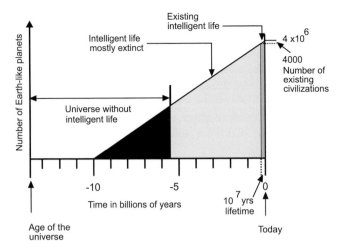

Fig. 10.2. The number of Earth-like planets and of past and present intelligent societies

(solid) do not carry intelligent life. Afterwards, additional intelligent societies formed but also, due to our assumed value of L, many of them died. This leaves an estimated 4000 ($= 4 \times 10^6 \times 10^7/10^{10}$) societies that formed mainly in the past $L = 10$ million years (dark gray) and still exist, while the previous ones (light gray) would be mostly extinct.

This means that of the roughly 2 million intelligent societies that went through our stage of development, 99.8% are already assumed to have perished. For the value of $L = 10^8$ years, we would have 40 000 surviving societies, with 98% of the created intelligent societies already being extinct. To assume a value for L between 10^7 and 10^8 years, therefore, would be a pessimistic view, because most of the created intelligent civilizations would already presumed to be dead.

Taking a definite value for L appears to contradict our assumption that there is no specific law of nature that limits the lifetime of intelligent societies. Yet we may interpret that lifetime as the unintended result of the struggle for survival of intelligent societies. That is, L could be seen as a consequence of the perpetual accumulation of knowledge, which unfortunately also increases the risk and the capability for self-destruction. But even if there are only 4000 existing extraterrestrial intelligent societies in our galaxy, the remaining question is: Can we find them in the huge accumulation of 160 billion stars?

10.5 Distances to the Extraterrestrial Societies

Before addressing the problem of how to find these intelligent societies, we have to discuss the distances in the galaxy. As shown in Fig. 10.3, our Milky Way has roughly the shape of a disk, with a radius of 50 000 Ly and a thickness of 1300 Ly, in which the Sun (dot) orbits at a distance of 26 000 Ly from the center (McNamara et al. 2000). If 4 million life-carrying Earth-like planets (dots) were distributed at random over this vast volume of space, then the average distance l between two planets would be about 170 Ly, while the mean distance between the 2 million planets, where in the course of galactic history intelligent life has presumably formed, would be about 210 Ly. As we have assumed that most of these civilizations have already perished, the average distance to the surviving 4000 intelligent societies would be about 1700 Ly.

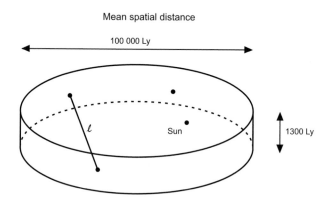

Fig. 10.3. The mean distance between life-carrying planets

These estimates have important consequences for our search efforts. In Chap. 8 we have discussed how, in the not too distant future, we will be able to directly observe terrestrial planets and detect traces of life by observing, for example, H_2O – and particularly O_3, in their atmosphere (Fig. 8.13). Detecting such life-carrying planets at the relatively small distance of about 170 Ly would constitute a huge advance in our understanding of how and under what conditions life forms.

Moreover, as the demise of an intelligent civilization is not expected to be fatal to all forms of life on a planet, it might well be possible that some eukaryotic cells or primitive multicellular life forms, perhaps surviving deep in the rocks of the planetary crust, would come forward and rebuild a situation similar to Earth 1 billion years ago, complete with an oxygen atmosphere. By the incessant working of evolutionary forces, intelligent life might then in due course appear for a second time, another billion years later. Because of this, the number of 4000 existing civilizations could be an underestimate – possibly even a big underestimate – and the nearest intelligent civilization may exist much closer than 1700 Ly. From a discrepancy between the age of the parent star and that of the planetary atmosphere, we might even be able to judge, from great distances, whether such life-carrying planets have undergone violent disturbances in the past. In any case, as about 2 million planets are supposed to have suffered a catastrophic demise of their intelligent society, an interstellar trip to such a planet at a distance of about 210 Ly would certainly be very interesting, as one could study then in detail the archaeology of this lost advanced civilization.

10.6 SETI, the Search for Extraterrestrial Intelligent Life

As, from our above discussions, it appears to be quite reasonable to expect the existence of a sizeable number of intelligent civilizations in our galaxy, is there a way to detect them? Let us discuss the search efforts for extraterrestrial intelligence (SETI). The only practical way by which we can obtain knowledge about our surrounding universe is by *electromagnetic radiation.* Except for material inscribed on matter (Rose and Wright 2004), other messengers from the universe do not deliver the huge amount of information that is transmitted in the electromagnetic spectrum. Such alternate messengers are *cosmic rays*, charged particles, primarily protons and electrons that are produced in supernova explosions or in outbursts on the Sun, *gravitational waves*, which are generated by the collapse of massive stars or star mergers, and *neutrinos*, generated by nuclear processes in the core of stars. Today, the entire electromagnetic spectrum, ranging from gamma rays and X-rays over the ultraviolet-, visible-, and infrared light range, up to microwaves and radio waves, can be observed either from the Earth's surface or from space.

There are parts of this spectrum that are particularly suitable for interstellar communication. The higher the frequency of the electromagnetic waves, the more easily they are absorbed by the interstellar gas and dust. In visible light most of the galaxy is heavily obscured, whereas in infrared light star formation regions or the galactic center can be observed much better. The galaxy is essentially transparent in the microwave region. Here the range of frequencies from 1 to 100 GHz (30 cm to 3 mm wavelength) shown in Fig. 10.4 presents a vital window, where the amount of background radiation is at a minimum and where it might be particularly easy to carry out interstellar communication (Oliver 1977).

For frequencies below 1 GHz, as seen in Fig. 10.4 (left side of the figure), the galactic synchrotron radiation gives an increasingly larger background contribution the smaller the galactic latitude angle $|b|$ is against the galactic plane. This synchrotron radiation is produced by high-energy cosmic ray electrons, which are forced to orbit around magnetic fields which lie in the galactic plane. There is also a high-frequency limit because of the spontaneous emission of photons in galactic clouds of various temperatures (right side of the figure). Figure 10.4 shows the microwave window outside the Earth's atmosphere. When observed from the Earth's surface, the spectrum from 20 to 1000 GHz is contaminated by H_2O and O_2 emission from the terrestrial atmosphere. This leaves a microwave window from 1 to 20 GHz from the

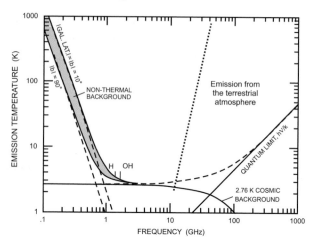

Fig. 10.4. The microwave window in terms of an emission temperature, for outside the Earth's atmosphere. The more intense the background radiation of the galaxy, the greater is the emission temperature. On the ground the window is narrowed to the *left of the dotted line* because of emission by H$_2$O and O$_2$ molecular bands in the atmosphere (modified after Oliver 1977)

ground with a very low background, called the *"water hole"*, although atmospheric water is no longer a limitation when observing from space. But even in the center of the "water hole" in Fig. 10.4 there is still noise from the 3 K cosmic background radiation, which is the remnant of the Big Bang fireball (Chap. 1). It is fortunate that in the "water hole" the very important 1.4024 GHz (21 cm) neutral hydrogen radio line is located, which because of the low galactic background noise has allowed us to map our galaxy and determine its spiral structure.

10.6.1 Radio and Optical Searches for Extraterrestrial Civilizations

The search for extraterrestrial intelligent life, SETI, began in the 1950s when Cocconi and Morrison (1959) suggested that microwave radio signals might be used to communicate between the stars. As already mentioned in Chap. 5, this suggestion was taken up by F. Drake, at the National Radio Astronomy Observatory (NRAO) in Green Bank, West Virginia, who over four months in project *Ozma* directed a 25 m radio telescope for six hours every day toward the two G-stars τ Ceti and ϵ Eridani, to search for regularly patterned pulses indicating intelligent civilizations. While Drake did not detect any signal from extraterrestrials, project Ozma spurred on the interest of others in the astronomical community, most immediately Russian colleagues such as I.S. Shklovskii and N. Kardashev. Initiated by the summer study project *Cyclops* at the NASA Ames Research Center, which looked for the best way to detect radio signals from extraterrestrials, the META project of Harvard University (now continued as BETA and run by the Planetary Society), the University of California's SERENDIP project, and an observing program at the Ohio State University were developed in the 1970s. The latter now cancelled program became famous because of the "wow!" signal, detected in 1977, which had the appearance of an extraterrestrial signal, but was only seen briefly and did never repeat.

META I (the Million-channel Extraterrestrial Assay) was a search program which, from 1985 to 1995, used the 26 m steerable radio telescope at Harvard, Massachusetts. Its counterpart, META II, with a 30 m antenna, at the Argentinian Institute for Radio Astronomy near Buenos Aires, provided coverage of the southern sky. META I and II monitored 8.4 million radio channels at once with a spectral resolution of 0.05 Hz, and reached a combined sky coverage of 93%. After five years of observations from the Northern Hemisphere and the recording of 6×10^{13} different signals, META I found 34 "alerts". Unfortunately, the data were insufficient to determine their real origin. Interestingly, the observed signals seemed to cluster near the galactic plane, where the largest numbers of Milky Way stars are located. After three years of observations, META II found 19 signals with similar characteristics.

BETA (the Billion-channel Extraterrestrial Assay), the follow-up program of META I (see SETI 2005), began observations in October 1995. It broke down in 1999, but is scheduled to resume operation in 2006. It employs a new strategy of rapid and automatic observation of candidate events, a better discrimination of terrestrial interference, and a much greater frequency coverage. With a 240-million-channel spectrometer, the output of which feeds an array of programmable "feature recognizers", BETA searches the full "water hole" of 1.4–1.7 GHz. The antenna incorporates two (east west) feed horns and a third low-gain terrestrial one. When, as a consequence of the Earth's rotation, a suspicious celestial signal is first seen in the east, then in the west, but not in the terrestrial horn (detection there is attributed to terrestrial sources), it triggers the antenna to jump to the west, forcing the source to move through the detection sequence again. If the signal is confirmed, the antenna will break off its general survey and start detailed tracking of the newly discovered source.

SERENDIP (Search for Extraterrestrial Radio Emissions from Nearby Developed Intelligent Populations) is the University of California at Berkeley SETI Program (see SETI 2005). The idea of the project is to "piggyback" alongside simultaneously conducted conventional radio astronomy observations. It uses the 300 m dish at the Arecibo Observatory in Puerto Rico (see Fig. 10.5), the largest radio telescope in the world. From 1992 to 1996 the spectrum analyzer, SERENDIP III, working essentially on a full-time basis, examined 4.2 million channels every 1.7 seconds in a 12 MHz-wide band centered at 1.429 GHz. This is only a small piece of the electromagnetic spectrum, but it is by far the largest segment ever examined so comprehensively. SERENDIP IV, the next-generation instrument, examines 168 million chan-

Fig. 10.5. The Arecibo Observatory, Puerto Rico (courtesy of NAIC)

nels every 1.7 seconds in a 100 MHz band centered at 1.42 GHz. SERENDIP signals that peak significantly above the background noise are run through a series of algorithms designed to reject terrestrial sources.

Since 1998, the Southern SERENDIP project at the University of Western Sydney, Macarthur, Australia, has operated on the same principle, by piggybacking onto the conventional radio astronomy observations at CSIRO's 64 m Parkes radio telescope, the largest radio astronomy telescope in the Southern Hemisphere. Southern SERENDIP currently scans 8 million radio channels every 1.7 seconds, and this will be increased to 58 million channels in the near future. Figure 10.6a shows a typical recording of 2.5 MHz bandwidth, from 1.4180 to 1.4205 GHz. Because of its appearance, this diagram is called a *waterfall plot*. On the horizontal axis one plots the 4.2 million channels that cover the 2.5 MHz, and on the vertical axis the observation time. Each horizontal line represents a momentary reading of the 4.2 million channels. A dot represents a "hit" where the signal strength is at least 12 times greater than the mean of the output of the 8000 adjacent channels. Most of these "hits" are receiver noise. The 2200 lines of observation represent about one hour of observation. This waterfall mode of recording is seen better in the older plot from the early 1970s shown in Fig. 10.6b, where the slanted track is due to a pulsar. Its frequency behavior is due to the fact that both the observer and the pulsar are moving in a nonuniform manner. The vertical tracks of Fig. 10.6a are due to interference by terrestrial radio sources. The SETI observers look for slanted tracks which indicate a celestial radio source. Comparison of the two panels of Fig. 10.6 shows the large increase in frequency and time resolution in 25 years.

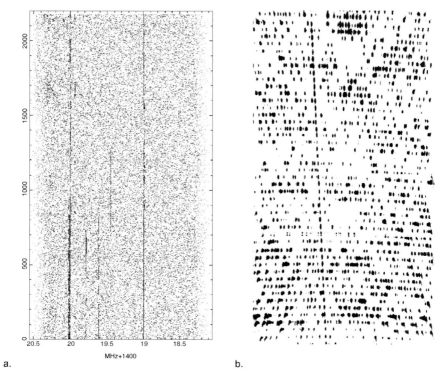

a.   b.

Fig. 10.6. Waterfall recordings. **a.** SERENDIP. **b.** Older observation from the 1970s with a pulsar (see SETI 2005)

In 1992, NASA started its own SETI program, called the High-Resolution Microwave Survey (HRMS). Unfortunately, after a year, as a result of intense lobbying by scientists with a strong conviction that there can be no close-by extraterrestrial intelligent societies, Congress cancelled funding. Fortunately, however, private means were found and the venture, renamed *Project Phoenix*, operated by the SETI Institute (see SETI 2005) from 1995 until its termination in 2004. It searched the frequency range between 1.0 and 3.2 GHz with 1 Hz resolution, using 28 million channels at a time. Rather than trying to scan the entire sky, the project targeted approximately 1000 nearby Sun-like stars. After its start at the Parkes radio telescope, it later moved to the newly upgraded Arecibo Observatory. After completing more than half of their targets, they have not yet found evidence of extraterrestrial transmissions.

Due to the large computational effort, to sift through millions of frequencies and events to isolate a few "alerts", a new strategy launched in 1999 is to tap the unused computer power of thousands of home PCs by inviting their owners to participate in the project SETI@Home (see SETI 2005). In this project, instead of running a screen saver, the PC analyses SETI data.

Here computer users from around the world are able to participate in a major scientific experiment.

Because of the limited time available on the Arecibo Observatory it was long felt desirable to have a radio telescope that would dedicate its entire time to SETI. This dream could be realized by generous private donations by Paul Allen (co-founder of Microsoft) and Nathan Myhrvold (Chief Technologist of Microsoft) that allowed the construction of a new instrument called the *Allen Telescope Array*. This telescope of which the first test runs are expected in 2005–2006 envisions an array of 350 commercial 6-m radio dishes with liquid air-cooled feeds, built on the grounds of the existing Hat Creek Observatory, located in the Cascades near Lassen Peak, 467 km north of San Francisco, and run by the Radio Astronomy Lab of the University of California, Berkeley and the SETI-Institute. With a collecting area comparable to the Arecibo Observatory this instrument will be simultaneously used for both SETI and radio astronomy research.

Finally, based on the idea that extraterrestrial societies may communicate with light pulses and that with present technology it is possible to produce laser pulses that outshine the visible spectrum of the Sun by 4 orders of magnitude there are several new *optical SETI Projects* at Harvard, Princeton, Berkeley, and elsewhere (see SETI 2005). Using both dedicated and piggy-back instruments, the aim is to detect laser pulses with time scales of a few nanoseconds (billionth of a second) emitted by extraterrestrial societies. The Harvard detectors, for example, were directed at some 13 000 Sun-like stars, and made some 16 000 observations totaling nearly 2400 h during five years of operation (Howard et al. 2004).

However, from the null result in all these intense efforts one must conclude that the attempt to identify extraterrestrial intelligent civilizations in our galactic neighborhood by their radio or light emissions have so far failed. This should not deter us from working hard to improve our technical equipment and search strategies because, as Giuseppe Cocconi and Philip Morrison put it in their 1959 article: "if we never search, the chance of success is zero".

10.6.2 Possible Contact in the not too Distant Future

The current SETI programs can detect 3 kW (a small radio station) at a distance of 100 Ly if its transmission is beamed toward Earth. Therefore, to detect a similar signal from an extraterrestrial civilization at a distance of 1700 Ly would be beyond our present technical capabilities. In addition to the signal strength, there is another important point that must be considered when trying to detect radio transmissions from extraterrestrial civilizations: the age distribution of the emitting societies. From the above estimate, that about 2 million intelligent societies have formed in the past 5 billion years, and assuming that star formation was a continuously ongoing process over all this time, one finds an average interval of 2500 years between the birth of intelligent societies. This birth interval means that the societies that are closest to our present state are either 2500 years more advanced or lag 2500 years behind. In the latter case, they would not yet be able to communicate.

If the next more advanced society, our closest temporal neighbor, also started to communicate with a profusion of radio waves like us, would there be a way detecting it? The difficulty here is the great size of our galaxy. Because the creation of the planet that carries the next society was a chance event, the likelihood that it formed nearby is small. Statistically, one must assume that this next society is located roughly a galactic radius, 50 000 Ly, away from us, and because of the finite speed of light, their first radio signals will not reach us until 47 500 years in the future. This shows that with our present receivers we will not be able to detect the initial radio transmissions from our closest temporal neighbors, but only signals from older more advanced societies that live closer.

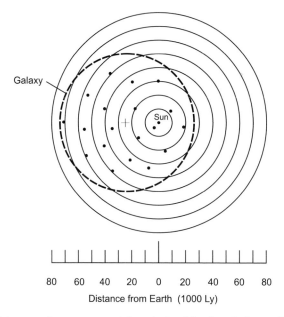

Fig. 10.7. Distances of extraterrestrial societies (*dots*) emitting radio signals. The *dashed circle* marks the galactic disk, the cross its center and the *dot* marked Sun is the position of the Earth. *Solid circles* indicate distances from which travel times of radio waves to the Earth are multiples of 10 000 years

If we assume that as soon as an extraterrestrial society becomes communicating, it starts with an outburst of radio emission then, in principle, we should be able to detect societies that are up to 76 000 years older (see Fig. 10.7). These signals would come to us from the opposite edge of the galaxy and would have spent all this time traveling the 76 000 Ly toward Earth. Figure 10.7 shows our galaxy (dashed) from above the galactic plane and the Earth (dot marked Sun) at a distance of 26 000 Ly from the galactic center (cross). Also shown are circles (solid) that mark positions from which one has the same travel times of radio waves (in multiples of 10 000 years) to Earth. Possible radio-emitting societies somewhere in the galaxy are shown as dots. If intelligent societies are born every 2500 years, with the above mentioned maximum travel time, initial radio bursts should be observable from at most 30 (= 76 000/2500) older societies. Yet this number is probably too high, because the density of stars is greater in the inner parts of the galaxy (see Fig. 10.7) where the star formation rate was higher. A guess is that there may be about 25 societies from which the initial radio outburst could be observable.

But for how long will intelligent civilizations emit radio waves, and would these advanced societies eventually stop this conspicuous emission voluntarily? If we assume that radio emissions continue at least over a thousand years before they are stopped, then the probability of detection until another society commences radio emission is 0.4 (= 1000/2500), which means, that we should presently be able to detect the initial radio emissions of about 10(= 25 × 0.4) extraterrestrial societies from our entire Milky Way, which were roughly in a technological state like ours when the signals were sent off. For shorter radio emission times, we would detect even fewer societies. Yet with such small numbers one always has the chance that because of an unequal distribution in birth times and distances one might have no detection at all. On the other hand, if the extraterrestrials do not hesitate using radio waves for their entire lifetimes, we should detect at least 4000 radio sources from extraterrestrial societies.

As mentioned above, there is also the problem of detecting such relatively weak radio sources at the present time over large galactic distances. Here, luckily, our rapidly advancing radio technology will greatly improve this situation in the near future. One day, with still more advanced technology, we might even detect initial radio transmissions from other galaxies. This indicates that extraterrestrial intelligent civilizations will not be hidden from us forever and that SETI searches will eventually be successful. However, because these radio communications would originate from civilizations that lived many thousands of years ago in our Milky Way, and 2–40 million years ago in nearby galaxies, there will clearly be no chance of a two-way conversation.

10.7 The Fermi Paradox: Where are the Extraterrestrials?

Detecting radio transmissions from intelligent civilizations is one form of establishing contact. Another would be an actual visit by extraterrestrials here on Earth. For the 4000 intelligent high-technology societies that may presently exist, with lifetimes of the order of 10 million years, a visit to Earth should not be impossible. But why is there no historical evidence of such an encounter? Here, an even more formidable problem poses itself: As mentioned above, over the past 5 billion years possibly 2 million intelligent technological societies have formed, of which some may have lived close by. Why do we not find a single artifact left by such extraterrestrial visitors or a single trace that unambiguously proves a visit here on Earth? Despite universal fame and instant fortune awaiting the finder, nobody has ever come forward with verifiable alien evidence collected somewhere or obtained from UFO encounters (Chap. 8). This poses the big problem, "If there are extraterrestrials, where are they?", the so-called *Fermi paradox*, named after the famous physicist and Nobel laureate Enrico Fermi who, in 1950, 10 years before the modern era of SETI, asked this question at a luncheon at Los Alamos (Dick 1996).

There are many answers to explain the Fermi paradox. The simplest is that extraterrestrials do not (or presently not) exist. Less simple ones are that the technical obstacles for a visit are too great, that they are nearby but undetected, or that they are not sufficiently interested in a visit. Let us look at these answers in detail.

10.7.1 They do not Exist

An unspectacular answer to the Fermi paradox is that the precise conditions for intelligent life are so unique that they have occurred only once in our galaxy. A somewhat weaker answer is that presently extraterrestrial intelligent societies do no longer exist. There are two problems with these explanations.

First, in our estimate of the factors for the Drake formula we found that it is very likely that extraterrestrial life exists in our galaxy, but that the greatest uncertainty is the lifetime of intelligent technological societies L (Table 10.1). Yet with our estimate that 2 million intelligent civilizations existed at distances of as little as 210 Ly, even societies with low life expectancies should be able to manage a trip or send an unmanned probe. This suggests that nonexistence, even under extreme assumptions for L, is an unlikely answer to the Fermi paradox.

Second, there is an even more fundamental consideration. To picture mankind and its place in the universe as unique has always turned out to be wrong. This was the case with the idea that the Earth was at the center of the universe. Supported by the powerful authority of Aristotle, this view had been a dogma for 1900 years up to the time of Copernicus in 1530, who established the heliocentric system. But our unique position was also shattered when, in 1924, Edwin Hubble discovered that our galaxy is only one of many in the universe, and in 1927 when Jan Oort found that the Sun orbits around the galactic center. The conviction that we cannot be in a unique place in the universe is now so profoundly established in astronomy that any world model that astrophysicists produce is required to satisfy the so-called *cosmological principle*, which states that the universe must look similar from every point inside it. The idea that mankind is unique in the universe or in our galaxy therefore simply does not fit into modern scientific thought.

10.7.2 Technically, a Visit is not Possible

Another frequently given explanation for the Fermi paradox is that the Milky Way is very large, and because spacecraft must travel well below the speed of light an expedition over great galactic distances is unrealistic because it would take too long. This thinking appears to be valid for low life expectancy civilizations, and is also correct for travel between galaxies, where for example a trip at half the speed of light to the Andromeda nebula would take more than 4 million years.

Aside of severe energy problems with propulsion, the concept of a spacecraft traveling close to the speed of light is not very realistic because of the interstellar material in its flight path. At very high speed such material would become extremely dangerous, as its impact energy would rapidly increase with the speed of the ship. This restricts the travelling speed to values that are low enough, that obstacles can be located and evaded. As already mentioned in Chap. 9, a velocity of 20% of the light speed, $c_L/20$, might be a realistic speed for interstellar travel. Here, distances of a few hundred light years would require travel times of thousands of years. This could pose problems for life forms with short individual life spans like ours, although many different travel schemes have been envisioned for the future, such as sending entire populations on a migrating space colony, or using suspended animation, or even disabling the aging process.

Life spans would not be a problem for unmanned expeditions, which could travel for hundreds and even thousands of years, sending back their reports at light speed using radio waves. As speculated in Chap. 9, interstellar travel might be accomplished with much less difficulty by employing intelligent nonorganic life and miniaturization. For androids, a thousand times smaller than us, the energy, cost, and time requirements would be greatly reduced. It is interesting to speculate that one day interstellar travel at light speed may indeed be possible by first sending a robot factory to the target planet at sub-lightspeed, where it could assemble a body from construction plans sent at the velocity of light and afterwards load the brain with software also sent at light speed. An even wilder speculation is that faster-than-light travel might one day be possible, covering large distances by tunneling through so-called "worm holes" in space-time to other parts of our universe, or even to other universes (Kaku 1995, Al-Khalili 1999).

The extraterrestrials would also have no problem locating us among the 160 billion stars of our galaxy. In Chaps. 4 and 8 we have outlined how our great advances in planet search strategies and instrumentation should enable us, in about a century, to find essentially all the terrestrial planets in our galactic neighborhood where atmospheric H_2O and O_3 absorptions indicate the presence of life. We therefore can expect that, long before their trips to distant stellar systems, information about the presence of life on the target planets would already be known to the extraterrestrials. Moreover, from the age of the parent star, they might even have a rough idea of the expected evolutionary state of the organisms on that planet. We can thus conclude that for societies which are thousands, millions, or billions of years older than us, it would be highly unlikely that technical obstacles would prevent them from visiting Earth.

10.7.3 They are Nearby, but have not been Detected

An intriguing suggestion to answer the Fermi paradox is that extraterrestrial beings are already here, and we have so far not detected them. This very speculative idea appears to be supported by claimed UFO sightings (see Chap. 8), and also by mythological stories. As has already been discussed above, the problem with this explanation is that so far nobody has found a single alien artifact, or produced a scientifically documented recording of an encounter with an extraterrestrial being. If contacts with UFOs are so frequent, why is there always a credibility discussion associated with these events, or a conspiracy legend, without any good reason why such encounters should be suppressed? Another strong argument against the reality of UFOs is the fact that these sightings invariably involve aliens that look quite similar to us and that operate ships and gadgets on a level of contemporary technology, while from our discussion above about the likely nature of advanced extraterrestrials, we can be quite certain that in their present state they would look very different and operate at a technological level far beyond our comprehension. So we must conclude that there presently is no evidence that extraterrestrials are here on Earth.

10.7.4 They are not Interested in Us

This view explains the Fermi paradox as being due to the fact that we are not interesting enough for the attention of extraterrestrial societies, and that therefore there is no reason for them to undertake space travel for a great number of years in order to visit us. Would we be sufficiently interested to go on a space expedition lasting several 1000 years to study single-cell life forms on other Earth-like planets? If not, what could be considered interesting enough for a long space expedition within our galaxy?

We can expect that advanced extraterrestrials would have long ago unraveled how stars function, how planets form and black holes are created. Yet even planets with primitive life, like Earth for over 3 billion years, may not really be worth an extended trip. There would be nothing to learn that could not have already been studied on other closer stellar systems. Here remote observations or visits to their own planetary neighborhood would give them all the information they might need.

This shows that the development of the highest life forms – the primate evolution, the evolution of man, and particularly human history – are the only really interesting objects of study in our galaxy which perhaps merit the effort of a lengthy trip. The very rapid evolution of our society over the past few thousand years cannot be observed from great distances without a consequently large time delay. Compared to the leisurely development of human culture over millions of years starting from *Homo habilis* it is the very rapid technological development in our times, where the transition from organic to nonorganic life might happen, almost a "mental big bang", which should be intensely interesting to extraterrestrial beings. This is even more so because the development of an intelligent society is unrepeatably individualistic, depending on prominent persons and unique historical developments as well as unpredictable systems of thought, but it might also rapidly develop into a distinct final catastrophe. Moreover, as was discussed earlier, this development would occur only very infrequently in our galaxy and at vastly different spatial locations.

For these reasons, the far-fetched speculation that we may presently be under observation by extraterrestrials, makes some sense. But then again the question remains: "Why do these visitors not reveal themselves?"

10.8 The Zoo Hypothesis

A possible answer to this question is the *zoo-hypothesis*. First proposed by John Ball (1973) and Carl Sagan (1973), it explains the lack of contact with extraterrestrial beings by their conscious decision not to reveal themselves. While at first sight this idea appears quite extravagant, it makes considerable sense on closer inspection.

Imagine what it would mean to have contact with an advanced civilization. Our history would suffer an irreversible bend, with drastic and fundamental changes – much more so than the history of the native Americans, annihilated when Columbus and Cortez arrived. It would be a catastrophe, a culture shock, and essentially an irresponsible act on behalf of the extraterrestrials. The individualistic evolution based on our managers, politicians, artists, scientists and leaders in all fields would suddenly be contaminated with elements of a foreign evolution from a very different stage of development. We could never again pursue our own destiny and follow the unique and individualistic expression of life on planet Earth.

It is precisely this lack of discipline that cannot be expected from the extraterrestrials, because if they were capable of such irresponsible acts, they would have never survived thousands, millions, or billions of years without falling victim to the dangers of such behavior. If highly advanced extraterrestrial civilizations – very much older than our own – exist, it could only be because they have learned to act responsibly. From this, the essence of the zoo hypothesis can be deduced: under no circumstances will they disturb or contact us, nor will they allow us to trace them by radio waves, through artifacts, or by direct contact. This also means that there is no possibility that UFOs are extraterrestrial vehicles, because in no way would they reveal themselves.

Fortunately, we will not be kept from contact with extraterrestrial intelligent life forever. As argued above, by detecting their initial radio bursts we will learn about the existence of civilizations close to our own level of development, and as discussed in Chap. 8, we should soon find less developed life forms. Also, one day, man will attempt interstellar missions to search for intelligent societies, living or extinct ones. To discover extinct advanced civilizations would be exciting but also risky, because of their influence on our society. For more advanced living civilizations, on the other hand, it is clear that if they do not want contact here on Earth, they would not want it on their home planets either. Using their superior technology, they would deflect us.

But what about contact with less developed societies? Here the time factor comes into play: not only will considerable time be needed to develop the technology to travel over interstellar distances, but it would also take at least 3600 years to send an unmanned probe at one 20th of light speed to the next life-carrying habitable planet (at an estimated distance of 170 Ly) and receive back its report by radio wave transmission. Over the intervening years, however, it is quite certain that, because of the rapid advances of our knowledge and technology, we will already have been forced to change our society into a responsible civilization in order to avoid the danger of self-destruction. This means that although we can be quite certain that we will discover extraterrestrial intelligent societies in our galaxy and other galaxies, the possibility of ever directly interacting with them is very bleak, because even for a contact with a less advanced society, the basic idea behind the zoo hypothesis, of acting responsibly, will then already have become the guiding principle for our own behavior.

References

Chapter 1

Allègre, C.J., Manhès, G., Göpel, C. 1995: The age of the Earth, Geochim. Cosmochim. Acta 59, 1445

Amelin, Y., Krot, A.N., Hutcheon, I.D., Ulyanov, A.A. 2002: Lead isotopic ages of chondrules and Calcium-Aluminum-rich Inclusions, Science 297, 1678

Bernasconi, P.A. 1996: Grids of pre-main sequence stellar models. The accretion scenario at Z = 0.001 and Z = 0.020, Astron. Astrophys. Suppl. 120, 57

Bloecker, T. 1995: Stellar evolution of low- and intermediate-mass stars. II. Post-AGB evolution, Astron. Astrophys. 299, 755

Bowring, S.A., Williams, I.S. 1999: Priscoan (4.00–4.03 Ga) orthogneisses from northwestern Canada, Contrib. Mineral. Petrol. 134, 3

Bressan, A., Fagotto, F., Bertelli, G., Chiosi, C. 1993: Evolutionary sequences of stellar models with new radiative opacities. II. Z = 0.02, Astron. Astrophys. Suppl. 100, 647

Carney, B.W., de Almeida, M.L.T., Seitzer, P. 2005: Elemental abundance ratios in stars of the outer galactic disk II. Field red giants, Astrophys. J. 130, 1111

Cavosie, A.J., Valley, J.W., Wilde, S.A., E.I.M.F. 2005: Magmatic $\delta^{18}O$ in 4400-3900 Ma detrital zircons: A record of the alteration and recycling of crust in the Early Archean, Earth Planet. Sci. Letters 235, 663

Cox, A.N., Ed. 2000: Allen's Astrophysical Quantities, 4th edn., Berlin Heidelberg: Springer

Ferreras, I., Melchiorri, A., Silk, J., 2001: Setting new constraints on the age of the Universe, Mon. Not. R. Astron. Soc. 327, L47

Gonzalez, G., Brownlee, D., Ward, P. 2001: The galactic habitable zone, galactic chemical evolution, Icarus 152, 185

Harrison, T.M. et al. 2005: Heterogeneous Hadean Hafnium: Evidence of Continental Crust at 4.4 to 4.5 Ga, Science 310, 1947

Hartmann, L. 1998: Accretion Processes in Star Formation, Cambridge: Cambridge University Press

Holleman, A.F., Wiberg, E. 1995: Lehrbuch der Anorganischen Chemie, 101st edn., Berlin: Walter de Gruyter

Jacobsen, S.B. 2005: The Hf-W isotopic system and the origin of the Earth and Moon, Ann. Rev. Earth Planet. Sci. 33, 531

Jenkins, A. et al. 1998: Evolution of structure in cold dark matter universes, Astrophys. J. 499, 20

Kleine, T., Palme, H., Mezger, K., Halliday, A.N. 2005: Hf-W Chronometry of Lunar Metals and the Age and Early Differentiation of the Moon, Science 310, 1671

Lineweaver, C.H. 2001: An estimate of the age distribution of terrestrial planets in the universe: Quantifying metallicity as a selection effect, Icarus 151, 307

Lineweaver, C.H., Fenner, Y., Gibson, B.K. 2004: The galactic habitable zone and the age distribution of complex life in the milky way, Science 303, 59

Salaris, M., Weiss, A., Percival, S.M. 2004: The age of the oldest open clusters, Astron. Astrophys. 414, 163

Sandage, A., Lubin, L.M., VandenBerg, D.A. 2003: The age of the oldest stars in the local galactic disk from Hipparcos parallaxes of G and K subgiants, Publ. Astron. Soc. Pacific 115, 1187

Valley, J.W. 2005: A cool early Earth, Scientific American, Oct., 58

Wilde, S.A., Valley, J.W., Peck, W.H., Graham, C.M. 2001: Evidence from detrital zircons for the existence of continental crust and oceans on the Earth 4.4 Gyr ago, Nature 409, 175

Wootten, H.A. 2002: http://www.cv.nrao.edu/ awootten/allmols.html Mojzsis S.J., Harrison T.M., Pidgeon R.T. 2001: Oxygen-isotope evidence from ancient

Chapter 2

Alibert, Y., Mordasini, C., Benz, W., Winisdoerffer, C. 2005: Models of giant planet formation with migration and disc evolution, Astronomy & Astrophys. 434, 343, see also: Alibert, Y., Mousis, O., Mordasini, C., Benz, W. 2005: New Jupiter and Saturn formation models meet observations, Astrophys. J. 626, L57

Armitage, P.J. 2003: A reduced efficiency of terrestrial planet formation following giant planet migration, Astrophys. J. 582, L47

Baker, J., Bizzarro, M., Wittig, N., Connelly, J., Henning, H. 2005: Early planetesimal melting from an age of 4.5662 Gyr for differentiated meteorites, Nature 436, 1127

Briceño, C. et al. 2001: The CIDA-QUEST large scale survey of Orion OB1: Evidence for rapid disk dissipation in a dispersed stellar population, Science 291, 93

Comets 2005: http://cfa-www.harvard.edu/iau/Ephemerides/Comets/
http://cfa-www.harvard.edu/icq/ICQComref.html
http://www.johnstonsarchive.net/astro/sslist.html

Delsemme, A. 1997: The origin of the atmosphere and of the oceans, in: Comets and the Origin and Evolution of Life, P.J. Thomas, C.F. Chyba, C.P. McKay, Eds., Berlin Heidelberg: Springer, p. 29

Feigelson, E.D., Garmire, G.P., Pravdo, S.H. 2002: Magnetic Flaring in the Pre-Main-Sequence Sun and Implications for the Early Solar System, Astrophys. J. 572, 335

Haisch, K.E., Lada, E.A., Lada C.J. 2001: Disk frequencies and lifetimes of young clusters, Astrophys. J. 553, L153

Hughes, D. 1992: Where planets boldly grow, New Scientist, Dec. 12, 29

Ida, S., Lin, D.N.C. 2004: Toward a deterministic model of planetary formation. I. A desert in the mass and semimajor axis distributions of extrasolar planets, Astrophys. J. 604, 388

Ida, S., Lin, D.N.C. 2005: Toward a deterministic model of planetary formation. III. Mass distribution of short-period planets around stars of various masses, Astrophys. J. 626, 1045

Jacobsen, S.B. 2005: See Chapter 1

Kitamura, Y., Saito, M., Kawabe, R., Sunada, K. 1997: NMA imaging of envelopes and disks around low mass protostars and T-Tauri stars, IAU Symp. 182, 381

Kowal, C.T. 1988: Asteroids, their Nature and Utilization, Chichester: Ellis Horwood

Lissauer, J.J. 1993: Planet formation, Ann. Rev. Astr. Astrophys. 31, 129

Miura, H., Nakamoto, T. 2005: A shock wave heating model for chondrule formation

Minor Planets 2005: Minor Planet Center at the Harvard-Smithsonian Center for Astrophysics, Cambridge MA, European Asteroid Research Node (EARN) and other archives:

http://cfa-www.harvard.edu/iau/mpc.html

http://earn.dlr.de/

http://cfa-www.harvard.edu/iau/lists/MPLists.html

http://earn.dlr.de/nea/

http://www.ipa.nw.ru/PAGE/DEPFUND/LSBSS/englenam.htm

http://neo.jpl.nasa.gov/stats/

http://en.wikipedia.org/wiki/Near-Earth_object

http://en.wikipedia.org/wiki/Near-Earth_asteroid

Murray, N., Hansen, B., Holman, M., Tremaine, S. 1998: Migrating planets, Science 279, 69

Nakamoto, T., Hayashi, M.R., Kita, N.T., Tachibana, S. 2005: Generation of chondrule forming shock waves in solar nebula by x-ray flares, 36th Annual Lunar and Planetary Science Conference, 1256

Palme, H., Jones, A. 2005: Solar system abundances of the elements, in: Meteorites, Comets, and Planets, Treatise in Geochemistry Vol 1, A.M. Davies Ed., Elsevier, London, p. 41

Pollack, J.B. et al. 1996: Formation of giant planets by concurrent accretion

Rodgers, S.D., Charnley, S.B. 2002: A model of the chemistry in cometary comae: deuterated molecules, Monthly Not. Roy. Astr. Soc. 330, 660

Schneider, J. 2005: Extra-solar Planets Catalog:

http://www.obspm.fr/encycl/catalog.html

http://cfa-www.harvard.edu/planets/

Shu, F.H., Shang, H., Lee, T. 1996: Toward an astrophysical theory of chondrites

Trilling, D.E. et al. 1998: Orbital evolution and migration of giant planets: Modeling extrasolar planets, Astrophys. J. 500, 428

Trieloff, M., Palme, H. 2005: The origin of solids in the early solar system, in: Planet formation – theory, observations, and experiments, H. Klahr, W. Brandner Eds., Cambridge University Press, in press

Wetherill, G.W. 1986: Accumulation of the terrestrial planets and implications concerning lunar origin, in: Origin of the Moon, W.K. Hartmann, R.J. Phillips, G.J. Taylor, Eds., Houston: Lunar Planet. Inst., p. 519. See also Weissman, P.R. 1989: The impact history of the solar system: Implications for the origin of atmospheres, in: Origin and Evolution of Planetary and Satellite Atmospheres, S.K. Atreya, J.B. Pollack, M.S. Matthews, Eds., Tucson: University of Arizona Press, p. 230

Wetherill, G.W. 1990: Formation of the Earth, Ann. Rev. Earth Planet. Sci. 18, 205

Chapter 3

Agee, C.B. 2004: Earth science: Hot metal, Nature 429, 33

Agnor, C.B., Canup, R.M., Levison, H.F. 1999: On the Character and Consequences of Large Impacts in the Late Stage of Terrestrial Planet Formation, Icarus 142, 219

Armstrong, R.L. 1981: Radiogenic isotopes: The case for crustal recycling on a near-steady-state no-continental-growth Earth, Phil. Trans. Roy. Soc. London, A301, 443

Benz, W., Cameron, A.G.W., Melosh, H.J. 1989: The origin of the Moon and the single-impact hypothesis III, Icarus 81, 113. See also Cameron, A.G.W. 1997: The origin of the Moon and the single-impact hypothesis V, Icarus 126, 126

Best, M.G. 2003: Igneous and metamorphic petrology, Malden MA, Oxford UK: Blackwell Science

Brandes, J.A. et al. 1998: Abiotic nitrogen reduction on the early Earth, Nature 395, 365

Canup, R.M. 2004: Simulations of a late lunar-forming impact, Icarus 168, 433, see also: Palme, H. 2004: The giant impact formation of the Moon, Science 204, 977

Carr, M.H., Saunders, R.S., Strom, R.G., Wilhelms, D.E. 1984: The Geology of the Terrestrial Planets, NASA SP-469. See also Weissman, P.R. 1989: The impact history of the solar system: Implications for the origin of atmospheres, in: Origin and Evolution of Planetary and Satellite Atmospheres, S.K. Atreya, J.B. Pollack, M.S. Matthews, Eds., Tucson: University of Arizona Press, p. 230

Cavosie, A.J., Valley, J.W., Wilde, S.A., E.I.M.F. 2005: see Chapter 1

Cohen, B.A., Swindle, T.D., Kring, D.A. 2000: Support for the lunar cataclysm hypothesis from lunar meteorite impact melt ages, Science 290, 1754. See also Kerr, R.A. 2000: Beating up on a young Earth, and possibly life, Science 290, 1677

Deming, D. 2002: Origin of the oceans and continents: A unified theory of the Earth, International Geology Review 44, 137

Garnero, E.J. 2004: A new paradigm for Earth's core-mantle boundary, Science 304, 834

Harrison, T.M. et al. 2005: see Chapter 1

Hart, S.R., Zindler, A. 1986: In search of a bulk-earth composition, Chemical Geology 57, 247

Hartmann, W.K., Ryder, G., Dones, L., Greenspon, D. 2000: The time-dependent intense bombardment of the primordial Earth/Moon system, in: Origin of the Earth and Moon, Canup, R.M., Righter, K., eds., Tucson: Univ. of Arizona Press, p. 493

Helffrich, G.R., Wood, B.J. 2001: The Earth's mantle, Nature 412, 501

Ida, S., Canup, R.M., Stewart, G.R. 1997: Lunar accretion from an impact-generated disk, Nature 389, 353

Jacobsen, S.B. 2005: See Chapter 1

Kasting, J.F. 1993: Earth's early atmosphere, Science 259, 920

Kasting, J.F., Brown, L.L. 1998 (repr. 2000): The early atmosphere as a source of biogenic compounds, in: The Molecular Origins of Life. Assembling Pieces of the Puzzle, A. Brack, Ed., Cambridge: Cambridge University Press, p. 35

Kleine, T., Palme, H., Mezger, K., Halliday, A.N. 2005: see Chapter 1

Lunine, J.I. 1999: Earth, Evolution of a Habitable World, Cambridge: Cambridge University Press, Chap. 16.

McDonough, W. 2003: Compositional models for the core, in: The mantle and core, Treatise on geochemistry Vol. 2, R.W. Carlson Ed., Elsevier, London

Miller, S.L. 1998 (repr. 2000): The endogenous synthesis of organic compounds, in: The Molecular Origins of Life, A. Brack, Ed., Cambridge: Cambridge University Press, p. 59

Montelli, R. et al. 2004: Finite-frequency tomography reveals a variety of plumes in the mantle, Science 303, 338

Müller R.D. et al. 1997: Digital isochrons of the world's ocean floor, Journal Geophys. Res. Solid Earth 102, 3211

Nisbet, E.G., Sleep, N.H. 2001: The habitat and nature of early life, Nature 409, 1083

Palme, H., O'Neill, H.St.C. 2003: Cosmochemical estimates of bulk composition, in: The mantle and core, Treatise on geochemistry Vol. 2, R.W. Carlson Ed., Elsevier, London

Press, F., Siever, R., Grotzinger, J., Jordan, T.H. 2004: Understanding Earth 4th edn., W.H. Freeman and Co., New York

Regenauer-Lieb, K., Yuen, D.A., Branlund, J. 2001: The initiation of subduction: Criticality by addition of water?, Science 294, 578

Rowley, D.B., Currie, B.S. 2006: Palaeo-altimetry of the late Eocene to Miocene Lunpola basin, central Tibet, Nature 439, 677

Sandwell, D.T., Schubert, G. 1992: Flexural ridges, trenches, and outer rises around coronae on Venus, Journal Geophys. Res. 97, no. E10, p. 16069

Skinner, B.J., Porter, S.C., Park, J. 2004: Dynamic Earth, An introduction to physical geology 5th edn., John Wiley, Somerset NJ, USA

Scotese, C.R. 2002: http://www.scotese.com/earth.htm

Strom, R.G. et al. 2005: The origin of planetary impactors in the inner solar system, Science 309, 1847

Taylor, S.R., McLennan, S.M. 1995: The geochemical evolution of the continental crust, Reviews of Geophys. 33, 241

Valley, J.W. 2005: see Chapter 1

Valley, J.W., Peck, W.H., King, E.M., Wilde, S.A. 2002: A cool early Earth, Geology 30, 351

van der Hilst, R.D., Widiyantoro, S., Engdahl, E.R. 1997: Evidence for deep mantle circulation from global tomography, Nature 386, 578

van Thienen, P., Vlaar, N.J., van den Berg, A.P. 2005: Assessment of the cooling capacity of plate tectonics and flood volcanism in the evolution of Earth, Mars and Venus, Physics Earth Planetary Interiors 150, 287

Walzer, U., Hendel, R., Baumgardner, J. 2003: Variation of non-dimensional numbers and a thermal evolution model of the Earth's mantle, in: High performance computing in science and engineering '02, Springer Verlag, Berlin Heidelberg, p. 89
http://www2.uni-jena.de/chemie/geowiss/geodyn/poster2.html

Walzer, U., Hendel, R., Baumgardner, J. 2004: The effects of a variation of the radial viscosity profile on mantle evolution, Tectonophysics 384, 55

White, W.M. 2003: Geochemistry. An Online Textbook, Cornell Univ.
http://www.geo.cornell.edu/geology/classes/Chapters/Chapter11.pdf

Wilde, S.A., Valley, J.W., Peck, W.H., Graham, C.M. 2001: see Chapter 1

Zahnle, K.J., Sleep, N.H. 1997: Impacts and the early evolution of life, in: Comets and the Origin and Evolution of Life, P.J. Thomas, C.F. Chyba, C.P. McKay, Eds., Berlin Heidelberg: Springer, p. 175

Chapter 4

ALMA 2005: ALMA observatory:
   http://www.mma.nrao.edu/info/
Beaulieu, J.-P. et al. 2006: Discovery of a cool planet of 5.5 Earth masses through gravitational microlensing, Nature 439, 437, see also: Queloz, D. 2006: Light through a gravitational lens, Nature 439, 400
CNES missions 2005: COROT:
   http://smsc.cnes.fr/COROT/ and http://corot.oamp.fr
   http://sci.esa.int/
ESA missions 2005:
   http://sci.esa.int/
Marcy, G.W., Butler, R.P. 1998: Detection of extrasolar giant planets, Ann. Rev. Astr. Astrophys. 36, 57
NASA missions 2005: Origins, Kepler, TPF, LF, PI:
   http://origins.jpl.nasa.gov/missions/missions.html#2005
   http://planetquest.jpl.nasa.gov/Kepler/kepler_index.html
   http://www.kepler.arc.nasa.gov/index.html
   http://planetquest.jpl.nasa.gov/TPF/tpf_index.html
   http://tpf.jpl.nasa.gov/
Sargent, A.I., Beckwith, S.V.W. 1993: The search for forming planetary systems, Physics Today, Apr., 22. See also Beckwith, S.V.W., Sargent, A.I. 1996: Circumstellar disks and the search for neighbouring planetary systems, Nature 383, 139
Schneider, J. 2005: see Chapter 2
Udalski, A. et al. 2005: A jovian mass planet in microlensing event OGLE-2005-BLG-071, Astrophys. J. 628, L109

Chapter 5

Allen, C.W. 1973: Astrophysical Quantities, 3rd edn., London: Athlone Press
Bodiselitsch, B., Koeberl, C., Master, S., Reimold, W.U. 2005: Estimating duration and intensity of neoproterozoic snowball glaciations from Ir anomalies, Science 308, 239
Boynton, W.V. et al. 2002: Distribution of hydrogen in the near surface of Mars: Evidence for subsurface ice deposits, Science 297, 81
Bressan, A., Fagotto, F., Bertelli, G., Chiosi, C. 1993: see Chapter 1
Cameron, A.G.W. 1963: Communicating with intelligent life on other worlds, Sky & Telescope 26, 258. See also Goldsmith, D. 1980: The Quest for Extraterrestrial Life, Mill Valley CA: University Science Books, p. 132
Clefs CEA 49, 2004: Le soleil a rendez-vous avec la terre, in: Le soleil et la terre, http://www.cea.fr/fr/Publications/clefs2.asp?id=49
Corsetti, F.A., Awramik, S.M., Pierce, D. 2003: A complex microbiota from snowball Earth times: Microfossils from the neoproterozoic Kingston Peak Formation, Death Valley, USA

Goldsmith, D., Owen, T. 1993: The Search for Life in the Universe, 2nd edn., Reading, MA: Addison-Wesley

Hart, M.H. 1978: The Evolution of the atmosphere of the Earth, Icarus 33, 23

Hoffman, P.F., Schrag, D.P. 2002: The snowball Earth hypothesis: testing the limits of global change, Terra Nova 14, 129, see also: Hoffman, P.F., Kaufman, A.J., Halverson, G.P., Schrag, D.P. 1998: A Neoproterozoic snowball Earth, Science 281, 1342

Kasting, J.F., Whitmire, D.P., Reynolds, R.T. 1993: Habitable zones around main sequence stars, Icarus 101, 108

Kopp, R.E., Kirschvink, J.L., Hilburn, I.A., Nash, C.Z. 2005: The Paleoproterozoic snowball Earth: A climate disaster triggered by the evolution of oxygenic photosynthesis, Proc. Natl. Acad. Sci. USA 102, 11131

Landolt-Börnstein, 1982: New Series, K.-H. Hellwege, Ed., Group VI, Vol. 2b, Berlin Heidelberg: Springer, pp. 31, 453

Laskar, J., Robutel, P. 1993: The chaotic obliquity of the planets, Nature 361, 608

Mars missions 2005: see Chapter 8

Méra, D., Chabrier, G., Schaeffer, R. 1998: Towards a consistent model of the Galaxy. II. Derivation of the model, Astron. Astrophys. 330, 953

Narbonne, G.M. 2005: The Edicara biota: Neoproterozoic Origin of animals and their ecosystems, Ann. Rev. Earth Planetary Sci. 33, 421

Olcott, A.N., et al. 2005: Biomarker evidence for photosynthesis during neoproterozoic glaciation, Science Express, 29. Sept., see also: Science 309, 2127

Pierrehumbert, R.T. 2004: High levels of atmospheric carbon dioxide necessary for the termination of global glaciation, Nature 429, 646

Rood, R.T., Trefil, J.S. 1981: Are we Alone? The Possibility of Extraterrestrial Civilizations, New York: Charles Scribner's Sons

Sagan, C. 1963: Direct contact among galactic civilizations by relativistic interstellar spaceflight, Planet. Space Sci. 11, 485. See also Goldsmith, D. 1980: The Quest for Extraterrestrial Life, Mill Valley CA: University Science Books, p. 205

Trainer, M.G. et al. 1994: Haze aerosols in the atmosphere of early Earth: Manna from heaven, Astrobiology, 4, 409

Ulmschneider, P. 2002: Intelligent Life in the Universe, From Common Origins to the Future of Humanity, 1st edn., Berlin Heidelberg: Springer

Williams, D.M. 1998: The stability of habitable planetary environments, Ph.D. thesis, Pennsylvania State University

Chapter 6

Alberts, B., Bray, D., Lewis, J., Raff, M., Roberts, K., Watson, J.D. 1994: Molecular Biology of the Cell, 3rd edn., New York London: Garland

Barghoorn, E.S. 1971: The oldest fossils, Scientific American, May, 30

Bertone, P. et al. 2004: Global identification of human transcribed sequences with genome tiling arrays, Science 306, 2160

Botta, O. 2004: The chemistry of the origins of life, in: Astrobiology: Future perspectives, Astrophysics and Space Sci. Library 305, P. Ehrenfreund et al. Eds., Dordrecht, Boston, London: Kluwer, p. 359

Cairns-Smith, A.G. 1982: Genetic Takeover and the Mineral Origins of Life, Cambridge: Cambridge University Press

Claverie, J.-M. 2005: Fewer genes, more noncoding RNA, Science 309, 1529

Dauphas, N. et al. 2004: Clues from Fe isotope variations on the origin of Early Archean BIFs from Greenland, Science 306, 2077

de Duve, C. 1991: Blueprint for a Cell: The Nature and Origin of Life, Burlington, NC: Neil Patterson, Carolina Biol. Supply Co.

de Duve, C. 1998 (repr. 2000): Clues from present-day biology: the thioester world, in: The Molecular Origins of Life, A. Brack, Ed., Cambridge: Cambridge University Press, p. 219

Doolittle, W.F. 1999: Phylogenetic classification and the universal tree, Science 284, 2124

Eigen, M., Winkler-Oswatitsch, R. 1981: Transfer RNA, an early gene?, Naturwiss. 68, 282

Ferris, J.P. 1998 (repr. 2000): Catalyzed RNA synthesis for the RNA world, in: The Molecular Origins of Life, A. Brack, Ed., Cambridge: Cambridge University Press, p. 255

Gavin, A.-C. et al. 2002: Functional organization of the yeast proteome by systematic analysis of protein complexes, Nature 415, 141. See also Kumar, A., Snyder, M. 2002: Protein complexes take the bait, Nature 415, 123

Giovannoni, S.J. et al. 2005: Genome streamlining in a cosmopolitan oceanic bacterium, Science 309, 1242

GOLD-EBI 2005: Genome Online Database, EMBL-European Bioinformatics institute and TIGR Database:
http://www.genomesonline.org/
http://www.ebi.ac.uk/genomes/
http://www.tigr.org/tdb/mdb/mdbcomplete.html

Green, N.P.O, Stout, G.W., Taylor, D.J., Soper, R. 1993: Biological Science 1. Organisms, Energy and Environment, 2nd edn., Cambridge: Cambridge University Press

Hart, H., Hart, D.J., Craine, L.E. 1995: Organic Chemistry, a Short Course, 9th edn., Boston: Houghton Mifflin

Hazen, R.M. 2001: Life's rocky start, Scientific American, Apr., 63

Holmes, B. 2005: Alive! The race to create life from scratch, New Scientist 12. Feb., 28

HPRD 2005: Human Protein Reference Database,
http://www.hprd.org/

Hutchinson III, C.A. et al. 1999: Global transposon mutagenesis and a minimal mycoplasma genome, Science 286, 2165

IHGSC 2004: International Human Genome Sequencing Consortium, Nature 431, 931

Kasting, J.F. 1993: see Chapter 3

Koonin, E.V. 2003: Comparative genomics, minimal gene-sets and the Last Universal Common Ancestor, Nature Reviews Microbiology 1, 127

Kunin, V. et al. 2005: The net of life: reconstructing the microbial phylogenetic network, Genome Research 15, 954

Johnston, W.K. et al. 2001: RNA-catalyzed RNA polymerization: Accurate and general RNA-templated primer extension, Science 292, 1319

Martin, W., Russell, M.J. 2003: On the origins of cells: a hypothesis for the evolutionary transitions from abiotic geochemistry to chemoautotrophic prokaryotes, and from prokaryotes to nucleated cells, Phil. Trans. R. Soc. Lond. B, 358, 59 see also: http://en.wikipedia.org/wiki/Origin_of_life

Mellersh, A. 1993: Origins of life and evolution of the biosphere 23, 261. See also Franklin, C. 1993: Did life have a simple start?, New Scientist, Oct. 2, 13

Miller, S.L. 1998: see Chapter 3

Miller, S.L., Orgel, L.E. 1974: The Origins of Life on the Earth, Englewood Cliffs, NJ: Prentice Hall

Mojzsis, S.J. et al. 1996: Evidence for life on Earth before 3,800 million years ago, Nature 384, 55

Moorbath, S. 2005: Palaeobiology: Dating earliest life, Nature 434, 155

Mushegian, A.R., Koonin, E.V. 1996: A minimum gene set for cellular life derived by comparison of complete bacterial genomes, Proc. Natl. Acad. Sci. USA 93, 10268

Orgel, L.E. 1998: The origin of life − a review of facts and speculations, Trends Biochem. Sci. 23, 491

Orgel, L.E. 2000: Self-organizing biochemical cycles, Proc. Natl. Acad. Sci. USA 97, 12503

Rasmussen, B. 2000: Filamentous microfossils in a 3,235-million-year-old volcanogenic massive sulphide deposit, Nature 405, 676

Robertson, M.P., Miller, S.L. 1995: An efficient prebiotic synthesis of cytosine and uracil, Nature 375, 772

Rosing, M.T. 1999: ^{13}C-Depleted carbon microparticles in >3700-Ma sea-floor sedimentary rocks from west Greenland, Science 283, 674

Rosing, M.T., Frei, R. 2004: U-rich archaean sea-floor sediments from Greenland – indications of >3700-Ma oxygenic photosynthesis, Earth Planetary Science Letters 217, 237

Schidlowski, M. 1988: A 3,800-million-year isotopic record of life from carbon in sedimentary rocks, Nature 333, 313

Schopf, J.W. 1992: The oldest fossils and what they mean, in: Major Events in the History of Life, J.W. Schopf, Ed., Boston: Jones and Bartlett, p. 29. See also Jakosky, B. 1998: The Search for Life on Other Planets, Cambridge: Cambridge University Press, p. 69

Schopf, J.W. et al. 2002: Raman-laser imagery of Earth's earliest fossils, Nature 416, 73

Schulze-Makuch, D., Irwin, L.N. 2004: Life in the universe, Expectations and constraints, Berlin Heidelberg: Springer

Schwartz, A.W. 1998 (repr. 2000): Origins of the RNA world, in: The Molecular Origins of Life, A. Brack, Ed., Cambridge: Cambridge University Press, p. 237

Smith, J.V. et al. 1999: Biochemical evolution III: Polymerization on organophilic silica-rich surfaces, crystal-chemical modeling, formation of first cells, and geological clues, Proc. Natl. Acad. Sci. USA 96, 3479

Tian, F., Toon, O.B., Pavlov, A.A., De Sterck, H. 2005: A hydrogen-rich early Earth atmosphere, Science 308, 1014

Tomita, M. 2001: Whole-cell simulation: a grand challenge of the 21st century, Trends in Biotechnology 19, 205. See also Periwal, V., Szallasi, Z. 2002: Trading "wet-work" for network, Nature Biotechnology 20, 345, and Endy, D., Brent, R. 2001: Modelling cellular behaviour, Nature 409, 391, as well as Tomita, M. et al. 1999: E-CELL: software environment for whole-cell simulation, Bioinformatics 15, 72

Tomita, M. 2005: E-Cell project, http://www.e-cell.org/

Wächtershäuser, G. 1988: Before enzymes and templates: Theory of surface metabolism, Microbiol. Rev. 52, 452

Wächtershäuser, G. 1998 (repr. 2000): Origin of life in an iron−sulfur world, in: The Molecular Origins of Life, A. Brack, Ed., Cambridge: Cambridge University Press, p. 206

Westall, F. 2004: Early life on Earth: The ancient fossil record, in: Astrobiology: Future perspectives, Astrophysics and Space Sci. Library 305, P. Ehrenfreund et al. Eds., Dordrecht, Boston, London: Kluwer, p. 287

Westphal, S.P. 2003: Ultra-long DNA is a geneticist's dream, New Scientist 22. Nov., 8

Whitehouse, M.J., Kamber, B.S., Fedo, C.M., Lepland, A. 2005: Integrated Pb- and S-isotope investigation of sulphide minerals from the early Archaean of southwest Greenland, Chemical Geology 222, 112

Woese, C.R. 1987: Bacterial evolution, Microbiol. Rev. 51, 221

Woese, C.R. 1998: The universal ancestor, Proc. Natl. Acad. Sci. USA 95, 6854

Chapter 7

Ambrose, S.H. 2001: Paleolithic technology and human evolution, Science 291, 1748

Arendt, D., Wittbrodt, J. 2001: Reconstructing the eyes of urbilateria, Philos. Trans. Roy. Soc. B, 346, 1545, see also: http://en.wikipedia.org/wiki/Eye

Balter, M. 2001: In search for the first Europeans, Science 291, 1722

Becker, L. et al. 2004: Bedout: A possible end-Permian impact crater offshore of northwestern Australia, Science 304, 1469, see also: Science 306, pages 609−613

Bekker, A. et al. 2004: Dating the rise of atmospheric oxygen, Nature 427, 117

Bennett, J., Shostak, S., Jakosky, B. 2003: Life in the universe, San Francisco, Boston, New York: Addison Wesley

Brain, C.K., Sillen, A. 1988: Evidence from the Swartkrans cave for the earliest use of fire, Nature 336, 464

Brunet, M. et al. 2002: A new hominid from the Upper Miocene of Chad, Central Africa, Nature 418, 145. See also Wood, B. 2002: Hominid revelations from Chad, Nature 418, 133

Campbell, N.A. 1996: Biology, 4th edn., Menlo Park CA: Benjamin/Cummings

Canfield, D.E., 2005: The early history of atmospheric oxygen, Ann. Rev. Earth Planet. Sci. 33, 1

Claeys, P. 1996: Chicxulub, le cratère idéal. Le chaînon manquant identifié dans le golfe du Mexique, La Recherche, Dec., 60. See also Morgan J. et al. 1997: Size and morphology of the Chicxulub impact crater, Nature 390, 472

Clark, J.A. 2004: From fins to fingers, Science 304, 57

Condie, K.C., Sloan, R.E. 1998: Origin and Evolution of Earth, Principles of Historical Geology, Upper Saddle River, NJ: Prentice Hall

Conway Morris, S. 1999: The Crucible of Creation. The Burgess Shale and the Rise of Animals, Oxford: Oxford University Press

de Lumley, H. 2001: http://www.culture.gouv.fr/culture/arcnat/tautavel/en/homme.htm

de Duve, C. 1996: The birth of complex cells, Scientific American, Apr., 38

Emery, N.J., Clayton, N.S. 2004: The mentality of crows: convergent evolution of intelligence in corvids and apes, Science 306, 1903

Falkowski, P.G. et al. 2005: The rise of oxygen over the past 205 million years and the evolution of large placental mammals, Science 309, 2202

Foley, J. 2005: Fossil hominids, the evidence for human evolution,
http://www.talkorigins.org/faqs/homs/index.html
http://www.talkorigins.org/faqs/homs/recent.html

Gavin, A.-C. et al. 2002: see Chapter 6

Gilbert, S.F. 1997: Developmental Biology, 5th edn., Sunderland, MA: Sinauer Associates, p. 18

Gore, R. 1997: The dawn of humans, expanding worlds, Natl. Geogr. 191, (May) 84

Gould, S.J. 1989: Wonderful Life: the Burgess Shale and the Nature of History, New York: Norton

Graham, J.B., Aguilar, N.M., Dudley, R., Gans, C. 1995: Implications of the late Palaeozoic oxygen pulse for physiology and evolution, Nature 375, 117

Haile-Selassie, Y. 2001: Late Miocene hominids from the Middle Awash, Ethiopia, Nature 412, 178. See also WoldeGabriel, G. et al. 2001: Geology and palaeontology of the Late Miocene Middle Awash valley, Afar rift, Ethiopia, Nature 412, 175, and: Haile-Selassie, Y., Suwa, G., White, T.D. 2004: Late Miocene teeth from Middle Awash, Ethiopia, and early hominid dental evolution, Science 303, 1503

Holden, C. 1998: No last word on language origins, Science 282, 1455

HOX 2005: Master genes, Homeobox (HOX) genes, Popovici, C. et al. 2001: Homeobox gene clusters and the human paralogy map, FEBS Letters 491, 237, see also: Luke, G.N. et al. 2003: Dispersal of NK homeobox gene clusters in amphioxus and humans, Proc. Natl. Acad. Sci. USA 100, 5292
http://en.wikipedia.org/wiki/Homeobox
http://www.hhmi.org/genesweshare/b120.html
http://www.dnaftb.org/dnaftb/37/concept/index.html

Kaplan, R.W. 1972: Der Ursprung des Lebens, Stuttgart: Georg Thieme

Kelley, J. 1992: Evolution of apes, in: The Cambridge Encyclopedia of Human Evolution, S. Jones, R. Martin, D. Pilbeam, Eds., Cambridge: Cambridge University Press, p. 223

Kenrick, P. 2003: Fishing for the first plants, Nature 425, 248

Kenrick, P., Davis, P. 2004: Fossil Plants, Washington: Smithsonian Books

Kerp, H. 2005: The Rhynie Chert and its flora,
http://www.uni-muenster.de/GeoPalaeontologie/Palaeo/
Palbot/erhynie.html

Leakey, M.D. 1979: Footprints in the ashes of time, Natl. Geogr. 155, (Apr.) 446. See also Leakey, M.D., Hay, R.L. 1979: Pliocene footprints in the Laetolil beds at Laetoli, northern Tanzania, Nature 278, 317

Lewin, R. 1989: Human Evolution. An illustrated introduction, 2nd edn., Cambridge, MA: Blackwell Sci.

Martin, R.D. 1990: Primate Origins and Evolution. A Phylogenetic Reconstruction, London: Chapman and Hall

Mayr, E. 1988: Toward a New Philosophy of Biology. Observations of an Evolutionist, Cambridge, MA: The Belknap Press of Harvard University Press

Mayr, E. 2000: Darwin's influence on modern thought, Scientific American, July, 67

McDougall, I., Brown, F.H., Fleagle, J.G. 2005: Stratigraphic placement and age of modern humans from Kibish, Ethiopia, Nature 433, 733

Mellars, P. 2004: Neanderthals and the modern human colonization of Europe, Nature 432, 461

Mojzsis, S.J. et al. 1996: see Chapter 6

Moyà-Solà, S., et al. 2004: Pierolapithecus catalaunicus, a new middle Miocene great ape from Spain, Science 306, 1339

Mundil, R., Ludwig, K.R., Metcalfe, I., Renne, P.R. 2004: Age and timing of the Permian mass extinctions: U/Pb dating of closed-system zircons, Science 305, 1760, see also: Bowring, S.A. et al. 1998: U/Pb zircon geochronology and tempo of the end-Permian mass extinction, Science 280, 1039

Norell, M.A., Xu, X. 2005: Feathered Dinosaurs, Ann. Rev. Earth Planet. Sci. 33, 277

Novikoff, A.B., Holtzman, E. 1976: Cells and organelles, 2nd edn., New York: Holt, Rinehart and Winston, p. 334

Padian, K., Chiappe, L.M. 1998: The origin of birds and their flight, Scientific American, Feb., 28

Pringle, H. 1998: The slow birth of agriculture, Science 282, 1446

Raven, P.H., Evert, R.F., Eichhorn, S.E. 2003: Biology of Plants, 6th edn., New York: W.H. Freeman

Rhode, R.A. 2005: Geowhen database of the stratigraphy organization, http://www.stratigraphy.org/geowhen/timelinestages.html

Rocchia, R. 1996: Naissance d'une théorie. D'une anomalie en iridium à la catastrophe cosmique, La Recherche, Dec., 53

Romer, A.S. 1974: Vertebrate Paleontology, 3rd edn., Chicago: University Chicago Press

Rouxel, O.J., Bekker, A., Edwards, K.J. 2005: Iron isotope constraints on the Archean and Paleoproterozoic ocean redox state, Science 307, 1088

Schaarschmidt, F. 1968: Paläobotanik I, Mannheim: Bibliographisches Institut

Serre, D. et al. 2004: No evidence of Neandertal mtDNA contribution to early modern humans, PLoS Biology 2, 313

Sibley, C.G. 1992: DNA-DNA hybridisation in the study of primate evolution, in: The Cambridge Encyclopedia of Human Evolution, S. Jones, R. Martin, D. Pilbeam, Eds., Cambridge: Cambridge University Press, p. 313. See also Pilbeam, D., Human origins and evolution, in: Origins. The Darwin College Lectures, A.C. Fabian, Ed., Cambridge: Cambridge Univ. Press, p. 89

Simons, E. 1992: The fossil history of primates, in: The Cambridge Encyclopedia of Human Evolution, S. Jones, R. Martin, D. Pilbeam, Eds., Cambridge: Cambridge University Press, p. 199

Skelton, P.W., Spicer, R.A., Kelley, S.P., Gilmour, I. 2003: The Cretaceous World Cambridge Univ. Press, Cambridge UK

Smit, J. 1996: Un épisode tragique: L'océan folamour. La brutalité des extinctions marines suggère une réduction soudaine de la luminosité, La Recherche, Dec., 62

Steiper, M.E., Young, N.M., Sukarna, T.Y. 2004: Genomic data support the hominoid slowdown and an early Oligocene estimate for the hominoid-cercopithecoid divergence, Proc. Natl. Acad. Sci. USA 101, 17021

Swisher III, C.C. et al. 1996: Latest Homo erectus of Java: Potential contemporaneity with Homo sapiens in southeast Asia, Science 274, 1870

Tattersall, I. 1997: Out of Africa. Again ··· and again? Scientific American, Apr., 46

Taylor, D.J., Green, N.P.O., Stout, G.W., Soper, R. 1997: Biological Sciences 2. Systems, Maintenance and Change, 3rd edn., Cambridge: Cambridge University Press

Thieme, H. 1997: Lower Palaeolithic hunting spears from Germany, Nature 385, 807

Tomita, M. 2001, 2005: see Chapter 6

Wallace, A.R. 1904: Man's Place in the Universe: A Study of the Results of Scientific Research in Relation to the Unity or Plurality of Worlds, London: Chapman and Hall

Ward, P.D. et al. 2005: Abrupt and gradual extinction among late Permian land vertebrates in the Karoo basin, South Africa, Science 307, 709

Wellman, C.H., Osterloff, P.L., Mohiuddin, U. 2003: Fragments of the earliest land plants, Nature 425, 282

Wood, B. 1994: The oldest hominid yet, Nature 371, 280

Zachos, J. et al. 2001: Trends, rhythms, and aberrations in global climate 65 Ma to present, Science 292, 686

Chapter 8

Angel, J.R.P., Woolf, N.J., 1996: Searching for life on other planets, Scientific American, Apr., 46. See also Beichman, C.A. 1996: A Road Map for the Exploration of Neighboring Planetary Systems, JPL Publ. 96-22, p. 4–6

Baalke, R. 2005: Mars meteorites:
http://www.jpl.nasa.gov/snc/

Boynton, W.V. et al. 2002: see Chapter 5

Buseck, P.R. et al. 2001: Magnetite morphology and life on Mars, Proc. Natl. Acad. Sci. USA 98, 13490

Condon, E.U. 1969: Final Report of the Scientific Study of Unidentified Flying Objects. E.U. Condon, scientific director. D.S. Gillmor, Ed., New York: Dutton. See also Dick, S.J. 1996: The Biological Universe. The Twentieth-Century Extraterrestrial Life Debate and the Limits of Science, Cambridge: Cambridge University Press, p. 290f

Head III, J.W. et al. 1999: Possible ancient oceans on Mars: Evidence from Mars Orbiter Laser Altimeter data, Science 286, 2134

Horowitz, N.H. 1977: The search for life on Mars, Scientific American, Nov., 52

Hynek, J.A. 1972: The UFO Experience: A Scientific Inquiry, Chicago: H. Regnery Co.. See also Hynek, J.A. 1998: The UFO Experience. A Scientific Inquiry, New York: Marlowe and Co.

Kasting, J.F. 1997: Habitable zones around low mass stars and the search for extraterrestrial life, Origins of Life and Evolution of the Biosphere 27, 291

Kasting, J.F., Brown, L.L. 1998: see Chapter 3

Kerr, R.A. 1998: Surveyor shows the flat face of Mars, Science 279, 1634

Lorenz, R. 1997: Death of a watery world, New Scientist, Sept. 20, 34

Lovelock, J.E. 1965: A physical basis for life detection experiments, Nature 207, 568

Lowell, P. 1909: Mars et ses canaux, Paris: Mercure de France

LP missions 2005: Lunar and Planetary Missions:
  http://nssdc.gsfc.nasa.gov/planetary/chrono.html
  http://nssdc.gsfc.nasa.gov/planetary/prop_missions.html
  http://nssdc.gsfc.nasa.gov/planetary/projects.html
  http://sci.esa.int/home/ourmissions/index.cfm
  http://www.nasa.gov/missions/solarsystem/explore_main.html
Mars missions 2005: Past, present and future missions:
  http://mars.jpl.nasa.gov/missions/log/
  http://nssdc.gsfc.nasa.gov/planetary/chrono.html
  http://mars.jpl.nasa.gov/mgs/
  http://marsrovers.jpl.nasa.gov/home/index.html
  http://marsprogram.jpl.nasa.gov/odyssey/
  http://sci.esa.int/marsexpress/
  http://www.esa.int/SPECIALS/Mars_Express/SEMGKA808BE_0.html
  http://www.esa.int/SPECIALS/Mars_Express/SEMVZF77ESD_0.html
  http://mars.jpl.nasa.gov/mro/
McKay, D.S. et al. 1996: Search for past life on Mars: Possible relic biogenic activity
  in Martian meteorite ALH84001, Science 273, 924. See also Kerr, R.A. 1996:
  Ancient life on Mars?, Science 273, 864
McKay, D.S. et al. 1997: No 'nanofossils' in martian meteorite, reply, Nature 390,
  455. See also the contribution on p. 454 by Bradley, J.P., Harvey, R.P., McSween
  Jr, H.Y.
Menzel, D.H. 1972: UFO's-the modern myth, in: UFO's: A Scientific Debate, C.
  Sagan, T. Page, Eds., Ithaca NY: Cornell University Press, repr. 1996, New York:
  Barnes & Noble Books, p. 123
Picardi, G. et al. 2005: Radar soundings of the subsurface of Mars, Science 310,
  1925
Priscu, J. 2001: http://www.homepage.montana.edu/~lkbonney/
Sagan, C. 1977: Reducing greenhouses and the temperature history of Earth and
  Mars, Nature 269, 224. See also Sagan, C. 1977: The long-range martian forecast,
  Sky & Telescope 54, 468
Sagan, C. et al. 1993: A search for life on Earth from the Galileo spacecraft, Science
  365, 715
Smith, D.E. et al. 2001: MOLA Science Team home page:
  http://ltpwww.gsfc.nasa.gov/tharsis/mola.html
Turtle, E.P., Pierazzo, E. 2001: Thickness of a Europan ice shell from impact crater
  simulations, Science 294, 1326

Chapter 9

Asimov, I. 1982: Evidence, in: The Complete Robot, London: Grafton Books, p. 518
Asimov, I. 1991: Foundation's Edge, New York: Bantam Books
Becker, L. et al. 2004: see Chapter 7
Billingham, J., Gilbreath, W., O'Leary, B. 1979: Space Resources and Space Set-
  tlements, NASA SP-428. See also:
  http://lifesci3.arc.nasa.gov/SpaceSettlement/spaceres/index.html
  http://lifesci3.arc.nasa.gov/SpaceSettlement/
  http://www.ssi.org/

Burrows, A. 2000: Supernova explosions in the universe, Nature 403, 727. See also van Paradijs, J. 1999: From gamma-ray bursts to supernovae, Science 286, 693

Bussey, D.B.J. et al. 2005: Planetary science: Constant illumination at the lunar north pole, Nature 434, 842, see also: McConnochie, T.H. et al. 2002: A Search for water ice at the lunar poles with Clementine images, Icarus 156, 335, and: Hodges, R.R. 2002: Reanalysis of Lunar Prospector neutron spectrometer observations over the lunar poles, Journal Geophys. Res. Planets 107, 8

de Lange, T. 1998: Telomeres and senescence: Ending the debate, Science 279, 334

Dragicevich, P.M., Blair, D.G., Burman, R.R. 1999: Why are supernovae in our galaxy so frequent?, Mon. Not. R. Astron. Soc. 302, 693. See also Tammann, G.A., Löffler, W., Schröder, A. 1994: The galactic supernova rate, Astrophys. J. Suppl. 92, 487, and: van den Bergh, S. 1994: Astronomical catastrophes in Earth history, Publ. Astr. Soc. Pacific 106, 689

Gavin, A.-C. et al. 2002: see Chapter 6

Grice, K. et al. 2005: Photic zone euxinia during the Permian-Triassic superanoxic event, Science 307, 706
http://www.space.com/scienceastronomy/helium3_000630.html

Hartmann, W.K., Miller, R., Lee, P. 1984: Out of the Cradle. Exploring the Frontiers beyond Earth, New York: Workman

Hamilton, C.J. 2001: Terrestrial Impact Craters:
http://www.solarviews.com/eng/tercrate.htm

He3-mining 2005: http://www.asi.org/adb/02/09/he3-intro.html
http://fti.neep.wisc.edu/pdf/wcsar9311-2.pdf
http://fti.neep.wisc.edu/pdf/wcsar9201-2.pdf

Henry, T.J. 2002: 100 nearest stars, CHARA, Georgia State University:
http://joy.chara.gsu.edu/RECONS/TOP100.htm
See also Cox A.N., Ed. 2000 see Chapter 1

IMP 2005: Impact crater locations:
http://www.unb.ca/passc/ImpactDatabase/
http://www.solarviews.com/eng/tercrate.htm
http://www.lpi.usra.edu/publications/slidesets/craters.html

ISS 2005: International Space Station,
http://spaceflight.nasa.gov/station/
http://spaceflight.nasa.gov/
http://spaceflight.nasa.gov/gallery/
http://www.esa.int/esaHS/iss.html
http://www.nasa.gov/mission_pages/station/main/index.html
http://science.nasa.gov/temp/StationLoc.html
http://en.wikipedia.org/wiki/International_Space_Station

ITER 2005: International Thermonuclear Experimental Reactor,
http://www.iter.org/index.htm, http://en.wikipedia.org/wiki/ITER

Jaffe, R.L., Busza, W., Wilczek, F., Sandweiss, J. 2000: Review of speculative "disaster scenarios" at RHIC, Rev. Modern Physics 72, 1125, see also: Dar, A., De Rujula, A., Heinz, U. 1999: Will relativistic heavy-ion colliders destroy our planet? Physics Letters B 470, 142

Jin, Y.G. et al. 2000: Pattern of marine mass extinction near the Permian-Triassic boundary in south China, Science 289, 432

Lewis, J.S. 1997: Mining the Sky, Reading MA: Helix Books, Addison-Wesley

LP missions 2005: see Chapter 8

Minor Planets 2005: see Chapter 2

NEA Search 2003: Study to Determine the Feasibility of Extending the Search for Near-Earth Objects to Smaller Limiting Diameters, http://neo.jpl.nasa.gov/neo/neoreport030825.pdf

O'Neill, G.K. 1974: The colonization of space, Physics Today, Sept., 32. See also O'Neill, G.K. 1974: A lagrangian community?, Nature 250, 636

O'Neill, G.K. 1989: The High Frontier. Human Colonies in Space, Princeton: Space Studies Institute Press

Payne, J.L. et al. 2004: Large perturbations of the carbon cycle during recovery from the end-Permian extinction, Science 305, 506

Sepkoski, J.J. 1995: in: Global Events and Event Stratigraphy, O.H. Walliser, Ed., Berlin Heidelberg: Springer, p. 35. See also Marshall, C.R. 1998: Mass extinction probed, Nature 392, 17

Space Policy 2005: US Space policy announcement 2003
http://www.nasa.gov/missions/solarsystem/explore_main.html
http://en.wikipedia.org/wiki/Vision_for_Space_Exploration

Spaceguard Survey 2005: Search to detect dangerous Near-Earth Objects,
http://en.wikipedia.org/wiki/Spaceguard
http://neo.jpl.nasa.gov/stats/
http://earn.dlr.de/nea/
http://impact.arc.nasa.gov/
http://cfa-www.harvard.edu/iau/lists/Dangerous.html
http://spaceguard.dlr.de/default_eng.htm
http://www.spaceguarduk.com/

Tomita, M. 2001, 2005: see Chapter 6

Zeck, G., Fromherz, P. 2001: Noninvasive neuroelectronic interfacing with synaptically connected snail neurons immobilized on a semiconductor chip, Proc. Natl. Acad. Sci. USA 98, 10457

Chapter 10

Al-Khalili, J.S. 1999: Black Holes, Wormholes and Time Machines, Bristol: Institute of Physics Publishing

Ball, J.A. 1973: The zoo hypothesis, Icarus 19, 347. See also Goldsmith, D. 1980: The Quest for Extraterrestrial Life, Mill Valley, CA: University Science Books, p. 241

Cameron, A.G.W. 1963: see Chapter 5

Cocconi, G., Morrison, P. 1959: Searching for interstellar communications, Nature 184, 844. See also Goldsmith, D. 1980: The Quest for Extraterrestrial Life, Mill Valley, CA: University Science Books, p. 102

de Duve, C. 1991: see Chapter 6

Dick, S.J. 1996: The Biological Universe. The Twentieth-Century Extraterrestrial Life Debate and the Limits of Science, Cambridge: Cambridge University Press

Goldsmith, D., Owen, T. 1993: see Chapter 5

Howard, A.W. et al. 2004: Search for nanosecond optical pulses from nearby solar-type stars, Astrophys. J. 613, 1270

Kaku, M. 1995: Hyperspace, a scientific odyssey through parallel universes, time warps, and the tenth dimension, Oxford New York: Oxford University Press

Kardashev, N.S. 1964: Transmission of information by extraterrestrial civilizations, Sov. Astron. 8, 217. See also Goldsmith, D. 1980: The Quest for Extraterrestrial Life, Mill Valley, CA: University Science Books, p. 136

McNamara, D.H., Madsen, J.B., Barnes, J., Ericksen, B.F. 2000: The Distance to the galactic center, Publ. Astr. Soc. Pacific 112, 202

Oliver, B.M. 1977: The rationale for a preferred frequency band: The water hole, in: The Search for Extraterrestrial Intelligence SETI, P. Morrison, J. Billingham, J. Wolfe, Eds., NASA SP 419, 63

Rose, C., Wright, G. 2004: Inscribed matter as an energy efficient means of communication with an extraterrestrial civilization, Nature 431, 47

Rood, R.T., Trefil, J.S. 1981: see Chapter 5

Sagan, C. 1963: see Chapter 5

Sagan, C. 1973: On the detectivity of advanced galactic civilizations, Icarus 19, 350. See also Goldsmith, D. 1980: The Quest for Extraterrestrial Life, Mill Valley, CA: University Science Books, p. 140

SETI 2005: Berkeley, Harvard, SETI@Home, SETI institute homepages:
http://seti.ssl.berkeley.edu/serendip/serendip.html
http://seti.harvard.edu/seti/
http://setiathome.ssl.berkeley.edu/
http://www.seti.org/

Author Index

Agee, C.B.   45, 71, 282
Agnor, C.B.   40, 282
Aguilar, N.M.   289
Al-Khalili, J.S.   274, 294
Alberts, B.   119, 123, 124, 126, 130, 285
Alibert, Y.   24, 26, 280
Allègre, C.J.   17, 279
Allen, C.W.   104, 284
Ambrose, S.H.   191, 288
Amelin, Y.   17, 279
Angel, J.R.P.   291
Angel, R.P.   215
Arendt, D.   195, 288
Aristotle   31, 117, 258, 273
Armitage, P.J.   26, 280
Armstrong, R.L.   70, 282
Asimov, I.   240, 242, 292
Atreya, S.K.   281
Awramik, S.M.   284

Baalke, R.   208, 209, 291
Baker, J.   37, 280
Ball, J.A.   276, 294
Balter, M.   186, 288
Barghoorn, E.S.   136, 285
Barnes, J.   295
Baumgardner, J.   283
Beaulieu, J.-P.   75, 82, 284
Bechman, C.A.   291
Becker, L.   173, 244, 288
Beckwith, S.V.W.   79, 284
Bekker, A.   153, 288, 290
Bennett, J.   150, 288
Benz, W.   40, 280, 282
Bernasconi, P.A.   9, 279
Bertelli, G.   279, 284
Bertone, P.   134, 285

Best, M.G.   63, 282
Billingham, J.   231, 292, 295
Bizzarro, M.   280
Blair, D.G.   293
Bloecker, T.   9, 279
Boccaccio, G.   243
Bode, J.E.   29
Bodiselitsch, B.   108, 284
Botta, O.   144, 285
Bowring, S.A.   18, 279, 290
Boynton, W.V.   102, 212, 284
Brack, A.   282, 283, 286, 287
Bradley, J.P.   292
Brain, C.K.   186, 288
Brandes, J.A.   48, 282
Branlund, J.   283
Bray, D.   285
Brent, R.   287
Bressan, A.   9, 98, 279, 284
Breuil, H.   196
Briceño, C.   27, 280
Brown, F.H.   289
Brown, L.L.   48, 215, 282
Brownlee, D.   279
Brunet, M.   183, 288
Burman, R.R.   293
Burrows, A.   249, 293
Buseck, P.R.   208, 291
Bussey, D.B.J.   225, 293
Busza, W.   293
Butler, R.P.   78, 284

Cairns-Smith, A.G.   146, 285
Cameron, A.G.W.   111, 261, 282, 284, 294
Campbell, N.A.   160–162, 171, 172, 288
Canfield, D.E.   153, 288

Canup, R.M.   41, 282
Carney, B.W.   15, 279
Carr, M.H.   45, 282
Cavosie, A.J.   18, 70, 279
Chabrier, G.   285
Charnley, S.B.   33, 281
Chiappe, L.M.   173, 290
Chiosi, C.   279, 284
Chyba, C.F.   280, 284
Claeys, P.   174, 288
Clark, J.A.   172, 288
Claverie, J.-M.   135, 286
Clayton, N.S.   198, 288
Cocconi, G.   266, 270, 294
Cohen, B.A.   45, 282
Condie, K.C.   177, 288
Condon, E.U.   216, 291
Connelly, J.   280
Conway Morris, S.   198, 288
Copernicus, N.   3, 273
Corsetti, F.A.   108, 284
Cox, A.N.   13, 279, 293
Craine, L.E.   286
Currie, B.S.   67, 283

Dar, A.   293
Darwin, C.   149, 196
Dauphas, N.   135, 286
Davis, P.   164, 289
de Almeida, M.L.T.   279
de Duve, C.   128, 132, 144, 151, 152,
    255, 286, 288, 294
de Lange, T.   239, 293
de Lumley, H.   186, 288
De Rujula, A.   293
De Sterck, H.   287
Delsemme, A.   33, 280
Deming, D.   70, 282
Dick, S.J.   273, 291, 294
Dones, L.   282
Doolittle, W.F.   142, 286
Dragicevich, P.M.   249, 293
Drake, F.   109, 110, 266
Dudley, R.   289

Edwards, K.J.   290
Eichhorn, S.E.   290
Eigen, M.   137, 286
Emery, N.J.   198, 288

Endy, D.   287
Engdahl, E.R.   283
Ericksen, B.F.   295
Evert, R.F.   290

Fabian, A.C.   290
Fagotto, F.   279, 284
Falkowski, P.G.   154, 289
Fedo, C.M.   288
Feigelson, E.D.   35, 280
Fenner, Y.   280
Fermi, E.   273
Ferreras, I.   3, 279
Ferris, J.P.   146, 286
Fleagle, J.G.   289
Foley, J.   183, 289
Franklin, C.   287
Frei, R.   136, 287
Fromherz, P.   241, 294

Göpel, C.   279
Gans, C.   289
Garmire, G.P.   280
Garnero, E.J.   63, 282
Gavin, A.-C.   133, 197, 238, 286, 289,
    293
Gibson, B.K.   280
Gilbert, S.F.   160, 289
Gilbreath, W.   292
Gillmor, D.S.   291
Gilmour, I.   290
Giovannoni, S.J.   141, 142, 286
Goldsmith, D.   111, 261, 284, 285,
    294, 295
Gonzalez, G.   15, 279
Gore, R.   184, 188, 191, 289
Gould, S.J.   198, 289
Graham, C.M.   280
Graham, J.B.   154, 289
Green, N.P.O.   120, 122, 286, 291
Greenspon, D.   282
Grice, K.   244, 293
Grotzinger, J.   283

Haile-Selassie, Y.   183, 289
Haisch, K.E.   27, 280
Halliday, A.N.   279
Halverson, G.P.   285
Hamilton, C.J.   246, 293
Hansen, B.   281

Harrison, T.M.   18, 70, 279
Hart, D.J.   286
Hart, H.   119, 121, 286
Hart, M.H.   106, 109, 285
Hart, S.R.   57, 282
Hartmann, L.   7, 279
Hartmann, W.K.   45, 227, 281, 282, 293
Harvey, R.P.   292
Hay, R.L.   289
Hayashi, M.R.   281
Hazen, R.M.   146, 286
Head III, J.W.   210, 291
Heinz, U.   293
Helffrich, G.R.   52, 282
Hellwege, K.-H.   285
Hendel, R.   283
Henning, H.   280
Henry, T.J.   237, 293
Hilburn, I.A.   285
Hodges, R.R.   293
Hoffman, P.F.   108, 285
Holden, C.   193, 289
Holleman, A.F.   13, 279
Holman, M.   281
Holmes, B   141
Holmes, B.   286
Holtzman, E.   157, 290
Horowitz, N.H.   206, 291
Howard, A.W.   270, 294
Hubble, E.   3, 273
Hughes, D.   24, 25, 280
Hutcheon, I.D.   279
Hutchinson III, C.A.   140, 286
Hynek, J.A.   217, 291

Ida, S.   24, 41, 280, 282
Irwin, L.N.   118, 287

Jacobsen, S.B.   18, 24, 40, 44, 279
Jaffe, R.L.   250, 293
Jakosky, B.   287, 288
Jenkins, A.   5, 279
Jin, Y.G.   244, 293
Johnston, W.K.   145, 286
Jones, A.   38, 281
Jones, S.   289, 290
Jordan, T.H.   283

Kaku, M.   274, 294

Kamber, B.S.   288
Kaplan, R.W.   158, 289
Kardashev, N.S.   259, 266, 295
Kasting, J.F.   48, 92, 100, 109, 144, 215, 282, 285, 286, 291
Kaufman, A.J.   285
Kawabe, R.   281
Kelley, J.   182, 183, 289
Kelley, S.P.   290
Kenrick, P.   164, 289
Kepler, J.   77
Kerp, H.   164, 289
Kerr, R.A.   213, 282, 291, 292
King, E.M.   283
Kirschvink, J.L.   285
Kita, N.T.   281
Kitamura, Y.   27, 281
Kleine, T.   18, 44, 279
Koeberl, C.   284
Koonin, E.V.   139, 140, 142, 286, 287
Kopp, R.E.   107, 285
Kowal, C.T.   30, 281
Kring, D.A.   282
Kroot, A.N.   279
Kuiper, G.P.   24, 31
Kumar, A.   286
Kunin, V.   142, 286

Löffler, W.   293
Lada C.J.   280
Lada, E.A.   280
Lamarck, J.B.   196
Laskar, J.   104, 285
Leakey, M.D.   185, 289
Lee, P.   293
Lepland, A.   288
Levison, H.F.   282
Lewin, R.   189, 289
Lewis, J.   285
Lewis, J.S.   230, 293
Lin, D.N.C.   24, 280
Lineweaver, C.H.   15, 280
Lissauer, J.J.   20, 25, 281
Lorenz, R.   210, 291
Lovelock, J.E.   214, 291
Lowell, P.   205, 291
Lubin, L.M.   280
Ludwig, K.R.   290
Luke, G.N.   289

Lunine, J.I.   66, 69, 283
Lyot. B.   83

Müller R.D.   60, 283
Méra, D.   111, 285
Madsen, J.B.   295
Manhès, G.   279
Marcy, G.W.   78, 284
Marshall, C.R.   294
Martin, R.D.   178, 179, 289, 290
Martin, W.   146, 287
Master, S.   284
Matthews, M.S.   281
Mayr, E.   196, 197, 289
McConnochie, T.H.   293
McDonough, W.   57, 283
McDougall, I.   186, 289
McKay, C.P.   280, 284
McKay, D.S.   208, 209, 292
McLennan, S.M.   70, 283
McNamara, D.H.   264, 295
McSween Jr, H.Y.   292
Melchiorri, A.   279
Mellars, P.   186, 290
Mellersh, A.   137, 138, 287
Melosh, H.J.   282
Menzel, D.H.   217, 292
Metcalfe, I.   290
Mezger, K.   279
Miller, R.   293
Miller, S.L.   48, 138, 139, 283, 287
Mohiuddin, U.   291
Mojzsis, S.J.   135, 153, 287, 290
Montelli, R.   63, 283
Moorbath, S.   136, 287
Mordasini, C.   280
Morgan, J.   288
Morrison, P.   266, 270, 294, 295
Mousis, O.   280
Moyà-Solà, S.   182, 290
Mundil, R.   174, 290
Murray, N.   25, 281
Mushegian, A.R.   139, 287

Nakamoto, T.   35, 36, 281
Narbonne, G.M.   107, 285
Nash, C.Z.   285
Nisbet, E.G.   45, 283
Norell, M.A.   173, 290

Novikoff, A.B.   157, 290

O'Leary, B.   292
O'Neill, G.K.   232–235, 252, 294
O'Neill, H.St.C.   57, 58, 283
Olcott, A.N.   108, 285
Oliver, B.M.   265, 266, 295
Oort, J.   32, 273
Oppenheimer, R.   250
Orgel, L.E.   139, 145, 146, 287
Osterloff, P.L.   291
Owen, T.   111, 261, 285, 294

Packard, N.   141
Padian, K.   173, 290
Page, T.   292
Palme, H.   34, 37, 38, 57, 58, 279, 281–283
Park, J.   283
Pavlov, A.A.   287
Payne, J.L.   244, 294
Peck, W.H.   280, 283
Percival, S.M.   280
Periwal, V.   287
Phillips, R.J.   281
Picardi, G.   213, 292
Pierazzo, E.   203, 292
Pierce, D.   284
Pierrehumbert, R.T.   109, 285
Pilbeam, D.   289, 290
Pollack, J.B.   281
Popovici, C.   289
Porter, S.C.   283
Pravdo, S.H.   280
Press, F.   48, 52, 54, 56, 69, 283
Pringle, H.   199, 290
Priscu, J.   204, 292
Ptolemy, C.   3

Queloz, D.   284

Raff, M.   285
Rasmussen, B.   146, 287
Rasmussen, S.   141
Raven, P.H.   164, 170, 290
Regenauer-Lieb, K.   65, 71, 283
Reimold, W.U.   284
Renne, P.R.   290
Reynolds, R.T.   285
Rhode, R.A.   162, 171, 177, 178, 290

Roberts, K.   285
Robertson, M.P.   139, 287
Robutel, P.   104, 285
Rocchia, R.   175, 176, 290
Rodgers, S.D.   33, 281
Romer, A.S.   172, 290
Rood, R.T.   111, 260–262, 285, 295
Rose, C.   265, 295
Rosing, M.T.   135, 136, 287
Rouxel, O.J.   154, 290
Rowley, D.B.   67, 283
Russell, M.J.   146, 287
Ryder, G.   282

Sagan, C.   111, 211, 214, 261, 276,
   285, 292, 295
Saito, M.   281
Salaris, M.   14, 280
Sandage, A.   14, 280
Sandweiss, J.   293
Sandwell, D.T.   72, 283
Sargent, A.I.   79, 284
Schaarschmidt, F.   165, 290
Schaeffer, R.   285
Schiaparelli, G.   204
Schidlowski, M.   135, 287
Schneider, J.   26, 74, 75, 281
Schopf, J.W.   135, 136, 287
Schröder, A.   293
Schrag, D.P.   108, 285
Schubert, G.   72, 283
Schulze-Makuch, D.   118, 287
Schwartz, A.W.   145, 287
Scotese, C.R.   67, 283
Seitzer, P.   279
Sepkoski, J.J.   243, 294
Serre, D.   186, 290
Shklovskii, I.S.   266
Shostak, S.   288
Sibley, C.G.   188, 290
Siever, R.   283
Silk, J.   279
Sillen, A.   186, 288
Simons, E.   181, 290
Skelton, P.W.   170, 290
Skinner, B.J.   48, 49, 283
Sleep, N.H.   42, 43, 45, 283, 284
Sloan, R.E.   177, 288
Smit, J.   176, 290

Smith, D.E.   212, 292
Smith, J.V.   146, 287
Snyder, M.   286
Soper, R.   286, 291
Spicer, R.A.   290
Steiper, M.E.   178, 290
Stewart, G.R.   282
Stout, G.W.   286, 291
Strom, R.G.   44, 283
Sukarna, T.Y.   290
Sunada, K.   281
Suwa, G.   289
Swindle, T.D.   282
Swisher III, C.C.   186, 290
Szallasi, Z.   287

Tachibana, S.   281
Tammann, G.A.   293
Tattersall, I.   185, 290
Taylor, D.J.   155, 190, 196, 286, 291
Taylor, G.J.   281
Taylor, S.R.   70, 283
Teilhard de Chardin, P.   196
Teller, E.   250
Thieme, H.   192, 291
Thomas, P.J.   280, 284
Tian, F.   48, 144, 287
Titius, J.D.   29
Tomita, M.   133, 197, 238, 287, 291,
   294
Toon, O.B.   287
Trainer, M.G.   107, 285
Trefil, J.S.   111, 260–262, 285, 295
Tremaine, S.   281
Trieloff, M.   34, 37, 38, 281
Trilling, D.E.   26, 281
Turtle, E.P.   203, 292

Udalski, A.   82, 284
Ulmschneider, P.   111, 285
Ulyanov, A.A.   279
Urey, H.   138, 139

Valley, J.W.   18, 45, 69, 70, 279, 280,
   283
van den Berg, A.P.   283
van den Bergh, S.   293
van der Hilst, R.D.   52, 283
van Paradijs, J.   293
van Thienen, P.   72, 283

VandenBerg, D.A.   280
Venter, J.C.   141
Vlaar, N.J.   283

Wächtershäuser, G.   146, 288
Wallace, A.R.   197, 291
Walzer, U.   62, 283
Ward, P.   279
Ward, P.D.   174, 291
Watson, J.D.   285
Weiss, A.   280
Weissman, P.R.   281
Wellman, C.H.   164, 291
Westall, F.   135, 288
Westphal, S.P.   141, 288
Wetherill, G.W.   22, 281
White, T.D.   289
White, W.M.   57, 58, 283
Whitehouse, M.J.   136, 288
Whitmire, D.P.   285
Wiberg, E.   13, 279
Widiyantoro, S.   283
Wilczek, F.   293
Wilde, S.A.   17, 70, 279, 280, 283

Williams, D.M.   102, 104, 105, 113, 285
Williams, I.S.   18, 279
Winisdoerffer, C.   280
Winkler-Oswatitsch, R.   137, 286
Wittbrodt, J.   195, 288
Wittig, N.   280
Woese, C.R.   130, 145, 288
WoldeGabriel, G.   289
Wolfe, J.   295
Wood, B.   183, 288, 291
Wood, B.J.   52, 282
Woolf, N.J.   215, 291
Wootten, H.A.   7, 280
Wright, G.   265, 295

Xu, X.   173, 290

Young, N.M.   290
Yuen, D.A.   283

Zachos, J.   180, 291
Zahnle, K.J.   42, 43, 284
Zeck, G.   241, 294
Zindler, A.   57, 282

Subject Index

Acasta gneisses   18
accretion disk   8, 19, 20, 24–27, 35, 36,
   39–41, 79, 80, 112, 197, 209
Aegyptopithecus   178, 181, 182
aging   238, 274
albedo   101, 102, 109
Allende meteorite   17, 34, 35
ALMA observatory   85
alpha Centauri   33, 236
alpha-helix   120, 122
amino acids   34, 119, 120, 124, 125,
   130, 131, 137, 138, 144, 175
ammonia   7, 47, 48, 138
amniotic egg   172
amphibians   130, 162, 163, 172, 177,
   195, 257
andesite   53, 54
androids   236, 237, 240, 241, 252, 258,
   259, 274
angiosperms   163, 168–171, 180, 195
apes   182, 187, 195
– great, 178, 179, 181–183, 187–190,
   195, 257
– lesser, 178, 179, 181, 182, 187, 188
apoptosis   159, 238
archaebacteria   131–133, 141, 143, 151
Archean era   46, 66, 69, 107, 108, 136,
   153
arthropods   162, 163, 171, 195
asteroids   24, 28–31, 33, 34, 39, 76,
   79, 112, 201, 224, 226, 230, 234, 236,
   243, 246, 247
– Amor, 28
– Apollo, 28
– Aten, 28
– NEA, 28, 30, 228–230, 233, 247
asthenosphere   61, 65
asymptotic giant branch   9, 12, 13, 37

ATP   119, 124, 127, 133, 145, 153, 154
Australopithecus   183, 185, 189, 192,
   257
– A. ramidus, 183
– A. ramidus kadabba, 183
– afarensis, 183–188, 192
– africanus, 183, 188
– anamensis, 183
– Sahelanthropus tchadensis, 183

backbone   162, 171, 172
bacteria   129–131, 133, 143, 152, 153,
   159, 195, 208, 214, 215, 238, 239,
   243, 244, 256
– aerobic, 154
– halophilic, 143
– methanogenic, 214, 215
– photosynthetic, 105, 135, 136, 152,
   153, 157
– thermophilic, 143, 203
banded iron formations BIFs   136, 153
Barringer crater   246
basalt   53–55, 59, 61, 63, 64, 68, 72,
   244
bases   121, 122, 134, 138, 139, 157
beta-sheet   120
big bang   3, 266
birds   130, 170, 173, 177, 180, 192,
   195, 199, 257, 262
black holes   12, 94, 250, 275
blastula stage   159, 160
bonds
– atomic, 118
– disulfide, 120
– hydrogen, 120, 122
– ionic, 118
– peptide, 119
– phosphodiester, 122

brown dwarfs 11, 73, 74
bryophytes 163, 164, 166–170
bubonic plague 243
buoyancy 61, 68, 163

CAIs 17, 34, 35, 37, 38
Callisto 24, 202, 204
Cambrian 67, 162, 197
Cambrian explosion 198
carbohydrates 119, 120, 127
carbon burning 12
carbon dioxide 47, 48, 99, 100, 102,
    105, 108, 109, 118, 138, 144, 153,
    166, 201, 206, 207, 210, 211, 215, 244
carbon monoxide 7, 21, 47, 48, 207,
    215
carbonate silicate cycle 99–101, 105,
    106
Carboniferous 107, 173
Carboniferous forest 167
catastrophic impact 247, 248
cell membrane 120, 121, 127, 144,
    145, 152
cell plasma 120, 126
cell specialization 158–160
cells 117, 119, 122, 125, 127, 130–132,
    142, 144, 146, 149, 152–154, 197, 208,
    238, 256, 257, 264, 275
– eukaryotic, 120, 122, 125, 127, 128,
    131–133, 136, 149, 151, 152, 154, 155,
    157–159, 199
– germinal, 159, 238
– prokaryotic, 122, 125, 127, 128, 131,
    134, 135, 139, 151, 153, 158, 197
– somatic, 159, 238
Cenozoic era 108
center of mass 61, 77, 78, 80, 95
Ceres 20–22, 28, 40–42, 57
Chandrasekhar limit 12
chemical composition 4, 12–14, 21,
    24, 29, 30, 33, 37, 38, 41, 48, 52, 54,
    56, 57, 71, 106, 212, 230
Chicxulub crater 43, 174, 246
chimpanzee 135, 182–185, 187, 189,
    190
chirality 119, 137, 145
chondrules 17, 34–36, 38
chordates 162, 163, 171, 195
chromosomes 122, 123, 127, 145, 239

CI-chondrites 34, 37, 38, 57
circumstellar disks 79, 80
codons 124–126, 137
coelom 160
Comet Halley 31, 32
Comet Hyakutake 32
comets 20, 24, 29, 31–33, 46–48, 80,
    83, 91, 173–176, 201, 226, 236, 243,
    246, 248
composition
– felsic, 53, 61
– mafic, 53, 61
compounds
– inorganic, 87, 91, 118, 241
– organic, 7, 31, 42, 48, 87, 88, 91, 118,
    119, 137, 138, 140, 143, 144, 146, 238,
    241
connected societies 241, 242, 253, 258
conquest of the solar system 221–223,
    231, 236, 242, 252
continental crust 17, 55, 61, 65, 66,
    69, 70
continuously habitable zone 97, 98,
    101, 109, 112, 113, 257, 262
convection 10, 26, 35, 46, 59–63, 66,
    69–72, 102
convergent boundaries 64, 65
core–mantle boundary 50, 61, 63
coronal mass ejections CMEs 35
cosmic background radiation 4, 266
cosmic rays 16, 208, 249, 251, 265
cosmological principle 273
craters impact 28, 44, 174, 203, 206,
    212, 213, 245–248
cratons 69
Cretaceous 60, 67, 68, 171, 173, 176,
    178–180, 244, 246
crossing over 156, 157
crossopterygian 172
cyanobacteria 107, 152, 153

dark age 4
Darwin's theory 87, 145, 147, 149,
    150, 196–198, 256, 257
Deccan Traps 64, 174, 244
deep-sea vents 47, 48, 55, 99, 100,
    133, 138, 143, 144, 146, 203
degeneracy 11–13
Deimos 205

deuterium burning   10, 73
deuterium main sequence   10
deuterostomes   160
Devonian   162, 165, 166, 171, 172, 195
Devonian explosion   166
Devonian forest   167
dinosaurs   173, 176, 195, 257, 262
diploid   155, 168, 169
DNA   117, 119, 121–125, 127, 129, 131, 132, 134, 137, 139, 141, 143, 145, 149, 154, 157, 159, 187, 188, 193, 237–239
DNA hybridization   187, 188
DNA-world   132, 145, 158
Drake formula   109–111, 255, 260–262, 273
dust   4, 7, 9, 14, 19, 20, 26–28, 31, 32, 36, 37, 39, 47, 48, 76, 79, 81, 108, 138, 143, 176, 197, 202, 206, 247, 265

E-Cell project   133
Earth   39
– core, 37, 46, 49, 50, 56–60, 71
– crust, 17, 49, 50, 52, 54–56, 58, 61
– magnetic field, 58, 59, 67, 69, 107
– magnetic poles, 59
– magnetic-field, 60
– mantle, 13, 14, 38, 45, 47, 49, 50, 52, 54, 56–58, 61–63, 66, 68, 71, 99–101, 118, 244
Earth crust   see continental crust, see oceanic crust
earthquakes   49–51, 55, 252
eccentricity   22, 23, 74, 96
ectoderm   160
Ediacaran   107, 158, 161
effective temperature   9, 90
Efremovka meteorite   17, 142
element abundances   4, 12, 13, 19, 38, 209
element production   12
Emperor seamounts   64
endoderm   160
endosymbionts   128, 132, 151, 152, 154, 256
environmental damage   222, 233, 234, 249, 250
enzymes   119, 127, 133, 144, 145, 152
Eocene   179

epsilon Eridani   266
escape speed   42, 92, 93, 228, 229
eubacteria   131–133, 141, 143, 151, 152
eukaryotes   120, 122, 125, 127, 128, 131–134, 136, 143, 149, 151–155, 157–159, 197, 199, 256, 257, 264
Europa   24, 92, 202–204
evolution
– biological, 109, 113, 131, 133, 137, 145, 149–152, 155, 157–159, 161, 162, 177–179, 181, 182, 187, 188, 190, 195–199, 256, 257, 259, 276
– chemical, 157
– convergence, 198
– cultural, 199, 221, 222, 228, 230, 234, 239–242, 252, 257, 276
– genetic, 157, 158
exons   125, 134, 154, 157
extraterrestrial intelligence   87, 94, 98, 109, 110, 117, 193, 194, 196, 197, 216, 217, 255, 257, 259–262, 270, 275–277
extraterrestrial life   9, 19, 87, 97, 109, 110, 117, 143, 149, 151, 201, 202, 204, 214, 216
extraterrestrial societies   221, 242, 253
– lifetime, 110, 111, 262, 263
– location, 263, 271, 274, 277
– number, 263, 272

Fermi paradox   255, 272–275
field stars   15
fireball   4, 266
fish   130, 134, 163, 171, 172, 176, 192, 195, 198, 257
– chondrichthyes, 162
– osteichthyes, 162
flares   35–37
fluid-induced melting   see wet melting
foraminifera   176, 177, 179, 180

Galactic Habitable Zone   15
galaxy clusters   4
galaxy formation   6
Galileo navigation system   223
gametophyte   168, 169
gamma ray bursts   243, 248, 249
Ganymede   24, 202–204
gastrulation   160, 194

genes   117, 126, 127, 134, 139,
    142–146, 154, 156–161, 199
genetic code   119, 124, 126, 131, 133,
    137, 138, 143
genome   133–135, 146, 156, 157, 159,
    187, 195, 199, 239
gibbon   179, 182, 187
Giotto   31, 32
glaciation irreversible   98, 101, 102,
    109, 211, 243, 245
Global Positioning System   223
Gondwana   68
gorilla   135, 182–185, 187, 189, 190,
    192
grains   20
– ice, 21
– iron, 21
– silicate, 21, 41
granite   53, 54
gravitational collapse   6, 7, 9, 10, 12,
    14, 19, 20, 265
gravitational waves   265
greenhouse effect   99, 100, 102, 105,
    109, 244
– moist, 100, 101, 109
– runaway, 98, 100, 245
greenhouse gases   99, 105
gymnosperms   163, 167, 169–171, 180

habitable zone   87–92, 94, 97, 98, 101,
    105, 109, 110, 112, 202, 204
Hadean era   45–48, 66, 69, 138, 153
half-life time   16
haploid   155, 156, 168, 169
Hayashi tracks   10, 11, 26
helium burning   11–13
helium main sequence   12
Hertzsprung–Russell diagram   9–11
Homo
– erectus, 184–188, 192, 240, 257, 262
– habilis, 185, 192, 240, 257, 259, 262
– heidelbergensis, 186
– rudolfensis, 185, 262
– sapiens, 184, 186–188, 193, 199, 262
– sapiens (archaic), 186
– sapiens (modern), 186
– sapiens neanderthalensis, 186
homunculus   189

hot spots   46, 60, 63, 64, 68, 70, 72,
    244
Hox genes   161
Hubble Space Telescope   79, 80, 85
hunting   192, 193, 199
– spears, 192, 241
Huronian glaciation   107, 245
hydrogen burning   10, 94
hydrothermal vents   see deep-sea
    vents

ice-formation boundary   21, 24, 25, 28,
    31, 33, 75, 91
impact events   21, 28, 33, 40–43, 47,
    61, 92, 93, 133, 142–144, 173–176,
    208, 225, 243, 244, 246–248
Insectivora   178–180
insects   134, 162, 170, 177, 180, 192
introns   125, 126, 134, 154, 157
Io   44, 202
iridium   108, 175, 176
Iridium communication system   223
iron   4, 12, 14, 34, 53, 57, 71, 102, 105,
    136, 146, 153, 211
iron core   21, 39, 40, 45, 71, 102, 202
iron–sulfur world   144, 146, 147
isotopes   10, 16, 17, 21, 39, 45, 135,
    175, 179, 180, 208
Isua Formation   18, 135, 136, 142, 157

jets   26, 27, 249
Juno   28
Jupiter   11, 24–26, 28–30, 32, 44, 71,
    73–75, 77, 78, 80, 81, 87, 91, 92, 96,
    102, 104, 109, 112, 113, 201, 202,
    204, 214, 262
Jurassic   60, 67, 68, 154, 173

K/T boundary event   43, 47, 173–177,
    195, 244, 246–248, 257
Kardashev societies   259
Kuiper belt   24, 31
Kuiper-belt objects   24, 27, 32, 39, 91,
    112, 236

Lagrange point   80, 84
language   190–194, 199
Last Universal Common Ancestor
    131–133, 137, 139, 141–143, 145, 146,
    151, 157, 158

Laurasia   68
lemurs   178, 179
lipids   119–121, 127, 128, 145
lithospheric plates   see tectonic plates
lorises   178, 179
Los Alamos Bug   141

magnetic field   see Earth, magnetic
    field
magnetic fields   6, 16, 20, 25–27, 35,
    202, 203, 252, 265
mammals   163, 170, 173, 177, 178,
    180, 182, 195, 199, 208, 257
Mars   22–24, 28–30, 33, 40, 57, 71,
    72, 89, 101, 102, 105, 109, 201, 202,
    204–207, 209–213, 225, 226, 231, 247
– life detection experiments, 206
– meteorites, 208–210
Mars missions   212, 225, 226
– Mariner 4 and 9, 205
– Mars Express, 102, 212
– Mars Global Surveyor, 210, 212
– Mars Odyssey, 102, 212
– Mars Reconnaissance Orbiter, 214
– Mars Scout, 214
– Opportunity, 213
– Pathfinder, 210
– Phoenix Mars Lander, 214
– Spirit, 213
– Viking 1 and 2, 205–207, 209, 214
mass drivers   227, 228
mass extinction events   173, 176,
    243–246, 249
mass spectrometer   16, 230
master genes   161, see HOX genes
matrix   34, 35, 38
Maxwellian velocity distribution   93,
    94
meiosis   152, 154–156
Mercury   22, 23, 71, 75, 96, 103, 144,
    201, 226, 247
mesoderm   160
Mesozoic era   108, 169
meteorites   16, 17, 30, 31, 33, 35, 37,
    46–48, 56, 174, 225, 246
– iron, 34, 35, 56
– stone, 34, 35, 57
– stone CI, 57
– stony-iron, 34, 35

methane   47, 48, 74, 99, 105–107, 138,
    154, 206, 214, 215
Mid-Atlantic Ridge   58, 64
minimal organism   141, see Last
    Universal Common Ancestor, 143,
    238
mining in space   227, 230, 231, 233,
    234
Miocene   60, 182, 183
mitochondria   127, 152, 154, 186
mitosis   152, 154, 155, 157
– anaphase, 155, 156
– cytokinesis, 155
– interphase, 154, 156
– metaphase, 155
– prophase, 154
– telophase, 155
Mohorovičić discontinuity   50
molecular clocks   129, 130, 143, 187
molecular clouds   6–9, 13, 14, 19, 20,
    24, 27, 39, 79, 85, 256
monkeys   182, 187, 188, 195
– new world, 178, 179
– old world, 178, 179, 181, 187
Moon   18, 19, 22, 23, 33, 40–44, 47,
    57, 201, 202, 224–229, 231–233, 242
– missions, see space missions
Moon station   226–228
multicellularity   149, 154, 158–161,
    194, 195, 238, 241, 256, 264
mycoplasmas   117, 127, 128, 134, 139,
    141, 238

NASA spaceguard survey   248, 252
Neptune   24, 29, 32, 44, 201
neutrinos   265
neutron stars   12, 75, 94
nuclear envelope   125, 127
nucleic acids   119–121, 127, 145
nucleosomes   122

oceanic crust   54, 55, 60, 61, 65, 68,
    69, 202
Oetzi   193
Oligocene   60, 179, 181, 182
olivine   31, 52–54, 71
Olympus Mons   206, 211, 212
Oort cloud   32, 33
orang-utan   182, 183, 187, 190

Ordovician   107, 162
organelles   120, 127, 128, 152, 154, 157–159
organic chemistry   118, 133, 137, 143, 149
organs   149, 154, 158–161, 196, 222, 238, 239, 256
Orion nebula   79, 80
orogeny   see mountain-building episode
oxygen catastrophe   153
ozone   76, 211, 214, 215, 264, 274

P/T boundary event   173, 174, 244, 246
Paleocene   60, 178, 179
Paleomap Project   66, 67
Paleozoic era   108
Pallas   28
Pangea   67–69
Pauli principle   11
Permian   107, 173, 174, 246
peroxisomes   127, 152, 154
Phanerozoic era   64, 107, 108, 153, 154, 244
phloem   167
Phobos   205
phosphoric acid   121, 122, 125, 145
photolysis of water   48, 100, 102, 106, 138, 211, 215
photosynthesis   48, 100, 105, 135, 136, 142, 147, 152, 153, 157, 163, 164, 166, 167, 203, 207, 214
phylogenetic tree   130–132, 142, 160, 162, 178, 183, 187
Pierolapithecus catalaunicus   182, 183
pillow basalt   55
placenta   173, 177, 178, 198
planet formation   3, 13, 14, 19, 25, 75, 91, 102, 112, 113, 137, 197, 262
planet migration   25, 112, 113
planet search
– direct methods, 76
– indirect methods, 76
– interferometric method, 84
– microlensing method, 81
– photometric method, 82
– proper motion method, 80
– radial velocity method, 78

planet-search missions   see space missions
planetary missions   see space missions
planetary nebula   9, 13
planetesimal disk   40, see acretion disk40
planetesimals   19–25, 27, 29, 31–34, 36–41, 44–47, 56, 57, 71, 102, 138, 142, 202
planets
– extrasolar, 73
– habitable, 87, 98, 101, 105, 109–111, 113, 214, 260, 277
– habitable, numbers, 260–262
– jovian, 91, 112, 113
– spin-axis variations, 96, 98, 103–105, 113
– spin-orbit resonance, 104
– terrestrial, 21–24, 26, 33, 45, 73, 76, 78, 91, 92, 99, 102, 105, 109, 112, 113, 118, 264
plate tectonics   39, 59, 60, 62–64, 67, 68, 71, 72, 100, 102, 105, 109, 202, 210, 211, 227
plate-tectonics   107
Pleistocene   107
Pliocene   179
plume convection   46, 63, 64, 68, 70, 72
Pluto   24, 91, 103, 226, 236
polypeptides   119, 144
Pongola glaciation   107, 108
predicting the future   221, 222, 252
primates   135, 178–182, 195, 199, 257, 276
primordial soup   48, 143, 146
Proconsul   178, 182
Progenote   133, 142, 143, 145, 157
projected mass   74, 77
prokaryotes   122, 125, 127, 128, 131, 134, 135, 139, 151, 153, 158, 197
proteins   119–122, 124, 125, 127, 131, 137, 144, 145
Proterozoic era   107–109, 153
protists   131, 158, 197
protoplanetary disk   27
protoplanets   23, 24, 85
protostars   8, 9, 19, 20, 25–27, 39, 73
pulsars   73, 75, 268, 269

purines   121, 122, 137, 139
pyrimidines   121, 137, 139
pyroxene   34, 52–54, 71

quartz   52–54, 118, 175

radiative tracks   10, 11
radio window   110, 266
radiometric clocks   15–17, 34
red-giant branch   9, 11, 12
reptiles   130, 163, 172, 173, 176, 177,
    192, 195, 198, 257
rhizome   164–166
Rhynie Cherts   165, 171
rhyolite   53–55
ribosomes   125, 126, 130
rifts   see spreading ridges
RNA   237
– heterogeneous nuclear, 125
– messenger, 119, 121, 122, 124–127,
    129, 131, 132, 137, 142–146
– ribosomal, 130
– transfer, 125
RNA-world   132, 133, 142–146, 157
Roche limit   97
rocks
– igneous, 34, 52–55
– metamorphic, 52, 66, 135
– sedimentary, 52, 65, 68, 135, 136, 175
Rodinia   69, 107

Sahelanthropus tchadensis   183
Saturn   24, 32, 44, 201, 202, 226
sea-floor spreading   see plate tectonics
seed plants   166–169
seedless vascular plants   163, 165–169
seismic tomography   51
seismic waves   49–51, 56
– P-wave, 49, 50
– S-wave, 49, 50
– shadow zone, 50
– surface waves, 49, 50
semimajor axis   22, 74, 75, 77, 78
sequencing   117, 129, 130, 133, 134,
    137, 151, 199
SETI   216, 255, 265–270, 272, 273
SETI projects
– BETA, 267
– Cyclops, 266

– HRMS, 269
– META I and II, 267
– optical, 270
– Ozma, 266
– Phoenix, 269
– SERENDIP, 267
– Home, 269
– southern SERENDIP, 268
Siberian Traps   64, 174
Silurian   107, 162, 165
solar radiative flux   89
space habitats
– ISS, 223–225, 229, 230
– O'Neill type, 232–235
– torus-type, 231–233
space missions
– Clementine, 225
– COROT, 82
– Darwin, 83
– ESA missions, 80, 83, 284
– Gaia, 80
– Galileo, 202–204
– Hipparcos, 80
– JWST, 85
– Kepler, 82
– LF, 84
– Lunar Prospector, 225
– Lunar/Planetary missions, 204, 225,
    226, 292
– NASA missions, 80, 82–85, 284
– PI, 84
– Pioneer 10 and 11, 236
– SIM, 80
– TPF-C, 83
– TPF-I, 84
– Voyager 1 and 2, 236
sporangia   164–166, 168, 169
sporophyte   168, 169
spreading ridges   55, 58–60, 64, 67
star clusters   8, 14, 15, 79
– globular, 14
– open, 14
star formation   4, 7, 8, 13, 80, 265,
    270, 272
stars
– dwarfs, 90
– giants, 90, 94
– lifetime, 90, 94, 97, 112

– luminosity, 90, 91, 97, 98, 100, 101, 106, 109
– main sequence, 90
– population I, 13
– population II, 4, 13
– population III, 4
– spectral types, 90, 91, 97
stellar evolution   19
– asymptotic giant branch, 12
– main sequence, 10, 90, 98, 112
– post-main sequence, 11, 45, 94, 97
– pre-main sequence, 8, 26
– red-giant stage, 97, 98
stellar wind   12, 13, 27, 31, 79
stereoscopic vision   181, 194
stomata   164, 165
stone tools   190–192
– Acheulean industry, 191
– Oldowan industry, 191
stromatolites   136, 143
sugars   120–122
supernovae   4, 7, 11, 12, 15, 75, 243, 248, 249, 256, 265

T-Tauri phase   24, 26, 27, 79
T-Tauri wind   26, 27
tau Ceti   236, 266
Tautavel man   186
tectonic plates   51, 55, 60, 63–65
teleological arguments   196, 197
telomeres   239
tephra   54
terrorism   251, 252
thermal pulsations   13
thermal speed   92, 93
Thioester-world   145
thrust faults   see convergent boundaries
tidal effects   24, 41, 47, 95–97, 102–104, 109, 112, 202
tidal heating   202
tidal lock radius   90, 96, 112
Titan   202

Titius–Bode law   29
tools
– fire, 186, 199, 241
– making, 186, 190–194, 199, 241
– using, 184, 188–190
transform fault   64, 65
Triassic   69, 154, 173, 174, 199, 246
Tunguska event   246

UFOs   216–218, 273, 275, 277
Uranus   24, 32, 40, 44, 103, 201

vacuoles   152, 157
Venus   22, 23, 71, 72, 89, 96, 100, 103, 201, 215, 226, 247
Vesta   28
volcanism   44, 46–48, 72, 99–102, 106, 138, 143, 174, 184, 202, 203, 206, 211, 212, 227, 243, 244, 252
– hot spot, 72
volcanoes   53–56, 63, 64, 70
– andesitic, 60, 65, 68
– rhyolitic, 55
– shield volcanoes, 53–55, 64, 72
– stratovolcanoes, 54, 55, 64
Volvox   159
von Neumann probes   236

war   244, 251
water hole   266
wet melting   54, 55, 65, 68
white dwarfs   9, 11–13, 94, 236, 256
Wilson cycle   69
worm holes   274

xenoliths   56
xylem   167

yield stress   62

zero-age main sequence   10
zodiacal light   79
zoo hypothesis   276

Printing: Krips bv, Meppel
Binding: Stürtz, Würzburg